BOSTON STUDIES IN THE PHILOSOPHY OF SCIENCE

VOLUME LVIII

PROGRESS AND RATIONALITY IN SCIENCE

SYNTHESE LIBRARY

VOLUME 125

BOSTON STUDIES IN THE PHILOSOPHY OF SCIENCE

EDITED BY ROBERT S. COHEN AND MARX W. WARTOFSKY

VOLUME LVIII

PROGRESS AND RATIONALITY IN SCIENCE

Edited by

GERARD RADNITZKY AND GUNNAR ANDERSSON

Universität Trier

D. REIDEL PUBLISHING COMPANY

DORDRECHT : HOLLAND / BOSTON : U.S.A. / LONDON : ENGLAND

Library of Congress Cataloging in Publication Data

Main entry under title:

Progress and rationality in science.
    (Boston studies in the philosophy of science; v. 58)
    Bibliography: p.
    Includes indexes.
    1. Science—Philosophy. I. Radnitzky, Gerard.
II. Andersson, Gunnar. III. Series.
Q174.B67 vol. 58 [Q175]   501's [501]   78-14305
ISBN-13: 978-90-277-0922-6   e-ISBN-13: 978-94-009-9866-7
DOI: 10.1007/978-94-009-9866-7

Published by D. Reidel Publishing Company,
P.O. Box 17, Dordrecht, Holland

Sold and distributed in the U.S.A., Canada, and Mexico
by D. Reidel Publishing Company, Inc.
Lincoln Building, 160 Old Derby Street, Hingham,
Mass. 02043, U.S.A.

EDITORIAL NOTE

The extraordinary Kronberg conference of 1975 brought the international discussion of the creative and critical work of the philosophers at LSE into sharp focus. We are pleased to bring the edited proceedings to public attention in this volume of papers that have been carefully revised in the light of the Kronberg conversations. Gratitude is due to Gerard Radnitzky and Gunnar Andersson for their efforts in organizing the conference, to Gerd Brand and his associates for their intellectual and financial support, and especially to our friends and colleagues at LSE for their good humor and incisive behavior under friendly fire.

Boston, October 1978   R.S.C.
                                                               M.W.W.

TABLE OF CONTENTS

PART III: THE LSE REPLY

PART IV: TWO BRIEF REJOINDERS

PREFACE

This collection of essays has evolved through the co-operative efforts, which began in the fall of 1974, of the participants in a workshop sponsored by the Fritz Thyssen Foundation. The idea of holding one or more small colloquia devoted to the topics of rational choice in science and scientific progress originated in a conversation in the summer of 1973 between one of the editors (GR) and the late Imre Lakatos. Unfortunately Lakatos himself was never able to see this project through, but his thought-provoking methodology of scientific research programmes was ably expounded and defended by his successors. Indeed, this volume continues and deepens the debate inaugurated in *Criticism and the Growth of Knowledge* (edited by Imre Lakatos and Alan Musgrave), a book which grew out of a conference held in 1965. That debate has continued during the years that have passed since that conference. The group of discussions about the place of rationality in science which have been held between those who emphasize the history of science (with Feyerabend and Kuhn as the most prominent exponents) and the critical rationalists (Popper and his followers), with Imre Lakatos defending a middle ground, these discussions were seen by almost all commentators as the most important event in the philosophy of science in the last decade. This problem area constituted the central theme of our Thyssen workshop.

The workshop operated in the following manner. A group of philosophers of science from the London School of Economics and Political Science, guided by John W. N. Watkins, prepared a position paper treating the central issues of the possibility of objective criteria for scientific progress, the ways of assessing a theory's cognitive merit and the status of such criteria. John W. N. Watkins was asked to present the basic tenets of falsificationism, so that his essay could be used for introducing students to the problems in this area of contemporary philosophy of science. J. Worrall, P. Urbach and E. Zahar assumed the task of presenting Lakatos's methodology of scientific research programmes. Thus in the first part of the book the question arises of how to evaluate the comparative achievements of these two methodologies. In early 1975 the LSE-position paper was

ix

G. Radnitzky and G. Andersson (eds.), Progress and Rationality in Science, ix–x.

distributed to the workshop's invited participants. They prepared critical commentary and often presented alternative positions. The workshop group held a plenary conference, organized by G. Radnitzky and G. Andersson, at the Schlosshotel Kronberg near Frankfurt, July 6–12, 1975. The participants then revised their contributions in the light of the critical discussion from the conference, sometimes making minor adjustments and sometimes virtually rewriting their essays. Most of the revisions were finished by 1977.

Next came the difficult task of selecting the papers to be published together as conference proceedings. We were often forced to exclude outstanding papers, for example because they did not bear directly on the LSE-position paper or because they were of a highly technical nature. The selected papers, chosen strictly according to the criterion of thematic unity, yielded two volumes. The first of these, the present book, poses the question of the objective criteria for scientific progress as its central concern. The second volume, which is being published under the title *The Structure and Development of Science* by D. Reidel Publ. Co., Dordrecht, deals with the philosophical presuppositions, the limits and the implications of scientific theorizing.

This volume opens with the LSE-position paper, and then follows the section of critical commentary and proposals of alternatives. Next come replies by J. W. N. Watkins and J. Worrall, as well as short rejoinders by P. Feyerabend and by K. Hübner. This structuring has made it seem most advantageous to pool the references for the LSE-position paper and the LSE-reply, and to place this unified bibliography at the end of the LSE-reply. The other papers all have separate references.

The editors acknowledge with gratitude the efforts and kind cooperation of all participants involved. Our thanks go above all to Prof. Dr. Gerd Brand, Director of the Fritz Thyssen Stiftung, who not only made the entire project possible through generous support but also gave us the benefit of his advice. Without him neither of the two volumes would ever have seen the light of day. Our thanks also go to Dipl. Ök. Klaus Pähler for compiling the book's subject index, and to Sieglinde Kordel for compiling the book's author index, to Barbara Hill for editorial assistance and to Dorothea Hill for struggling through a heavy load of secretarial responsibilities.

Trier, May 1978   THE EDITORS

INTRODUCTION

GERARD RADNITZKY AND GUNNAR ANDERSSON

OBJECTIVE CRITERIA OF SCIENTIFIC PROGRESS? INDUCTIVISM, FALSIFICATIONISM, AND RELATIVISM

Ideally science should tell us many interesting things about the world, and what it says should be certain. It should tell us much both in the sense of being precise and in the sense of giving us deep knowledge, knowledge of the underlying structures, of the innermost constitution of the world. At least so it was thought for a long time, from the Pre-socratics to recent times. J. W. N. Watkins calls this ideal of science 'the Bacon-Descartes Ideal' in his contribution 'The Popperian Approach to Scientific Knowledge', which opens the LSE-position paper at the very beginning of this volume.

Ideally, thus, scientific knowledge should be certain; scientific theories should be true. How can we know that scientific knowledge is true? In the empiricist tradition it has been held that this can be shown by recourse to our experience, with the help of observations and experiments. It was thought that certainty, truth, or at least, as the next best hope in this world, high 'probability' of scientific knowledge must be and can be demonstrated with the help of an inductive principle, i.e. a principle that says something about the general nature of the world and tells us that, roughly speaking, the future will be much like the present in the sense that observed regularities will hold also at a later date. But how can an inductive principle be justified? How could we know that it is true? Hume raised this question and answered it with a clear demonstration that an inductive principle cannot be justified. In European culture this was one of the factors leading to irrationalism. In reaction there were various more or less desperate attempts to save rationality, to restore the hope in the possibility of certainty – from transcendental philosophy to contemporary justificationalist approaches in the philosophy of science. However, the fact remained that an inductive principle warranting the truth or probability of scientific knowledge could not be justified. As John Watkins aptly put it, "Since Hume any serious philosophy of scientific knowledge has been bound to contain *some* pessimistic elements. 'Never glad confident morning again!'" (This volume p. 24.)

G. Radnitzky and G. Andersson (eds.), Progress and Rationality in Science, 3–19.
All Rights Reserved.

POPPER VS. INDUCTIVISM

According to Popper it is unrealistic to demand that science give us certainty. All of our scientific knowledge is in principle fallible and hypothetical. It cannot be justified by experience (nor by intellectual intuition) in the sense of being proven to be true. We cannot be certain of its truth because there is no indicator which could guarantee truth. However, our knowledge about empirical phenomena can be criticized and even falsified by experience. Popper abandons the ideal of certainty. Although we cannot justify knowledge, we can criticize it. Knowledge is not provable but is improvable, improvable through criticism. Instead of the requirement that scientific knowledge should be certain, he suggests the ideal that science should say much, describe and explain a wide range of phenomena, and that it should do so with theories which have been hardened in empirical criticism, i.e. controlled by experiments and observations. The hope of ever attaining certainty in science is a vain hope, but by means of criticism we can find the weak spots of our hypotheses and theories, and sometimes we succeed in improving them, refining them or replacing them by better ones. In this way we may come closer and closer to the truth in the sense that our theories can give us a better and better representation of the relevant aspects of reality. (Cf. Hans Albert's contribution.)

Popper's criticism of inductivism and his proposal of falsificationism as the basic method of science has generated one of the main intellectual confrontations in 20th-century philosophy of science. This introduction is not the place to weigh the balance of this controversy. However, the volume provides deep, probing exchanges if not a stock-taking.

Adolf Grünbaum deals with some inductivist themes in his contribution 'Popper vs Inductivism'. John Watkins comments, "Grünbaum is in the fortunate position of being a belligerent neutral in the battle between Popper and inductivism." (This volume, p. 369.) In the history of philosophy Bacon and J. S. Mill are often regarded as arch-inductivists. Grünbaum, however, shows that there were some falsificationist elements in their philosophy of science. This of course is good news for falsificationism. Adherents of this side in the controversy probably will think that if there were falsificationist elements in Bacon and J. S. Mill, so much the better for them. The other side will perhaps say that some attacks on inductivists have been attacks on strawmen.

Popper has argued that *repeated tests yield diminishing returns* and has shown how this naturally follows from a falsificationist position. Popper has argued that, provided Carnap's formalism is taken seriously, according to inductivism every test should count equally. This is counter-intuitive because in actual research repeated tests do yield diminishing returns. This theme is discussed by Musgrave, who thinks that there is a 'step function': when the results of tests are integrated into the background knowledge, further tests are of no value, i.e. the degree of severity of a test suddenly drops to nil. Musgrave thinks that there is no such thing as a gradual integration of results of tests into the background knowledge, but that at a certain point we decide to accept a particular piece of knowledge as unproblematic. We do so because we are convinced that for the moment there is no reason to doubt the correctness of the experimental results so that performing the test again would be a waste of time.

Grünbaum contends that at least for so-called Bayesian inductivism, the counter-intuitive conclusion that every test counts equally need not follow and hence the argument that there is a diminishing return from the repetition of a test cannot be directed against the Bayesian variant of inductivism. Grünbaum argues that Bayesian inductivism gives a solution to this problem which is as good as falsificationism's.

THE CHALLENGE TO POPPERIAN METHODOLOGY FROM THE KUHNIAN VIEW OF THE HISTORY OF SCIENCE

In 1962 Kuhn's *The Structure of Scientific Revolutions* was published. There Kuhn tried to show that in the history of science there emerge typical patterns. There are periods of 'normal science', in which basic theoretical assumptions are not questioned, but are only applied in order to produce scientific explanations and predictions. During periods of normal science, scientists are, so to speak, in a state of dogmatic slumber with respect to their theoretical frameworks, and hence they can concentrate their efforts on what Kuhn calls 'puzzle solving'. Incidentally, this gives a good example of the well-known phenomenon that a tradition itself can relieve its members from the burden of scrutinizing their basic assumptions.

Kuhn's view of science and in particular his idea of normal science contradict the Popperian view according to which criticism and especially falsification attempts are basic for successful research. For a falsificationist the dogmatic slumber of the 'normal' scientist would be a dangerous thing for science, endangering the growth of knowledge, impairing and eventually making impossible cognitive progress.

Kuhn tried to show that the falsificationist ideal of science is unrealistic. He maintained that his study of the history of science shows that every theory is infected with 'anomalies', but that scientists do not mind, at least not in the period of normal science. This theme was later taken up by Lakatos, who rhetorically claimed that theories are submerged in an ocean of 'anomalies' (or 'counter-examples').[1]

If 'anomaly' is construed as counter-example, this appears to have unpleasant consequences for falsificationism. Every scientific theory, submerged as it is in an 'ocean of anomalies', would have to be regarded as falsified! If the study of the history of science shows that this is actually the case, serious doubts will arise whether falsificationism can provide a realistic view of science. Can falsificationism maintain that there are any important differences between two falsified theories with respect to their achievements as representations of the world? Does it not follow from falsificationism that every falsified theory has to be rejected as false?

These and similar questions were basic in a series of debates which has been going on since 1962 and which usually is subsumed under the heading of the Kuhn-Popper controversy. The most important contribution to this debate so far is the well-known volume *Criticism and the Growth of Knowledge*, edited by Imre Lakatos and Alan Musgrave. These questions have received different answers from different authors. In Lakatos's and Musgrave's volume, Lakatos took Kuhn's challenge to falsificationism very seriously and tried to meet it. Lakatos found Kuhn's account of science to be subjectivistic and psychologistic. In his more rhetorical moments, Imre Lakatos thus labelled it 'mob-psychology', and accused Kuhn of irrationalism, while Kuhn denied his reproach of irrationalism and charged Lakatos with evading history. Incidentally the exchange between Kuhn and Lakatos illustrates how difficult it is to be engaged in an intensive polemic and to stick to one's view: in the various reformulations of their views, Kuhn and Lakatos came closer to each other than perhaps either of them would have liked to admit.

Kuhn had stressed that the belief of a scientist in a theory and the consensus among scientists are important factors for the acceptance of a theory. These are obviously socio-psychological factors dependent on historical contingencies. In his so-called methodology of scientific research programmes Lakatos tried to replace Kuhn's subjectivist and psychologistic criteria by objective criteria. The fight against subjectivism and relativism as well as the concomitant plea for objectivism is common to Popper and Lakatos. Musgrave certainly is right when he writes in his contribution that Lakatos was not an 'epistemological anarchist' (as Paul Feyerabend would have it). However, there are great differences between Popper and Lakatos. From Kuhn, Lakatos adopted the view that in research falsifications are not important, and that falsificationism is an unrealistic view of science.

How, then, should theories be appraised? Lakatos proposed that the 'evidential support' of a theory is decisive and that the unit of appraisal should be the 'scientific research programme', under which a series of theories is produced. In passing it may be remarked that for the inductivist, too, 'evidential support', interpreted as inductive support or degree of confirmation, is essential in the appraisal of theories. In his attempt to secure objective criteria for the appraisal of scientific research programmes Lakatos fell back upon an idea which has some similarity with the idea of inductive support. Herbert Feigl has even argued that Lakatos 'cannot help being a second level inductivist'.[2] Alan Musgrave asks the question, "Have we rescued Lakatos's methodology from the frying pan of anarchism, only to see it cast into hell-fire of inductivism?"[3] and he answers this question in the negative – but only after having given a 'rational reconstruction' of Lakatos's methodology. Musgrave there refers to Feyerabend's claim that Lakatos is an epistemological anarchist, 'a fellow anarchist in disguise' as Feyerabend puts it, and he polemicizes against Feyerabend's claim in such a way that Lakatos appears to become a falsificationist in disguise.

Lakatos's methodology of scientific research programmes is defended and elaborated by Zahar, Worrall, and Urbach in the first part of this volume. One of their main claims is that the concept of 'evidential support' in the methodology of scientific research programmes is in better agreement with the intuitive appraisals of great scientists than corresponding ideas of falsificationism are.[4] In falsificationism theories are accepted if they have withstood severe tests, because in this way they have been highly corroborated.

The main thrust of the Kuhnian challenge was the thesis that, as history shows, most if not all scientific theories have been refuted. Is this thesis historically correct? If one wished to deny this, one could firstly point out that a falsification is always relative to the continued acceptance of the falsifying experimental results. Popperians consider experimental results to be as susceptible to revision as general hypotheses. Secondly, one could point out that what is at stake is never a theory alone, but a theoretical system consisting of the theory and the auxiliary hypotheses necessary for deducing the prediction which has been falsified.

However, in the present volume the Kuhnian thesis is taken for granted – at least for the sake of argument. Alan Musgrave, e.g., declares that he will not dispute that theoretical systems are 'born refuted'. (This volume, p. 183.) J. W. N. Watkins holds that 'it may be questioned' whether we are comparing an unrefuted and corroborated theory with a refuted ('discorroborated') one, and he emphasizes that "no doubt, the actual situation in a scientific debate is often messy, with each of the competing theories attended by theoretical or empirical disadvantages". (This volume, p. 36.) Thus in the compass of the present volume it is assumed that there is a methodologically interesting sense of 'being refuted' in which more or less all important scientific theories have been refuted. The responsibility for this messy situation in the history of science lies with the 'anomalies' to which Kuhn has drawn attention. This is one of the ideas put forward in favour of Lakatos's concept of 'evidential support'.

Lakatos proposed a new unit of appraisal. Instead of emphasizing theories, as Popper did, or 'paradigms', as Kuhn did, he focussed on 'scientific research programmes' (a term which is close to Popper's 'metaphysical research programmes', as N. Koertge points out in her contribution). Having rejected the logical positivists' central question, "When is a theory to be accepted?" (when is it justified, in the sense of having been shown to be true or at least probabilified to a sufficient degree?), Kuhn's central question in the philosophy of science is, "*When should a paradigm be accepted in science?*" His answer consists in pointing out that there often is a consensus among scientists with respect to a particular paradigm. Thus he abandons the logical empiricists' criteria of degree of confirmation or inductive support – which at least were objective – and also the Popperian comparative degree of corroboration as a fallible indicator for the hypothesis that a particular theory is more truthlike than its competitor, which is also objective, for a subjectivistic consensus criterion. Kuhn characterizes the transition by a particular scientist from one paradigm to another as a 'conversion experience' which he views in analogy with a Gestalt switch. He is influenced by Gestalt psychology and above all by the later philosophy of Wittgenstein. Working under a paradigm is much like living in a particular form of life in the later Wittgenstein's sense. The paradigms are incommensurable – much like the monadic forms of life in the later Wittgenstein. They cannot be criticized except from 'within'. The concept of cognitive progress becomes problematic, especially that of the progress from one paradigm to another paradigm.

Lakatos's central question is, "*When is it rational to accept (reject) a research programme?*" In contradistinction to Kuhn, he would accept only objective reasons, objective criteria. What are the guiding ideas Lakatos uses in his appraisal of scientific research programmes? The key notions are increase in empirical content in the sense of successful predictions, and heuristic power. Under a scientific research programme a series of theories is developed. A scientific research programme is better than its rivals (hence to be preferred) if the sequence of theories produced under it shows a greater increase of content than the sequence of theories produced under the rival programme. Lakatos speaks of 'excess empirical content'[5] and explains it as the prediction of new, hitherto unexpected facts. The crucial idea for Lakatos is that of 'evidential support'. (This key concept is discussed below by Worrall, Urbach, Zahar, Musgrave, Koertge and Post.)

According to Popper the explanation of a known fact does not 'support' a theory. If a fact is known, it belongs to the background knowledge, and a fact belonging to the background knowledge cannot corroborate a theory. This follows from the definition of corroboration. The reason is that the investigation of a known fact cannot constitute a test of the theory in question since the outcome is known in advance of the 'test'. But according to the methodology of scientific research programmes the explanation of a known fact does indeed 'support' a theory. The adherents of the methodology of scientific research programmes regard this as an important advantage of their position over falsificationism.

A falsificationist could answer that although the explanation of a known fact by a particular theory certainly cannot increase its degree of corroboration, since the degree of corroboration is intended only as a measure of the severity of the tests a theory has withstood, for an observed fact to count as 'evidential support' for a theory that fact must not be explained 'ad hoc' with the help of that theory. A major problem for the methodology of scientific research programmes is *how to rule out ad hoc evidential support*. Following E. Zahar's proposal, Worrall suggests that those facts which have been used in the construction of the theory to be appraised cannot "support" the theory. (Incidentally, the term 'support' carries semantic overtones from justificationalism.) Worrall must show, and attempts to do this in his paper, that it can be objectively determined whether or not a fact has been used in the construction of a theory. If 'used' could not be objectively defined, then the immediate objection to the proposal would be that the issue of whether or not a fact has been used in the construction of a theory by a particular scientist is a historical contingency, and may depend on such chance factors as whether or not the scientist knew about this fact. (Cf. Koertge who criticizes Worrall's position paper on these lines.) Lakatos always fought subjectivism and psychologism and has warned against the degenerating problem shift which consists in shifting from the problem of theory appraisal to the problem of producer appraisal. Worrall maintains that it can be objectively determined whether a fact was used in constructing a theory. His main examples are facts used in order to determine the values of parameters. Zahar gives an example of such a determination of a parameter in his case study of Kaufmann's experiment.

The problem of 'evidential support' is also discussed, although from a more Popperian point of view, by Alan Musgrave. For Popper, of course, a 'fact' can corroborate a theory only if it does not follow from the background knowledge of that theory. If it followed from the background knowledge, we would not be entitled to speak of testing at all. Thus, if we construe 'evidential support' as Popperian corroboration, then only a real test can support a theory. The problem of determining the severity of a test is obviously connected with the earlier problem of ruling out 'evidential support' that is *ad hoc*. If a fact is used in an objective sense, e.g. in determining a parameter of a theory, then the same fact can only be explained *ad hoc* by that theory – a fact cannot be used twice. Thus the problem of explicating '*ad hoc*ness' is related to the problem of 'evidential support' and to the problem of *explicating the concept of 'background knowledge'*.

In an earlier publication, Alan Musgrave distinguished among three senses of 'background knowledge'.[6] (1) One can adopt a temporal view according to which everything known at a certain time constitutes the background knowledge for that time. Something is said to be 'known' if it has been severely tested or is considered unproblematic. (2) One can use a theoretical construal, sometimes also called the 'touchstone theory' construal: the background knowledge in the relevant sense is the best rival of the theory to be assessed. This sense is the one relevant when one is appraising the comparative achievements and merits of two competing theories. (3) There is also a heuristic interpretation, which is the one Worrall uses: a fact is part of the background knowledge if it has been used in the construction of a theory – if this fact is indispensable for formulating the theory. In his contribution Musgrave argues that (2) and (3) are complementary (this volume, p. 186), a view contested by Worrall, who interprets Musgrave's move as a 'conventionalist strategem'. (This volume, p. 332.)

Imre Lakatos had two main criteria for the appraisal of scientific research programmes. One we have already mentioned: increase in empirical content, at least some of which has stood up to (more or less severe) empirical testing. The other is the idea that a research programme is better than a competitor if it has more heuristic power, potential for creating 'evidential support', i.e. if it makes possible the construction of a sequence of theories with increasing 'evidential support', at least if viewed overall, in retrospect. Roughly, a research programme is fruitful if it can lead to a series of theories which produce novel knowledge, potential knowledge which when tested severely stands up to the test. It appears that heuristic power is scarcely more than the capacity to generate increase in successful predictions, content of empirical information that gets corroborated. In his contribution Urbach attempts to show that the heuristic power of a research programme can be objectively determined.

Both Urbach and Worrall are impressed by Paul Feyerabend's criticism that heuristic power cannot guarantee success. Feyerabend has often pointed out that a research programme which has been 'progressing' up to now, i.e. has made a good showing with respect to both heuristic power and empirical support (which boils down, as we have seen, to 'empirical support'), may begin 'degenerating'. Of course – Lakatos's criteria are intended only for use *ex post*. (It is much the same as with stocks: a stock which up to now has been an excellent investment might suddenly stop being profitable and may even turn into a catastrophic investment.) Lakatos's criteria for backward-looking assessments of profitability in terms of gains in novel knowledge could be used for forecasting future profitability only if one were willing to add the assumption that roughly the future will be like the past at least so far as this research programme is concerned. But then we would be confronted with the successor problem of defending such an inductivistic principle. (Cf. H. Feigl's characterization of Lakatos as a second-level inductivist.) Avoiding the insoluble problem of defending a principle of induction was one of the main tasks of Popper's methodology. Lakatos appears to have been ambivalent on this point. In later writings he became sceptical of the possibility of methodology's giving any advice with respect to strategies that might facilitate cognitive progress. (Of course advice is even more fallible than appraisal.[7]) Feyerabend, although he thought that "here poor Imre was too modest" (personal letter), was quick to point out that Lakatos at least originally had very much wanted to give advice also in matters of research-policy-making based upon the methodology of scientific research programmes, but that, unless one made the unthinkable inductivist move, such a promise would be overselling. Hence Feyerabend greeted his friend Lakatos as a fellow anarchist, but an anarchist in disguise. Musgrave, however, sees it in a different light: "What he [Lakatos] has done with his notion of 'heuristic power' is to give us a falsificationist account of what it is to develop a theory and defend it against criticism. He has given us, in a nutshell, a Popperian account of Kuhn's 'normal science'." (This volume, p. 190.)

PROBLEMS OF FALSIFICATIONISM

In order to make testable predictions we have to use auxiliary hypotheses. Now suppose that such a prediction turns out to be false. The theoretical system consisting of the theory under test plus the auxiliary hypotheses necessary for deducing the prediction has been falsified. Where to put the blame? The problem has become known in the literature as the *Duhem problem*. In this volume Elie Zahar and Noretta Koertge deal with it, in both cases by means of illustrative *case studies*.

In support of the hypothesis that in the history of science 'all theories are falsified' Kaufmann's experiment is often cited as an experiment falsifying the special theory of relativity.[8] Zahar makes a detailed case study of this experiment and shows, as was pointed out by Planck in 1906, that the alleged falsification *de facto* was no falsification at all because there were some independent reasons to doubt one of the auxiliary hypotheses involved. Earlier it was thought that Kaufmann's experiment was a crucial experiment between classical and relativity theory; it had the appearance of an *experimentum crucis* because a classical assumption had tacitly been used as an auxiliary hypothesis. Zahar shows in his paper that when the crucial auxiliary hypothesis had been modified appropriately, both the classical theory and relativity theory could explain Kaufmann's experiment, and that thus the experiment was not a crucial experiment at all. Zahar then discusses whether or not the modification of the auxiliary hypothesis was *ad hoc*. He comes to the conclusion that according to Popperian falsificationism it was, but that according to Zahar's own version of Lakatos's methodology of scientific research programmes and its definition of *ad hoc*ness (fact 'used' in the construction of the theory under appraisal) it was not *ad hoc*. His reason for regarding the modification of the auxiliary hypothesis as '*ad hoc*' is that it was made with a view to saving the theory from falsification. Here too, one gets the successor problem of explicating in objective terms 'modified with a view to saving the theory' because any recourse to the intentions of the particular scientists would steep us in psychologism, and would be irrelevant for methodology.

N. Koertge examines the Duhem problem with the help of a case study of Mendeleev's periodic system. She analyses two cases of falsification with which Mendeleev was confronted. In one of the cases Mendeleev remedied the situation by auxiliary hypotheses, in the other by modifying the theory, the periodic system. Thus Koertge's case study can illustrate the two main possible procedures and so put into focus the problem of "when to change the theory and when to change the auxiliary hypotheses" and of what good reasons can be mustered for each of the strategies. She gives a suggestion for which strategy is to be used with the help of Bayes's formula and of plausibility appraisals, in answer to the question, "When is it plausible that a 'fact' 'disconfirms' the theory and when plausible that it 'disconfirms' some of the auxiliary hypotheses?"

Musgrave's paper deals with the Duhem problem also. He argues that Duhem did not hold that every falsification could be explained away by adjusting the auxiliary hypotheses, but that Duhem argued that different strategies are possible. Which strategy is chosen will depend not only upon the sensitivity of the researcher in judging the type of research situation at hand, but also on his personality traits: cautious thinkers will prefer conservative strategies, restoring the balance by minor changes, while bold thinkers will tend to revise the basic theory, or to devise entirely new systems. (This volume, p. 187.) Where Popper has improved upon Duhem is in adding the requirement that modifications, be they minor or major, must not be *ad hoc*.

Lakatos's solution to the Duhem problem was the requirement that the arrow of the modus tollens of a falsification not be directed towards the 'hard core' of a research programme. *Prima facie*, this appears to amount to a rejection of falsification. However, Musgrave is able to show that, surprisingly enough, this is not the case. Lakatos, like Popper and Feyerabend, insists on pluralism, on the free market with full competition, in demanding that science contain competing research programmes. Hence an 'anomaly' or falsification can lead either to modification of the auxiliary hypotheses involved or to the replacement of the research programme by a new one. Thus the original problem of when to modify the one and when the other *returns* in an only slightly modified version: when should we retain the research programme ('save it' by introducing a new auxiliary assumption inspired by the research programme's positive heuristic) and when should we replace the programme by a new one? Musgrave argues that Lakatos has not been able to give us more than Popper has. He just rephrases the problem in terms of his own unit of appraisal, the research programme.

In view of this it appears reasonable to take a second look at Popper's old solution, at his guiding ideas for theory comparison, and in particular to examine how we could manoeuvre with them if, for the sake of argument, we have accepted the Kuhnian thesis that every scientific theory has been falsified. The key concepts relevant to this undertaking are those of *truthlikeness*, *corroboration*, and *content comparison*.

The central question is, "Which of two competing theories is better than the other and hence to be preferred?" To answer it we have to explicate the concept of 'better than' in such a way that the explicatum is fruitful for methodology and then we have to provide indicators (fallible, but objective) for ascertaining in a concrete case which of the two particular theories exemplifies the property of 'being better' as explicated. Lakatos merely changed the unit of appraisal from theories to scientific research programmes. The gist of his answer to the above question was that a research programme is better than its rival if it has more 'evidential support' and more heuristic power, i.e. power to create theories with 'evidential support'. Thus everything hinges, again, upon the idea of 'evidential support'. We also saw that Lakatos's answer was motivated by Kuhn's criticism of falsificationism as an unrealistic model for stylizing the history of science.

Let us recall what solution falsificationism offers to the problem of theory comparison. In her paper Koertge offers a list of criteria for appraisal before testing and appraisal after testing. Among the post-testing criteria are "minus-marks for being inconsistent with the experiment" and "plus-marks for having passed severe tests". (This volume, p. 271.) Popper has suggested degree of corroboration as an index of how severely a theory has been tested and how well it has passed these tests. Thus, degree of corroboration is a sort of *ex post* resume of how successful predictions have been up to the time of the evaluation. Interestingly, the degree of corroboration does not function as a means of appraising the comparative achievements of two competing theories if both theories are refuted and if moreover we accept a formalism according to which all refuted theories get the degree of -1. Therefore, if we accept both the Kuhnian thesis that most, if not all, scientific theories have in fact been refuted, then, if we also accept the formalism, we must conclude that most, if not all theories have the degree of corroboration -1, i.e. they are equally bad. If so, in most cases the concept of degree of corroboration would not enable us to give good reasons for preferring one theory to its rivals.

In his theory of *verisimilitude*, Popper explicates the intuitive idea of truthlikeness, the idea that a particular hypothesis may give a better representation of certain aspects of reality than some particular competing hypothesis. Popper allows that a theory can have false consequences, a certain falsity content. If two theories are both refuted in that each has at least one known false consequence, one of the pair may convey less false information and more true information than its competitor, and hence it seems that two false theories could

be compared with respect to their truthlikeness, as explicated by
Popper's formal concept of verisimilitude. Thus it appears that
Kuhn's challenge could be met with the help of Popper's formal
explicatum of truthlikeness.[9]

However, Popper's explicatum has come under criticism from D.
Miller and P. Tichý. G. Andersson in his contribution investigates
whether this criticism can be met, and he comes to the conclusion
that at the present stage of the discussion it cannot be. But Andersson
argues that the formal criticism of the theory of verisimilitude should
not prevent us from using the intuitive idea of truthlikeness. This idea
is (as H. Albert argues in his contribution) indispensable in any
language.[10] The criticism of Popper's theory of verisimilitude is only a
criticism of a particular formal definition used to explicate truthlikeness. If we assume that it will never be possible to give a definition
couched in an ideal language, in standard logical formalism, this still
does not justify any conclusion about the fruitfulness of the explicandum, the intuitive idea that one hypothesis can represent certain
aspects of reality more precisely than a certain rival hypothesis does.
Today many logicians of science are trying to give an adequate formal
definition of verisimilitude. D. Miller has recently criticized various
re-definitions of verisimilitude in a convincing way.[11] These investigations suggest that truthlikeness may not be explicable through
a definition couched in a formalized language.[12]

Watkins and Post try to give an alternative to 'corroboration' and
'verisimilitude'. Or, more accurately, they attempt to show that at
least the idea of comparing empirical content – a pre-testing concept –
can be rescued. (Empirical content, according to Popper's original
idea, is identical with degree of falsifiability.) Watkins relates empirical content to how many scientific questions a theory can answer. He
tries to show that it is possible to compare empirical contents, in
particular that a sort of set-theoretical comparison of empirical
contents is possible with the help of the idea of 'counterpart
consequences' (different answers to the same question).

Feyerabend and Hübner in contrast object that content comparison
is impossible because contents of theories are not subsets of each
other. This argument is also discussed by Grünbaum. Of course, if
different theoretical entities are presupposed by two theories, then the
theories see the world in different ways. Feyerabend's main argument
hinges on the thesis that different forms of life, even in science –

different ways of looking at the world – are not comparable. Feyerabend's position has affinities with both the later philosophy of Wittgenstein and with the approach of phenomenology. (The criticism of the formal definition of verisimilitude uses a similar idea: truth contents and falsity contents are not comparable.)

In this volume cognitive progress has been the center of concern: what we should mean by that term and how we can recognize whether or not in a concrete case there is such progress. The volume focusses on one dimension: the *epistemic dimension*, the appraisal of theories with respect to how correctly they represent certain aspects of reality. This is perhaps the main dimension and the one that has to be investigated before one can hope successfully to attack other dimensions. In the Popperian tradition the necessity of appraisals in other dimensions is clearly recognized, for example in Popper's thesis that progress essentially consists in going from problems to 'deeper' problems.

It seems justified to hope that in this volume the topical problems of Popperian, Lakatosian as well as logical empiricists' methodologies have been developed to a higher level of understanding, even if – as according to the Popperian view should happen – in solving one problem new ones are discovered.

NOTES

[1] (Lakatos, 1970) p. 133.
[2] (Feigl, 1971) p. 146.
[3] (Musgrave, 1976) p. 480.
[4] Of course these intuitive appraisals must be endorsed by the methodologists because otherwise there is the danger of a naturalistic fallacy, of concluding from facts to prescriptions. Cf. (Radnitzky, 1976), §1.14.
[5] (Lakatos, 1970) p. 116.
[6] (Musgrave, 1974).
[7] (Radnitzky, 1979), §0.2.
[8] (Feyerabend, 1975) p. 56, n. 6.
[9] (Popper, 1963) ch. 10, (Popper, 1972) chs. 2 and 9, and (Popper, 1976).
[10] (Radnitzky, 1978a).
[11] (Miller, 1976).
[12] In the opinion of one of the present writers the discussion of versimilitude has an aura of *déja vu*: its whole ambience is the same as that of the almost classical discussion of the explication of 'empirical significance' by logical empiricists; in Lakatos's terminology, it is monster-producing and monster-barring. If so, this episode may, once again, illustrate the truth of the (Hegelian) saying that from history one can only learn that it is not possible to learn from history. G. R.

BIBLIOGRAPHY

Andersson, G. and Radnitzky, G.: 'Le progrès de la connaissance: Où en sont les théories de la science?', *Archives de Philosophie* **39** (1976), 619–628.

Buck, R. and Cohen, R. S. (eds.): In *Memory of Rudolf Carnap. Boston Studies in the Philosophy of Science*, Vol. 8, D. Reidel, Dordrecht 1971.

Cappelletti, V. and Grmek, M. (eds.): *On Scientific Discovery. Boston Studies in the Philosophy of Science*, D. Reidel, Dordrecht, forthcoming 1979.

Cohen, R., Feyerabend, P., and Wartofsky, M. (eds.): *Essays in Memory of Imre Lakatos. Boston Studies in the Philosophy of Science*, Vol. 39, D. Reidel, Dordrecht 1976.

Feigl, H.: 'Research Programmes and Induction', in Buck and Cohen, 1976, pp. 147–150.

Feyerabend, P.: *Against Method. Outline of an Anarchistic Theory of Knowledge*, New Left Books, London 1975.

Feyerabend, P.: *Science in a Free Society*, Vol. II of *Against Method*, New Left Books, London 1978.

Feyerabend, P.: *Rationalism and the Rise of Science*, Cambridge University Press, London, forthcoming 1979.

Lakatos, I.: 'Falsification and the Methodology of Scientific Research Programmes', in Lakatos and Musgrave, 1970, pp. 91–195.

Lakatos, I. and Musgrave, A. (eds.): *Criticism and the Growth of Knowledge*, Cambridge University Press, Cambridge 1970.

Marek, G. and Zelger, I. (eds.): *Österreichische Philosophie und ihr Einfluß auf das analytische Denken der Gegenwart*, Conceptus-Verlag, Innsbruck 1978.

Miller, D.: 'Verisimilitude Redeflated', *British Journal for the Philosophy of Science* **27** (1976), 373–381.

Musgrave, A. E.: 'Kuhn's Second Thoughts', *British Journal for the Philosophy of Science* **22** (1971), 287–297.

Musgrave, A. E.: 'Falsification and its Critics', in Suppes *et al.*, 1973, pp. 393–406.

Musgrave, A. E.: 'Logical Versus Historical Theories of Confirmation', *British Journal for the Philosophy of Science* **25** (1974), 1–23.

Musgrave, A. E.: 'Method or Madness?', in Cohen *et al.*, 1976, pp. 457–491.

Popper, K.: *Conjectures and Refutations*, Routledge & Kegan Paul, London 1963.

Popper, K.: *Objective Knowledge. An Evolutionary Approach*, Oxford University Press, London 1972.

Popper, K.: ' A Note on Verisimilitude', *British Journal for the Philosophy of Science* **27** (1976), 147–159.

Radnitzky, G.: 'Popperian Philosophy of Science as an Antidote Against Relativism', in Cohen *et al.*, 1976, pp. 505–546.

Radnitzky, G.: 'Philosophie und Wissenschaftstheorie zwischen Wittgenstein und Popper', in Marek und Zelger, 1978, (1978a).

Radnitzky, G.: 'Die Sein-Sollen-Unterscheidung als Voraussetzung der liberalen Demokratie', in Salamun, 1978, (1978b).

Radnitzky, G.: 'Rationality and Progress in Research', in Cappelletti and Grmek, 1979.

Radnitzky, G. and Andersson, G. (eds.): *The Structure and Development of Science*, D. Reidel Publ. Co., Dordrecht, Holland, forthcoming 1979.

Rescher, N.: 'Some Issues Regarding the Completeness of Science and the Limits of Scientific Knowledge', in Radnitzky and Andersson, 1979.

Salamun, K. (ed.): *Sozialphilosophie als Aufklärung. Festschrift für Ernst Topitsch*, J. C. B. Mohr (Paul Siebeck), Tübingen 1978.

Stegmüller, W.: 'Accidental (Non-Substantial) Theory Change and Theory Dislodgement', in Radnitzky and Andersson, 1979.

Suppes, P., Henkin, L., Moisil, Gr., and Joja, A. (eds.): *Logic, Methodology and Philosophy of Science. Proceedings from the Fourth International Congress for Logic, Methodology and Philosophy of Science, Bucarest, 1971*, North-Holland Publishing Company, Amsterdam 1973.

PART I

THE LSE POSITION

JOHN WATKINS

THE POPPERIAN APPROACH TO SCIENTIFIC
KNOWLEDGE

1. THE BACON-DESCARTES IDEAL

Criteria for scientific progress?[1] The Popperian tradition is opposed
to *criterion-philosophies*.[2] If we had come brandishing criteria we
would immediately have been asked what their *authority* was. Since
we hold that there is no (extra-logical) certainty either inside or
outside science, we would have had to admit that they are fallible and
that, in the event of a clash between our 'criteria' and scientific
practice, it might be our 'criteria' that were wrong. Then why, we
might very properly have been asked, set up as law-givers to science
if you do not believe in your own law?

Yes, we have no criteria. What we do have is a corrigible and
revisable methodology of scientific appraisal. (That it is corrigible and
revisable is shown by the fact that it has been quite extensively
corrected and revised during the last four decades, both by Popper
himself, and by Lakatos and others.) This methodology has two
unusual features: when applied to exemplary cases of scientific progress, to cases where everyone would agree what the right appraisal
would be, this methodology does actually yield the right appraisal.
And when, in its latest version as the methodology of scientific
research programmes, it is applied to more obscure and messy situations, it is capable of coming up with appraisals which are at once
revealing and unequivocal.[3]

But there are other methodologies and other philosophies of
science; and some of these have been developed by men who had an
intimate understanding of important areas of science and its history;
Mach, Peirce, Poincaré, Meyerson, Duhem, Campbell, Bridgman,
Schlick and other members of the Vienna Circle, not to mention more
recent writers. Are there any general arguments for preferring ours to
theirs?

Lakatos in [1971] proposed a quasi-historical meta-method 'for the
evaluation of rival methodologies' (p. 122). (On the ground that 'one

A unified bibliography can be found on pp. 379–383.

G. Radnitzky and G. Andersson (eds.), Progress and Rationality in Science, 23–43.
All Rights Reserved.
Copyright © 1978 by John Watkins.

must always stop somewhere' he declined to offer a meta-meta-method for the evaluation of rival meta-methods [1974, p. 251].) But we might be accused of a kind of self-reinforcement if we invoked that here. Also, Lakatos's 'more sophisticated answer' was indeed rather sophisticated. Let us try an *un*sophisticated approach.

Since Hume, any serious philosophy of scientific knowledge has been bound to contain *some* pessimistic elements: "Never glad confident morning again". (It might be objected that Kant's heroic attempt to repair the damage done by Hume resulted in an essentially optimistic epistemology: after all, it gave us *episteme*. True; but things-in-themselves fell into unknowability and the world, in so far as we understand it *a priori*, turned out, disappointingly, to be a world that *we* manufacture. There *was* a deeply pessimistic side to Kant's Copernican Revolution.)

This suggests a possible thought-experiment which might help us in the assessment of rival methodologies. Let us try to think ourselves back to the glad confident morning of modern science, back to the Bacon-Descartes era. Let us pretend, for a moment, that we know nothing of Hume's (or anyone else's) logical undermining, that we are in a state of epistemological innocence and optimism. What then would we have wanted, ideally, from science (that is, from pure science: we are not considering technological applications)? What then would we have taken the ideal aim of science to be? Having arrived at a naively optimal ideal for science in this way, we can then reintroduce the well-known difficulties. The object of this intellectual exercise will be to enable us to pose the following question: which extant methodology, if any, captures most of our utopian dream without running into known difficulties?

So let us, as it were (this is being written on Christmas Eve), compose a letter to Santa Claus telling him what sort of science we would like to have.

Well, in the first place we will surely want it: (1) to consist of *truths*; and not just of trivial truths but of *deep* truths, truths which *explain* a lot. Indeed, while we are about it we may as well go the whole way and ask that science should get to the very bottom of things, unlock nature's innermost secrets, give us *ultimate* explanations.

That, presumably, would be our main message; but there are some further desiderata which we would want to add. (2) Suppose that science were actually to achieve all this, only we did not realise that it had done so. Surely, it would be even more satisfactory if our science were not only true (deep etc.) but *known* to be true. (3) Suppose that our science became vaguer as it went deeper. That would be most disappointing. We will want it to give us *exact* truths at all levels. (4) We will not want there to be anything loose or slipshod in the way our science connects its statements at different levels; we will want these connections to be strictly logical or *deductive*.

Such an ideal would not have seemed absurd in the era in which we are imagining ourselves. The dream of doing for the physical world what Euclid was taken to have done for space persisted well into the seventeenth century (remember Spinoza's *Ethics*). Indeed, Newton's system was seen by many as a partial fulfilment of that dream.[4] Bacon, with his idea of an ultimate alphabet of nature which can be discovered by methodical empirical interrogation, and Descartes with his idea of the deductive unfolding of the consequences of infallible first principles, both subscribed to something like our ideal. (Bacon came almost as close as Descartes did to accepting (4) above. Both men rejected the only *formalized* deductive logic available at the time, namely Aristotle's syllogistic. For Descartes, 'deduction' consisted of making very short steps, one at a time, each of which is intuitively valid. For Bacon, 'induction' likewise consisted of a series of short steps, and was supposed to 'lead to an inevitable conclusion': it too was intended to be truth-preserving.) Our naive ideal is really an amalgam of ideas of theirs. We might call it the Bacon-Descartes ideal.

Snapping ourselves back into the latter half of the twentieth century we see at once that most of the individual components of this ideal are separately unattainable while some are mutually antagonistic. In particular, the demand for *depth* and the demand for *certainty* pull in opposite directions. (The deeper we try to penetrate beneath the surfaces of familiar objects that we can see or touch, the more conjectural our ideas are likely to be.) This ideal is bi-polar: it has the aim of deep (or ultimate) explanation at one pole, that of certainty at the other. And this polarity was reflected in a polarization within late nineteenth and twentieth century philosophy of science.

After Hume – or rather, after the breakdown (not yet admitted by Whewell or even, it seems, by Meyerson) of Kant's attempt to answer Hume – no one could persist with a demand for absolute certainty in science, just as no one could persist with a demand for ultimate explanations. But the great majority of philosophers of science during this period seem to have been attracted by the certainty-pole or repelled by the depth-pole. Some Catholic thinkers may have been repelled by the depth-pole because they wished to preserve deep-fishing rights for theology. But in most cases, philosophers of science seem to have been repelled by depth because attracted by certainty.

An anti-depth theme has persisted through much philosophising about science since the last decades of the nineteenth century.[5]

Mach's sensationalism (like Berkeley's idealism before it) is an empiricist metaphysics that excludes any possibility of science penetrating beneath the phenomena. For Mach, a physical theory is a pedagogically convenient device for packaging large quantities of observational material into economical formulas. Poincaré's case is less clear-cut; but he did interpret the more deep-seated components of physical theories as conventions which acquire a kind of incor-rigibility at the cost of ceasing to describe the world. Duhem's main tendency (there may have been other tendencies in his thought) was to regard a physical theory as a piece of mathematical apparatus which 'represents' (but without explaining) a set of experimental laws. Experimental laws describe the world; a good theory achieves a 'natural classification' of experimental laws; but the theory does not describe the world: "Concerning the very nature of things, or the realities hidden under the phenomena we are studying, a theory conceived on the plan we have just drawn teaches us absolutely nothing" [1906, p. 21]. Bridgman's operationalism carried the implication that even the most theoretically recondite parts of physics, provided they are meaningful, actually concern things that we can manipulate. Ramsey won the gratitude of later empiricists by inventing a method for taking the theoretical content out of theories while leaving their empirical content intact.

The Vienna Circle's verification-principle was a modernised, meta-level version of the demand for certainty. And it was accompanied by something like a horror of depth. In his [1928] Carnap declared:

If a physical object were irreducible to sensory qualities... , this would mean that there are no perceptible indicators for it. Statements about it would be suspended in the void; in science, at least, there would be no room for it. (p. 92)

Soon after this Neurath persuaded Carnap to shift to a physicalist observation language. But Neurath's physicalism was little more than

Mach's monist sensationalism under a new flag: "Physicalism knows no 'depth', everything is on the 'surface' " [1931, p. 326; in the original this sentence was italicised]. This pronouncement by Neurath echoed an earlier pronouncement in the 'Weiner Kreis' manifesto: "In science there are no 'depths'; there is surface everywhere".

The majority response to the certainty-depth polarity was unambiguous: discard the idea of depth in favour of the best available approximation to certainty.

And the minority? Peirce, despite certain tendencies which anticipated operationalist and logical empiricist theories of meaning, was a fallibilist who regarded science as helping to provide a general cosmology.[6] Clifford, who took a cheerfully realistic view of atoms etc., rejected the possibility of certainty, emphasising that our 'knowledge' of the external world relies upon various interpretative principles, some at least of which are logically unjustifiable; but, he added, they are necessary for our survival, and our possession of them can be explained (not justified) in terms of evolutionary selection. According to Clifford, the idea of the uniformity of nature is such a principle: "Nature is selecting for survival those individuals and races who act as if she were uniform; and hence the gradual spread of that belief over the civilised world." [1886, p. 209].

Meyerson was a scientific realist; but we can hardly include him here; for he did not renounce certainty in favour of realism. He was also a Kantian of kinds who regarded the various conservation-principles of his day as possessing, if not actual certainty, at least a special kind of *plausibility*.

It seems that the first *full-fledged* opposition to the majority view was Popper's [1934], in which the ideal of certainty is discarded altogether, in favour of scientific depth, explanatory power, and severe testability. On Popper's view, not only does a theory remain conjectural forever, no matter how successfully it may have passed very exacting tests, but the 'basic' statements against which it is tested also have a conjectural status, their (provisional) acceptance being the result of a conventional agreement. There is no empirical certainty at any level of science.

As so far stated, the situation might appear to be this. The Bacon-Descartes ideal turned out to be bi-polar. One philosophical party – we might call it the 'verificationist-phenomenalist' party – was drawn by one pole, while the 'conjecturalist-realist' party was drawn by the other pole. Each party achieved an approximation to its preferred part of the original ideal (scientific certainty, ultimate explanations); and the honours are about even.

As we see it, however, the honours are by no means even. The sacrifice of depth *was to no avail*: it did not bring the realization of other parts of the Bacon-Descartes ideal any closer. Whereas the sacrifice of certainty, and of all substitutes for it (such as high probability), makes possible the realization of surprisingly much of other parts of the Bacon-Descartes ideal; or so I will argue.

Why was the sacrifice of depth to no avail? Let T be some empirically powerful and mathematically exact scientific theory; let E be all the evidence known at a certain time that is relevant to T; and assume that all this evidence is favourable. Then verificationists and quasi-verificationists want to regard T as in some sense established, or justified, or rendered probably true, by E.

Now from a realist point of view, T is, so to speak, enormously top-heavy relative to E. For one thing T is universal: it makes assertions about all spatio-temporal regions of the universe; whereas E consists of a finite number of singular observation-reports. Again T is, or at least its fundamental postulates are, mathematically *precise*, whereas E reports measurements that are accurate only within the limits of experimental error.[7] Finally, T postulates a *theoretical on-tology* concerning 'the realities hidden under the phenomena' (to use Duhem's phrase, quoted earlier); and there is no mention of this ontology in E. In short, T transcends E in three distinct ways.

Then the problem, for the verificationist or inductivist who wants to retain some substitute for the old ideal of certainty, is to re-interpret T in a way that cuts it down to a manageable size so that it is no longer so top-heavy relative to E.

His main move here, as we know, is to de-ontologize T by interpreting it in some phenomenalist or otherwise radically empiricist way, so that it speaks only about the surfaces of things. (He might try to achieve this by replacing T by its Ramsey-sentence.)

But this, by itself, eliminates only one of the three ways in which T transcends E. Mach's phenomenalist interpretation, for instance, did not deprive scientific laws of their universality and mathematical exactitude. Mach declared: "The laws of nature are equations between the measurable elements $\alpha \beta \gamma \delta \ldots \omega$ of phenomena" [1912, p. 605]. And such equations he supposed to be 'mathematically exact and generally valid' (p. 306). It would seem that, so long as T retains these two characteristics, the inductivist cannot justifiably hold that E logically implies a finite proportion of the content of T.

Hempel in [1945] introduced an ingenious idea for overcoming the problem of universality. But in making something easier for oneself in one direction one sometimes makes it harder in another direction; and so it was in this case: Hempel's solution of the universality-problem rendered the precision-problem intractable.

His idea was this: given a universal T and a finite E, E will be said to confirm T if E logically implies what T would have said if the world were assumed to consist just of those objects that are mentioned in E. This idea has the great advantage that it succeeds both in retaining the *universality* of T and in construing the confirmation of T by E as a *deductive* relation (by interposing between E and T the 'development', as Hempel called it, of T for E); so this idea satisfies desideratum (4) of our Bacon-Descartes ideal.

But it also has a great disadvantage. Let T be a theory whose development for E is mathematically precise. Then, given that there are limits to observational accuracy, there will be another possible theory, say T', whose development for E is equally precise and observationally indistinguishable from but logically inconsistent with that of T. Now E (which we assume to be self-consistent) obviously cannot logically imply *both* developments, since they are mutually inconsistent; nor can E favour *one* of them, since they are observationally indistinguishable. Hence E cannot confirm either of them. Thus theories of the kind that have been the glory of science at least since the days of Galileo and Kepler become unconfirmable, on this view. Hempel's [1945] theory of confirmation fails to satisfy desideratum (3) of our Bacon-Descartes ideal.

It has been more usual, within the logical empiricist movement to which Hempel belongs, to attempt to construct a confirmation-function with the help of probability logic. Popper's decisive negative results against this approach are now sufficiently recognised not to need to be rehearsed here.[8] They boil down to this: if 'probability' is interpreted in accordance with the probability calculus (as it was by Keynes, Reichenbach, and Carnap), then the probability of a hypothesis relative to given evidence varies inversely with the content of the hypothesis and more especially with the extent to which its content goes beyond the evidence. Whether or not a universal law-statement, which infinitely transcends any finite evidence statement, must always have a probability of zero, as both Carnap and Popper supposed,[9] to aim at high probability would be to aim at hypotheses that are highly *ad hoc* relative to the evidence.

Thus to interpret T in some de-ontologising way as a hypothesis which merely correlates and predicts observable phenomena, fails by itself to yield a substitute for certainty, such as inductive confirmation or high probability. Depth has been sacrificed without any corresponding gain.

Some philosophers, perceiving that it is not enough to take just the theoretical content out of scientific theories, have resorted to the desperate step of taking all the factual content out of them. (This group includes Wittgenstein, Schlick, Ryle, and various later writers.) In other words, they denied that a universal scientific hypothesis is a true-or-false description of the world and re-interpreted it as an inference-ticket that licenses us to infer new singular observation-statements from suitable sets of old ones.

Precious little of the Bacon-Descartes ideal survives in *that* view of science. (Incidentally, in allowing singular but not universal statements to be true-or-false, it involves a peculiar logic: for if the singular statement 'Fa' is true-or-false, then so, presumably, is its existential generalisation '$\exists x \, Fx$'; and if '$\exists x \, Fx$' is true-or-false, then so, presumably is its negation, namely the *universal* statement '$\forall x \sim Fx$'.)

The pattern of the foregoing development seems to have been this. Some approximation to scientific certainty was desired. But it was recognised that science, interpreted realistically, is enormously top-heavy relative to its empirical base: so it needed to be cut down to a manageable size so that it could then be adequately supported by its base. But attempts to cut it down either did not go far enough, or went too far and reduced theoretical science to a kind of inference licensing authority.

The situation changes dramatically if we abandon the inductivist hope of re-transmitting truth from verified conclusions to scientific premises, and focus instead on the re-transmission of falsity from falsified conclusions.

There are those for whom the words 'Leave all hope behind' are inscribed over Popper's epistemology. Well, it does mean abandoning all hope of any kind of cognitive (as opposed to merely psychological) certainty in science. But if we can reconcile ourselves to this loss we can retain much of the rest of the Bacon-Descartes ideal.

I will take the desiderata in that ideal in reverse order. First, with regard to (4), we can now afford to be thorough-going deductivists. Our scientific inferences will proceed either from conjectural premisses to a testable conclusion, or from a tested and refuted conclusion to the negation of the set of premises from which it was derived: in either case we will be proceeding non-ampliatively from a proposition with more content to one with less.

Next, with regard to (3), we can *welcome* exactness in science; for an exact hypothesis can be tested against inexact measurements. It will generally be the case that the more precise hypothesis is also the more testable. That is why we value precision We do not worship precision for its own sake. Indeed, there are some kinds of precision which we regard as pointless or even as unfortunate. For example, replacing the antecedent clause of a universal hypothesis by a more precise one will normally reduce testability.

Passing over desideratum (2), the demand for certainty, we come to desideratum (1). As to ultimate explanations: well, our view does not actually rule out the logical possibility that science may one day hit upon an explanatory theory which is in fact, though no one could know this, at once true and not logically derivable from any true unified theory that is more general than it. But what our view really does is to substitute for the utopian idea of ultimate explanations the realistic idea that science can arrive at deeper and deeper explanations.[10] Indeed, we can offer at least a partial methodological explication of the idea of scientific depth (see below). Just as the universality and precision of a theory T cease being a stumbling-block, so the ontological richness of T relative to evidence E ceases to be a stumbling-block once we scrap the idea of any sort of inductive inference from E to T, or any sort of probabilistic confirmation by E of T, and regard scientific theories as conjectures which are only testable *against* basis-statements. (The term 'basis-statement', signifying a statement belonging to the 'empirical basis' but inevitably theory-impregnated, is perhaps preferable to Popper's original term 'basic statement' which some people took to mean an ultimate, rock-bottom, incorrigible, empirical statement.) In short, the whole problem of the 'top-heaviness' of theoretical science relative to experimental evidence ceases to be a problem for a falsificationist philosophy of science.

What, finally, of the primary component of the Bacon-Descartes ideal, namely that science should give us *truths*? Well, we accept the Tarskian idea of absolute truth. But it is a concomitant of that idea that there cannot be a general criterion for truth: we have to be content with a *regulative* idea of truth. Unlike the 'inference-license' party, we do at least regard a universal hypothesis as true or false; and furthermore, unlike the 'verificationist-phenomenalist' party, we regard an ontologically rich scientific theory as true or false. Moreover, we *want* true theories. And we want highly testable theories because we hope that, if such theories are severely tested they have a high chance of being weeded out if they are in fact false. But as we said earlier, every philosophy of science after Hume has to contain *some* pessimistic element; and the pessimistic element in ours enters at this point. From the fact that a highly testable hypothesis has been severely tested and has triumphantly passed all tests we can make no inference to the truth, or likely truth, of the hypothesis. Both logic and the history of science (for instance, the supersession of Newtonian theory) oblige us to reject this statement by J. S. Mill: "if the conclusions to which the hypothesis leads are known truths, the hypothesis itself either must be, or at least is likely to be, true" (*Logic*, III, xiv, 4.).

This concludes one general argument for preferring this methodology and its appraisals to rival methodologies: although it of course discards some parts of the Bacon-Descartes ideal, it appears to preserve considerably more of it than do any of them. Another argument for it is that it avoids, as it were naturally and without contrivance, certain notorious difficulties (additional to those already touched upon) which have been found to afflict inductivist theories of confirmation. But before turning to this further argument we need to take a quick look at the Popperian theory of corroboration. John Worrall will argue in the next chapter that that theory, as Popper presented it, is in need of certain significant modifications and developments. However, these modifications do not affect its performance with respect to the above-mentioned difficulties. So in the next section I will operate with an unmodified version of this theory of corroboration.

2. POPPER'S THEORY OF CORROBORATION

Let T be, at a certain time, the best theory in its field. What would a new theory T' have to do to constitute a *clear-cut advance* over T? The *prior* conditions for such an advance are that T' should:

(1)   answer every empirical question that T can answer, and with at least equal precision;

(2)   answer some empirical questions to which T offers no answer, or no answer of the like precision;

(3)   make observationally discernible revisions to some of the answers yielded by T, and not only in areas, if any, where T has run into some empirical difficulty, but also in areas where T has so far been well corroborated.

In short, T' should both *go beyond* and also *revise* T at the empirical level.[11]

The relation that the empirical content T'_E of T' should have to the empirical content T_E of T may be represented diagrammatically thus:

$$T'_E$$

Here, α is the part of T'_E that goes beyond T_E, β is the part of T'_E that revises T_E, γ is the part of T_E that gets revised, and δ is their common empirical content.

The idea behind these desiderata is that T' should be severely testable, and far from *ad hoc*, with respect to T and the evidence that corroborated T.

The posterior condition for such a scientific advance is that T' should go on to be highly corroborated by the results of subsequent tests. According to Popper, a theory is corroborated when it passes a test, the more severe the test the higher the degree of corroboration. The severity of the test depends in turn on the novelty of the tested prediction relative to background knowledge. Background knowledge is, roughly, all that 'knowledge' which, whether or not it is true, is being taken as unproblematic in the testing of a particular theory (or theories). It will normally include the results of past tests in the domain of the theory now under test. Thus if T had passed various tests in the δ-area, merely repeating these, now as tests on T', could not provide much corroboration for T'. But tests in the α-area may provide striking corroborations; and so may tests in the β/γ-areas: if T' successfully challenges T in areas where T had hitherto been *empirically successful*, that will constitute a famous victory for T'.

One unsatisfactory consequence of this idea of corroboration ought to be mentioned here. Suppose that, before the advent of T', some evidence e had come to light (e.g. concerning the perihelion of Mercury) which tells against T; and suppose that T' (e.g. relativity theory) was not adjusted to fit e, but explains e in a, so to speak, unpremeditated way. Then e cannot be said to corroborate T' since e was already in background knowledge. A way (due to Zahar) of removing this consequence will be discussed in the next chapter.

I promised a (partial) methodological explication of the idea of an increase of theoretical *depth* in science. Suppose we start with two independent theories, T_1 and T_2. Now it would be possible to construct a theory that satisfied just conditions (1) and (2) above merely by conjoining T_1 and T_2. But let T' be a unified theory which also satisfies condition (3) above (as would be the case if T_1 and T_2 were, respectively, Galileo's laws of terrestrial motion and Kepler's laws of planetary motion, and T' were Newton's dynamics). Then the fact that T' revises T_1 and T_2 in the process of unifying and explaining them would be a strong indication, though not a guarantee, that it is deeper than them: for to effect this T' has presumably postulated principles underlying, and connecting, what had hitherto appeared to be two separate domains.

We can go a step further in the explication of increased depth. I have recently proposed a method for indentifying a scientific theory's metaphysical content, of M-component.[12] (Briefly: given a theory T, we construct its Ramsey-sentence T_R. The M-component of T is then the set of those consequences of T that are not consequences of T_R and are not empirically testable.) If the relation of T' to T satisfies the above conditions for high corroboration there are three possibilities concerning the relation between their respective M-components. (1) T' and T have the same M-component: it is only in its subsidiary assumptions that T' differs from T. In this case there is no increase in depth, only an increase in predictive scope. (2) At the opposite extreme is the case where the M-component of T' *repudiates* some or all of that of T: in the derivation from T' of a revised and expanded version of the empirical content of T, some or all of the latter's ontology is by-passed. This may be called a revolutionary scientific reduction. (3) Between these two extremes we have the case where the M-component of T' absorbs that of T: the theoretical entities postulated by T are at once endorsed and explained in terms of those postulated by T'. This may be called a non-revolutionary or radical scientific reduction: the ontology of T is shown as rooted in a deeper ontology. In the case both of revolutionary and of radical reductions there is an advance to a deeper scientific theory. It is worth mentioning that, given that we can detect an increase of depth by a consideration of the relation between the theories' M-components, we need no longer insist upon condition (3) above, which required the deeper theory to *revise* the empirical content of the theory (or theories) taken over by it. Thus we can now affirm that Maxwell's electromagnetic wave theory was deeper than, although it did not revise, Fresnel's wave theory of light. And we can now say that merely to conjoin two disjoint theories does not lead to greater depth because it involves no change of M-components.

So much concerning the way in which the depth-pole of the Bacon-Descartes ideal is both preserved, and given a partial explication, by this view of science. As to the certainty-pole: it deserves to be mentioned that, although certainty itself is totally renounced, *the role that certainty was supposed to perform* in science is largely preserved. No doubt, certainty was wanted for its own sake: ("I am not content with the probable"; wrote Pascal, "I seek the sure"). But it was also intended to serve as a guide in the making of *rational choices* among propositions. This is very clear in the case of the arch-infallibilist Descartes, whose first methodological rule was to *accept* nothing which he did not recognise as indubitably true. The

intuition that underlies most probabilistic theories of confirmation
was, presumably, that where certainty is unattainable our choices
among competing hypotheses should be guided by a preference for
the less uncertain over the more uncertain.[13]

Popper's theory of corroboration can likewise serve as a guide in
the making of rational choices among competing scientific theories. It
will not tell a scientist which theory, or research-programme, he ought
to *work on*. (This question will be discussed by Peter Urbach.) But it
may indicate which of two extant theories, laid on the table for our
examination and appraisal, constitutes the better (conjectural)
scientific explanation of a certain range of empirical 'facts'. To take a
clear-cut case: suppose that our T' has, so far, won only corroborations in the a-area and the b-area, while T has been discorroborated in the c-area. Then T' is obviously better than T on two
counts: it explains *more*; and the evidence to date tells in favour of it
and against its rival. Whether such nice, clear-cut cases actually occur
in the history of science may be questioned. No doubt, the actual
situation in a scientific debate is often messy, with each of the
competing theories attended by theoretical or empirical disadvantages. In such cases, attempts to discriminate between them in terms
of their relative degree of corroboration are likely to fail. A method of
rational evaluation in such cases will be discussed by John Worrall in
the next chapter.

3. AVOIDANCE (NOT EVASION) OF TWO 'PARADOXES'

Popper's theory of corroboration, the ground-plan of which was laid
down in his [1934], was not designed with an eye to the two
'paradoxes of confirmation' high-lighted, respectively, in Hempel's
[1945] and Goodman's [1947]. So these might be regarded as posing
something of a test for it. Indeed, one has often seen it claimed that
the Popperian theory is hit by these 'paradoxes'.[14] That it is hit by
Hempel's was rebutted, to the present writer's satisfaction, some
years ago.[15] But there is still some philosophical interest in considering, not merely whether as a matter of fact the Popperian theory
happens to avoid these 'paradoxes', but whether it possesses some
underlying characteristic than enables it to do so. I will argue that it
does: it avoids them because it is a non-inductivist theory of corroboration and they are 'paradoxes' of *inductivist* confirmation (an

inductivist theory of confirmation being one that starts from the assumption that evidence E confirms a hypothesis H if E is consistent with H and verifies part of H).[16]

Let us consider Hempel's 'paradox' first. We take a universal hypothesis H, say 'All ravens are black', and some suitable universe of discourse. We divide the latter, first into the class R of ravens and the class R' of non-ravens, and second into the class B of black objects and the class B' of non-black objects; which gives us four intersection classes:

(1) RB
(2) RB'
(3) $R'B$
(4) $R'B'$

Hempel showed that instances of (3) and (4) are on a par, confirmation-wise, with instances of (1): if $Ra \wedge Ba$ instantiates $\forall x(Rx \rightarrow Bx)$, so does $\sim Rb \wedge Bb$ instantiate the logically equivalent hypothesis $\forall x(\sim Rx \vee Bx)$. and so does $\sim Rc \wedge \sim Bc$ instantiate the logically equivalent $\forall x(\sim Bx \rightarrow \sim Rx)$. Thus we need consider only two classes here, namely (2), the class of potential *falsifiers* of H, and the complement of (2), or the union of (1) and (3) and (4), the class of potential *satisfiers* of H.

Now the great cleavage between a verificationist or inductivist theory of confirmation and a falsificationist and deductivist theory of corroboration is this: the former is obliged to regard *all* verifications of potential satisfiers of a hypothesis as confirming it; the latter regards only *some* non-verifications of potential falsifiers of a hypothesis (namely, those that result from genuine tests) as corroborating it. The latter is discriminating, not in the sense that it allows only type (1) instances to corroborate, but in the sense that it allows instances – whether of type (1), (3), or (4) – to corroborate only if there was reason to suppose beforehand that they would turn out, on investigation, to be type (2) instances.[17]

Notice, also, the respective responses of the two approaches to a series of hypotheses with *degenerating empirical content*. If H_1 is a relatively strong hypothesis which logically implies the empirically weaker H_2, then the class of potential satisfiers of H_1 is a proper sub-class of that of H_2; it is the weaker H_2 that has the greater potential for inductivist confirmation. Conversely, of course, the class of potential falsifiers of H_2 is a proper sub-class of that of H_1; it is the stronger H_1 that has the greater potential for corroboration by tests.

Now let us turn to the other famous 'paradox' of confirmation, due to Goodman. Here again our thesis will be that this is a symptom of a disease which is endemic only in inductivist theories of confirmation.

Goodman claimed that the old problem of induction has been dissolved, but only to be succeeded by the new problem of projection [1954, III]. Actually, this 'new' problem is really the old inductivist curve-fitting (or curve-projecting) problem, but posed now in connection with qualitative observations and generalisations therefrom. Let us first take a look at this older problem.

A down-to-earth experimentalist, Mr A, has the misfortune to be paired with a wily and perverse curve-fitter, Mr B. The results which Mr A is plotting on a graph, as he measures changes in one magnitude against changes in another, seem to him obviously to be falling in a straight line. But Mr B prefers oscillating curves. Where there is a comparatively wide gap between two experimental points, he is content with a comparatively small amplitude. But as the gaps narrow, so the amplitudes of his curve increase. Thus Mr A, after each successful attempt to refute Mr B's latest curve in favour of his own straight-line hypothesis, is confronted by an even more outrageous curve. A section of Mr B's latest curve might look like this:

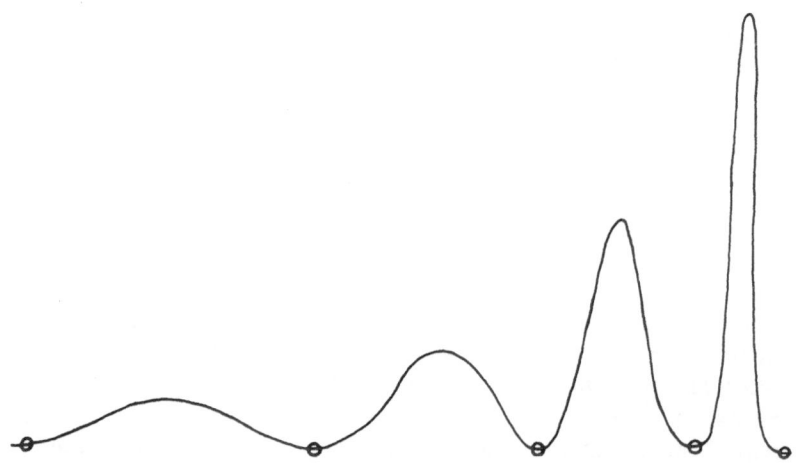

It would be a mistake to assume as a matter of course that it is Mr A who is in the right. In this connection my friend Peter Clark has observed:

Mr B's curve-fitting attitude cannot be excluded *a priori*, or by methodological fiat. A good example of the possible correctness of Mr B's attitude is provided by the curve of thermal equilibrium. In this case Mr A would no doubt draw a straight line corresponding to the constant temperature maintained by his gas sample, but Mr B would be quite right in supposing that the correct curve would consist of many sharp peaks and troughs of very short duration (10^{-11} s for a gas under ordinary conditions) interspersed with relatively rare peaks of great magnitude. Mr B, unlike Mr A would in this case be able to predict the existence of a novel physical phenomenon: the Brownian motion.

One might devise, in analogy with this, a dynamic version of Goodman's paradox: as soon as one Goodman-type hypothesis gets refuted it is to be succeeded by one that is even more gruesome. (Thus "All emeralds are grue" might be succeeded by "All emeralds are gredible", where 'gredible' applies to all things that are examined before time *t* and are green *or* are not examined before *t* and are both red and edible.)

As Poincaré had already put it: "every [experimental] proposition may be generalised in an infinite number of ways";[18] and this has very awkward implications for inductivist (or quasi-verificationist or probabilist) theories of confirmation, which it does not have for a testability theory of corroboration. If an experiment tests two mutually incompatible hypotheses equally severely, and if both pass this test, then each hypothesis may be corroborated as highly as it would have been if the other hypothesis had not been invented and it alone had been under test. Just because corroboration is non-inductive, does *not* confer an enhanced claim to *truth* upon corroborated hypotheses, the corroborating effect of a test does not have to be rationed out among competing hypotheses of which at most one can be true. Thus in making corroboration-appraisals there is no need to take into account possible hypotheses which would also have been corroborated if they had been proposed: we can concentrate upon the hypotheses actually before us.

But an inductivist theory of confirmation cannot so easily set aside such merely possible hypotheses. It interprets confirmation as partial verification (and verification as conclusive confirmation). We may think of a scale ranging from -1 (= falsified), through 0 (= neither confirmed nor disconfirmed), to $+1$ (= verified). If evidence *E* in-

ductively confirms hypothesis H, then E raises H to a higher position
on such a scale. But suppose that there is another hypothesis H',
inconsistent with H but enjoying the same relation to E that H has.
Then would it not be unreasonable to say that E raises only H?
Surely the confirming effect of E should be suitably apportioned
between H and H' or awarded to their disjunction?

But now suppose that over every hypothesis H that is actually
before us, and which we regard as confirmed by the actual evidence
E, there hovers a whole cloud of possible competing hypotheses H',
H'', ... , all having the same relation to E as H has. Will not any
confirming effect of E be completely dissipated as it disperses
through this cloud? We already knew that such a cloud exists when E
consists of measurements and H is a curve fitted over them. What
Goodman has taught us is that the cloud also exists when H is a
qualitative hypothesis like 'All emeralds are green.'

But it will be said, we may be able to discriminate between H, H',
H'', ... on grounds of *simplicity*. To quote Poincaré again: "The choice
can only be guided by considerations of simplicity" (*ibid.*, p. 146).
And Goodman in [1961] likewise appealed to simplicity as a guiding
consideration 'in the choice of hypotheses' (p. 151). But if simplicity is
to come to the rescue of inductivist confirmation theory, here, two
conditions will have to be met. First, a tolerably hard and objective
measure of comparative simplicity will be needed, so that it will not
be (in Boltzmann's words) "doubtful – a matter of taste, as it were
– which [hypothesis] we consider to be the simpler one" [1905, p.
247]. Second, if such a measure were forthcoming it would need to be
underwritten by a synthetic principle which somehow connects
greater simplicity with higher probability, or *a greater likelihood of
being true*. After all, simplicity is being invoked, here, to prevent the
dissipation of the confirming effect of E over the whole set of
competing hypotheses H, H', H'', ... The idea is that if H is the
simplest of these, then the confirming effect of E can be concentrated
upon it. It receives more 'verification' from E than the others do. And
to justify *this* sort of discrimination would require, not a methodological, but a *metaphysical* principle to the effect that, since nature is as
a matter of fact essentially simple, a relatively simple hypothesis is,
other things being equal, more likely to be true than a less simple one.
But is anyone still willing to affirm such a principle? Poincaré was
not.[19] Nor was Bridgman, who wrote in [1927]: "It seems to me that

as a matter of experimental fact there is no doubt that the universe at any definite level [of experimental knowledge] is on the average becoming increasingly complicated, and that the region of apparent simplicity continually recedes" (p. 205).

Concerning simplicity in Popper's methodology, I will say three things. First Popper's [1934] provided a comparative measure of empirical simplicity which is tolerably hard and objective.[20] Second, and more important, the reason for preferring simpler hypotheses is methodological and not metaphysical. Nature may be as complex as she pleases, but we should start with the simpler hypothesis because, if false, it should be easier to falsify. Third, and most important: whereas inductivist theories of confirmation require simplicity, in addition to logical consistency and empirical adequacy, as a third, independent criterion for the appraisal of hypotheses, a falsificationist theory of corroboration *dispenses with the need* for such a third requirement. It operates just with the deductivist idea of falsifiability (of which its idea of simplicity is a special application) and with the idea of empirical adequacy, which it interprets as withstanding attempted falsifications. Being non-inductive it does not have to invoke a special principle to prevent the corroborating effect of the result E of tests on a hypothesis H from being dissipated among grue-ish or other alternatives to H that might be regarded as also corroborated by E.

Two main arguments have been offered for taking this methodology and the appraisals it yields seriously: it seems to capture much more of the old Bacon-Descartes ideal than do alternative methodologies; and it avoids, easily and naturally, the well known difficulties that beset inductivist theories of confirmation. But to repeat: it is a corrigible and revisable methodology. And there are places where, as so far presented, it seems to call for some revision. One of these was mentioned earlier (p. 34 above). Another may be indicated now, with the help of an artificially simple example.

Let T' and T be two theories that are related to each other in the way depicted in the diagram on p. 33 above; and suppose that ten tests have been made on T', five of which were in the a-area (and hence not tests of T) and five of which were in the β/γ-area. Of the latter, four told in favour of T' and against T but one told in favour of T against T'. Of the former, all five corroborated T'.

Clearly, T' is in some sense a better theory than T. But how could this judgment be articulated within Popper's methodology? A few years ago he might have said that these test-results, together with the logical relations in which T' stands to T, at least permit us to conjecture that T' has more verisimilitude than T. But some recent results of Miller [1974] and Tichý [1974] show that such an appraisal is impermissible in the case of two false theories, which is presumably what we have here. Are we then reduced to saying that T and T' have the same degree of corroboration, namely -1? This question is taken up by John Worrall in the next chapter.

The London School of Economics

NOTES

[1] The LSE group had been invited to prepare a position paper on 'Criteria of Progress in the Natural Sciences'.

[2] See K. R. Popper [1962, pp. 373–4]. Popper also said: 'I do not propose any "criterion" for the choice of scientific hypotheses' [1963, p. 218n].

[3] See, for example, Urbach [1974].

[4] Laplace spoke of 'the perfection which [the human mind] has been able to give to astronomy' [1820, p. vii].

[5] Indeed, this theme appeared *within* science. My colleague Peter Clark writes: "A clear example of this preference for certainty over depth is provided by the argument which raged at the end of the century over the superiority of phenomenological thermodynamics over atomistic kinetic theory." The former 'started direct from a few very general empiricalfacts' (Planck [1897, p. viii]). For Planck, the great merit of thermodynamics was its empirical or phenomenological character and its compactness: it provided a wonderfully economical classification of experimental facts and laws. For Duhem and Mach it became *the* paradigm of good science: just because it achieved generality *without* postulating hypothetical entities (atoms), and had withstood all tests, it could claim empirical certainty. Ostwald, who declared that "everything we know of the world is transmitted to us through our sense organs ... Everything beyond that is subjective addition" [1895, p. 430], regarded (a reformulated version of) thermodynamics as the fulfilment of his anti-depth and pro-certainty ideal for science: "To relate realities, demonstrable and measurable quantities, to each other, so that, when some are given, the others may be deduced, *that* is the task of science. This task cannot be accomplished by hypostatizing some hypothetical picture, but only by proving relations of mutual interdependence between measurable quantities" (quoted in Post [1968, p. 6]).

Boltzmann had originally adhered to the other pole: "the bolder one is in transcending experience, the greater the chance to make really surprising discoveries ... the phenomenological account of physics therefore really has no reason to be so proud for sticking so closely to the facts." [1905, p. 8.] However, in order to make the kinetic theory conform with new experimental facts it became necessary to ascribe increasingly strange (and even contradictory) properties to the atom. Boltzmann eventually lost heart, and developed a methodology in which theories are mental pictures to be appraised according to their economy and simplicity. The ideal of phenomenological certainty had triumphed over the ideal of theoretical realism.

[6] On Peirce, see now Scheffler [1974].

[7] The importance of theoretical *exactness* as a stumbling block for inductivism was emphasised by Popper [1963, p. 186].

[8] Popper [1959, Chapter X and Appendix *ix] and [1963, pp. 228f. and 280f.].

[9] See Hintikka [1964] and Howson [1973].

[10] See Popper [1972, pp. 194f.].

[11] [Added in 1977] It turns out that these three conditions cannot be collectively satisfied. This serious difficulty is examined below in §§7-9 of my reply.

[12] Watkins [1975].

[13] See for example Robinson [1964], pp. 84f.

[14] See, for instance, Scheffler [1963], pp. 269f., and Goodman [1961], p. 150.

[15] In Watkins [1964].

[16] For instance, Carnap said of a quantitative inductive logic: "it tells him [the physicist] to what degree the hypothesis considered is supported by the observations; this is, so to speak, the degree of partial entailment or partial logical implication" [1950, p. 222].

[17] Agassi in [1959] gave a nice example of a type (4) instance providing corroboration. The hypothesis "All freely falling bodies fall with constant acceleration" is tested by dropping steel balls down a mine-shaft. They are observed to fall with *in*constant acceleration, and the hypothesis seems to be falsified. But then it is found that the mine-shaft passes through magnetic rock and that this accounts for their aberrant behaviour. The hypothesis has survived the test: these steel balls were *not* freely falling bodies after all.

[18] [1902], p. 130.

[19] "A century ago it was frankly confessed and proclaimed abroad that Nature loves simplicity; but Nature has proved the contrary since then on more than one occasion. We no longer confess this tendency, and we only keep of it what is indispensable, so that science may not become impossible." (*Ibid.*, p. 130.)

[20] See Chapter VII. It is an input-output measure: a hypothesis is simpler than another one if fewer initial conditions have to be put in for it to put out a prediction of a given precision, or if a more precise prediction is put out when a given set of initial conditions is put in. I say '*empirical* simplicity' because it is not a measure of the 'organic unity' of a theory.

The notes refer to the unified bibliography for Parts I and III on pp. 379-383.

JOHN WORRALL

THE WAYS IN WHICH THE METHODOLOGY
OF SCIENTIFIC RESEARCH PROGRAMMES
IMPROVES ON POPPER'S METHODOLOGY

1. INTRODUCTION

A theory is scientific rather than pseudoscientific if it is capable of receiving genuine 'support' from the 'facts'. One scientific theory is better than another rival theory if it is better supported by the facts than its rivals. Although some would reject the term 'support' and replace it by 'confirm' or 'corroborate', most recent attempts to provide an objective and generally applicable criterion of scientific merit have started essentially from these two assumptions. But when does a fact provide genuine support for a theory? And when do the facts support one theory better than another?

Popper's answers to these questions have developed over time in response to criticisms and open problems. But, rather than delve into the intricacies of this development, I shall take as definitive the answers developed in the preceding chapter. I shall (in Sections 2 and 3) contrast these Popperian answers with those provided by the methodology of scientific research programmes and I shall argue that these latter answers are better.[1]

The methodology of research programmes in providing new answers to these fundamental questions *corrects* Popper's methodology in two important ways. It also recognises, as we shall see, the programmatic aspect of scientific achievements and it imports this aspect into methodological appraisals. In these two ways it *goes beyond* Popper's methodology. I deal with this in Section 4.

Finally, having given an outline of the general features of this methodology, I consider (in Section 5) the account it gives of scientific revolutions. I shall try to show that it can explain the rationale of certain scientific developments about which Popper's theory of corroboration remains silent.[2]

2. WHEN DOES A FACT SUPPORT A THEORY?

According to Popper's theory of corroboration, that a description of a 'fact' is deducible from some theory (together with suitable auxiliary

A unified bibliography can be found on pp. 379–383.

G. Radnitzky and G. Andersson (eds.), Progress and Rationality in Science, 45–70.
All Rights Reserved.
Copyright © 1978 by D. Reidel Publishing Company, Dordrecht, Holland.

assumptions) does not guarantee that that 'fact' supports the theory.[3] In his approach an empirically accepted consequence of a theory supports the theory only if it describes the outcome of a 'severe test' of it. According to Popper, a test of a theory is *not* severe if the theory (together with 'background knowledge') predicts the same outcome as is predicted by 'background knowledge' alone.[4] Thus, for Popper empirical support is not a simple two-place relation between theory and evidence, but rather a three-place relation between theory, evidence and 'background knowledge'. Simplifying slightly, the relation holds whenever the evidence is implied by the theory but not by 'background knowledge'.[5]

Of what does 'background knowledge' consist? According to Popper it consists of all those statements provisionally accepted by the scientific community as unproblematic at the time of the test (though any part of it may come in for critical revision later).[6] With this characterisation of background knowledge, the Popperian account of empirical support says that a theory is supported by any fact which it describes correctly and which was first discovered as a result of testing this theory; and that a fact which was already known before the theory's proposal does not support it.[7]

One of the arguments in favour of this account of factual support with its strong temporal or historical element has been that it captures more of scientists' intuitive decisions about confirmation in particular cases. There are many cases, for example, in which each of two rival theories has had some factual statement as a consequence and yet most scientists have intuitively regarded the fact concerned as genuinely supporting only *one* of the theories. I agree that it is more successful in this respect than earlier atemporal theories of 'confirmation'.[8] But I shall argue that it is not so successful in this respect as the methodology of scientific research programmes. I shall now describe two cases where two theories had the same empirical consequence, acceptance of which was regarded as supporting only *one* of them; and I will show that the Popperian account sketched above captures scientists' intuitions in only one of the two cases.

The first case is one often cited by Lakatos. According to him,[9] the Cartesians kept managing to produce theories which 'explained' *post hoc* those facts with which their earlier theories had not dealt but which had been predicted by Newton's theory. Most scientists, however regarded these facts as supporting Newton's theory but not the Cartesian theories.

The second case concerns the perihelion of Mercury. The facts about the precession of Mercury's perihelion are regarded as strongly supporting the general theory of relativity but not as supporting Newton's theory – despite the fact that this phenomenon can now be fully explained classically by making suitable subsidiary assumptions.[10]

The Popperian corroboration theory can easily explain scientists' intuitive judgements in the first case. When Newton's theory was first proposed, it predicted certain hitherto unknown effects. The observation of these effects thus supported Newton's theory. Then, however, knowledge of these effects became part of 'background knowledge'. Thus no theory which was proposed subsequently could derive support from these facts. This applies in particular to the Cartesian theories mentioned.

However, in the case of the perihelion of Mercury, Popperian corroboration theory *fails* to capture scientists' intuitions about empirical support. For the facts about Mercury were known (*i.e.* were part of background knowledge) long before the proposal of either the general theory of relativity or the classical theory which explains them. And this means that the facts about Mercury's perihelion can, according to the Popperian account, support neither the modified classical theory *nor* the general theory of relativity. Yet these facts are widely acknowledged to constitute one of the few important pieces of evidence in favour of the latter theory.

Other cases can be cited in which the Popperian account runs counter to scientists' intuitions. For example, since the Michelson-Morley result had been known since about 1887 it could support *neither* Lorentz's 1904 theory of corresponding states *nor* Einstein's 1905 relativity theory. Similarly, on the Popperian account the fact that plane-polarised light twice internally reflected at a certain angle in a glass rhomb will emerge circularly polarised cannot support Fresnel's wave theory, since Fresnel had discovered it experimentally prior to proposing the theory which explained it.[11] Yet this was widely regarded as one of the most impressive pieces of support for Fresnel's wave theory.

Let us take a more detailed look at the important methodological features of examples of this sort. The precession of Mercury's perihelion can be explained using Newtonian theory by providing certain free parameters and then assigning particular values to them. This assignment in fact reflects some (so far non-independently testable) assumptions about the distribution of mass within the Sun.

Newtonian theory provides us with no independent way of assigning values to these parameters. The Newtonian explanation of Mercury's perihelion is actually arrived at by looking at the facts of Mercury's orbit and using them to work out what values those parameters *must* have in order for Newtonian theory to yield these facts. Of course once these parameter values have been filled in in this way, Newtonian theory does indeed entail the facts about Mercury's orbit but this is more like a consistency proof than a genuine prediction.[12] Mercury's perihelion is not regarded as supporting classical theory; but the reason behind this assessment is not simply that the facts about it were already known (included in 'background knowledge') but that they were known *and used in the construction of the theory.* On the other hand the general theory of relativity was arrived at through considerations which were quite independent of any facts about Mercury's orbit.[13] It nevertheless explains these facts and there seems no good reason not to regard them as supporting this theory.

The classical explanation (using the Lorentz-Fitzgerald contraction hypothesis) of the Michelson-Morley result will serve as a further example. The basis of the widespread opinion that the null result of this experiment, although entailed by the Lorentz-Fitzgerald hypothesis, does not support that hypothesis is the widespread belief that the hypothesis was *arrived at* in the following way: In response to the null result classical mechanics was provided with a new parameter reflecting the amount of contraction a rod undergoes in moving through the ether. The value of this parameter was fixed by working out precisely what value it needed to have for classical mechanics to explain the known experimental result.[14] Again the methodologically important feature of this account is not just that the result was known (the result was known before 1905 yet is generally taken to support relativity theory) but that it was known *and used in the construction of this theory* – a crucial component of this theory was 'read off' from the result.

Should we not rule, then, that of the empirically accepted logical consequences of a theory those, and only those, used in the construction of the theory fail to count in its support? It is precisely this suggestion that the methodology of scientific research programmes incorporates. This methodology embodies the simple rule that one can't use the same fact twice: once in the construction of a theory and then again in its support. But any fact which the theory explains but which it was not in this way pre-arranged to explain supports the theory *whether or not the fact was known prior to the theory's proposal.* Theoreticians should not automatically be penalised, as Popper's corroboration theory would penalise them, for the earlier successes of their experimentalist colleagues.,

This new account of the empirical support relation captures scientists' intuitions in all of the historical cases we have discussed. For example, the facts predicted by Newton's theory and subsequently accommodated within Cartesian theory provide the latter with no support, not because the facts were known prior to the construction of the Cartesian theory which explains them, but because they were *used* in the construction of that theory. The facts about circularly polarised light produced by total reflection of plane polarised light *do* support Fresnel's wave theory of reflection *despite* being known before the proposal of that theory because the theory was arrived at by considerations (of theoretical and mathematical kinds) completely independent of those facts.[15]

In the light of this new characterisation of empirical support, it is not difficult to see what was wrong with Popper's characterisation. Popper's introduction of 'background knowledge' into considerations of empirical support indicates that he had correctly identified the major problem with the requirement of simple testability.[16] This is that the requirement is too easy to satisfy: it is easy (and riskless) to make a theory 'testable' (to provide it with 'potential falsifiers') if one already knows how the tests will turn out (so that no one knows in advance that none of the 'potential falsifiers' are 'actual falsifiers'). Similarly, if a theory T is presently accepted and some new evidence e crops up which is not predicted by T, then it is generally trivially easy to use T and e to generate a new theory T' which does entail e. Popper's theory of corroboration excludes such cheap success. But it constitutes an *over*-solution of this problem – the proposed solution is too coarse-grained. There is no justification for regarding a fact as incapable of supporting any theory proposed subsequent to the fact's discovery. There is every justification for regarding a fact used in the construction of a theory as not capable of supporting the theory.[17]

Of course, if a fact was unknown at the time of the proposal of some theory then it could not have been used in the construction of the theory. Thus not being part of the theory's 'background knowledge' is a *sufficient* condition for the fact to support a theory which explains it. But it is not a necessary condition.[18] This is why Popper's account of corroboration captures scientists' intuitions in *some* of the examples cited; but when it does so it is for the wrong reason – the question of whether some fact was or was not known when some theory was proposed is *in itself* irrelevant to the question of whether or not the fact supports the theory.

The methodology of scientific research programmes regards a theory as supported by any fact, a 'correct' description of which it implies, provided the fact was not used in the construction of the theory. This seems a quite modest proposal and it seems to be the obvious solution to the problem posed by the ease with which *ad hoc* explanations of *given* facts may be generated. The proposal does, however, have the effect of bringing questions of how a theory was arrived at, questions of 'heuristic', into the methodological assessment of the empirical merits of a theory. This has certain consequences which appear unacceptable at first sight.[19]

Foremost amongst these seemingly unacceptable consequences is the following: according to this conception of empirical support it is possible for a theory arrived at in one way to be supported by a fact while the *same* theory arrived at in a different way is not supported by the *same* fact. This certainly sounds implausible and is in fact taken, in a recent article by Musgrave,[20] as a *reductio ad absurdum* of this whole approach to empirical support. The air of paradox about this consequence however, stems only from the fact that we are used to speaking of a fact supporting a theory whereas this new proposal speaks of a fact supporting a theory arrived at in a certain way. This new conception (like indeed the old Popperian one) makes empirical support a *three*-place, rather than a two-place, relation. Here two of the places are filled, as before, by a theory and a factual statement; but the third place is filled, not by background knowledge, but by the set of those factual statements used in the construction of the theory. The relation holds if and only if the factual statement is implied by the theory but is not a member of the set of factual statements used in the construction of the theory. It is possible, therefore, for the relation to hold for a given factual statement and a given theory constructed in one way, but not for the *same* fact and the *same* theory constructed in another way. (Just as, in the Popperian account, a theory may be supported by a fact in one historical situation whereas the same fact fails to support the same theory in another historical situation – because 'background knowledge' has changed.)[21] And this, far from being unacceptable, is precisely the consequence we want, for we have seen that the main problem which this new approach to empirical support was meant to solve is that posed by cases in which each of two theories implies a correct description of some fact which we intuitively want to regard as supporting only one of them. We have seen that this arises when one theory is nicely adjusted to imply some of its rival's empirical content. But if it is possible for one theory to 'explain' *post hoc* some of its rival's content in this way, it may be possible for it to explain *all* of it. In that case the two theories would become, if not identical, at least empirically equivalent. Yet we should regard one of them as supported by the facts it predicted and the other (despite its empirical equivalence to the first) as not supported by the facts it was adjusted to fit. The only extra difficulty in the case where a theory becomes identical *via* such *ad hoc* adjustments to its rival is one of formulation. To avoid paradox we now have to make the three-place character of the empirical support relation explicit by speaking of evidence supporting a *theory as arrived at in a certain way* (or as supporting a theory together with a heuristic) rather than as simply supporting a theory.

Perhaps some doubts will remain about this proposal: does it not, for instance, make empirical support a 'person relative' affair?[22] This is a very reasonable fear which is encouraged by Zahar's occasional use of such formulations as 'a theory is not supported by those facts it was *devised* to explain'. Such phrases make it sound as if two scientists might introduce the same theory, which, however, is supported differently according to which scientist proposed it, since each of them introduced it to explain different facts. But it is not a person-relative, but a *heuristic*-relative affair; and the heuristic considerations which led to the construction of a theory can be objectively specified as we shall see from some examples *below*.[23]

I shall return to the question of heuristics later.[24] For the moment, having answered the question "When does a fact support a theory?", I turn to the question "When is one theory better supported by the facts than another?".

3. ARE THERE ANY SITUATIONS IN SCIENCE WHICH ARE NOT 'MESSY'?

In the previous chapter Watkins claims that Popper's theory of corroboration can provide principles for discriminating between theories in certain 'clear-cut' cases, while for other 'messy' cases (in which, for example, both theories are refuted) other principles of theory-comparison may have to be invoked. The trouble with this suggestion is that, when looked at in any detail, most cases in the history of science turn out to be 'messy'. Even the best scientific theories (like Newton's) were inconsistent with accepted experimental results when first proposed, and remained inconsistent with experimental results (though not necessarily the same ones) until they were replaced. Thus it turns out that nearly all theories during the whole history of science have had the lowest possible degree of Popperian corroboration (*minus* one). This means that scientists' choices between theories cannot be accounted for on the basis of Popperian degree of corroboration appraisals. For example, in the 1830s say, both the latest wave theory of light and the latest corpuscular theory of light were refuted. Thus both had Popperian degree of corroboration *minus* 1, yet the wave theory was more or less universally regarded as vastly superior to its rival.

The methodology of scientific research programmes by contrast can easily accommodate this particular intuitive appraisal and other similar ones. According to this methodology one theory is better than its rival if it is supported by more facts than its rival (in the sense of empirical support outlined in Section 2) – and this condition may hold even if both theories are refuted. This, as Lakatos pointed out, transfers the methodological spotlight away from refutations and focusses it on verifications of excess content (*i.e.* that part of a theory's empirical content which it does not share with its rival).

That refutations do not play such a big role in the development of science had of course been forcefully pointed out by Kuhn.[25] He showed that the typical response of a theoretician to an inconsistency between the theories he accepts and accepted experimental results is *not* to regard the theory as ruled out by the experimental result but rather to regard the result as an 'anomaly' which he hopes and expects will be 'dissolved' by further research.

There is a well known *epistemological* rationale for scientists not getting too excited about refutations and for holding on to a theory despite clashes between it and the experimental evidence. This rationale, which was already in large part provided by Duhem, is taken into full account by the methodology of research programmes. Admitting the fallibility and 'theory laden' character of 'observation statements' (as they are usually considered) the point may be expressed as follows.

Assume that some observational consequence has been drawn from a theory and that the best available observational techniques indicate that this consequence is false. In the derivation of the testable consequences at least some statements of 'initial conditions' will have been assumed as premises in addition to the theory itself. There are then three distinct possible explanations of this clash between theory and experiment, only one of which is the falsity of the theory itself. The other two possibilities are that one of the statements of initial conditions is false, and that the ascription of falsity to the observational consequence is itself false. Both the ascription of truth to the statements of initial conditions and of falsity to the observational consequences are after all, decisions based on observational theories.[26] These ascriptions may be incorrect because the theories on which they are based are false.

But can't we at least narrow down the possibilities of resolving a clash between theory and experiment by requiring that the statement describing the experimental outcome be of such low level that we can hardly suppose that the truth-value we ascribed to it is wrong? For example, we might require, as Poincaré did, that our observation statements be about 'meter readings' – concurrences of needles with points on a scale. There are two points to be made about this suggestion. The first is that our inability to conceive of the possibility of our having made false ascriptions of truth values to such sentences may be due only to our lack of imagination.[27] The second, and more important, point is that if we adopt this suggestion we shall pay for the increased likelihood (in some informal sense) of our decision about the observation statement being practically incorrigible, by having to add extra premises to the theory in order for statements of the required level to be derivable. We may, for example, be able to derive a statement about a body's *temperature* just from some theory of heat and suitable initial conditions; but in order to derive predictions about, say, heights of mercury columns in tubes we shall have to add all sorts of auxiliary assumptions about the thermal properties of glass and mercury, about coefficients of expansion and so on. Similarly we may be able to derive some prediction about the position of a planet just from some astronomical theory and initial conditions; but to derive predictions about, say, the position of a spot on a photographic plate exposed in a camera in a suitably inclined telescope we shall have to invoke extra premises about the chemical properties of photographic emulsion, about optics, about the working of the telescope, about atmospheric refraction and so on. In other words in meeting the requirement that our observation statements be very low level we shall simply articulate and add as extra premises in our deductive test-structure those observational theories on which we had hitherto implicitly relied. But then of course even if we deduced an experimental consequence whose falsity we regarded as beyond dispute, it is perfectly possible to blame, not the theory under test, but one of the necessary auxiliary assumptions.

But then clashes between theory and experiment constitute not experimental disproofs of theories, nor even straightforward inconsistencies between the theory under test and experimental results. Rather such clashes constitute inconsistencies between experimental results and a whole group of theories. It clearly may therefore be reasonable to regard such a clash as not particularly endangering the theory under test. For one may expect the 'anomaly' to be 'dissolved'. This simply means that one expects that some change will be made to the auxiliary assumptions so that consistency with the experimental result is restored.

Indeed there are plenty of well-known examples from the history of science in which such expectations were satisfied. Some of the predictions of Newton's theory were refuted by Flamsteed's data, but this clash was resolved not by giving up the theory but by Newton's 'revising' Flamsteed's data by providing him with a new theory of atmospheric refraction. The facts about the orbit of the planet Uranus refuted the Newtonian prediction, but this resulted not in the rejection of Newton's theory but in the rejection of an auxiliary assumption about the total force acting on Uranus.

The methodology of scientific research programmes can deal very simply with these examples and with the Duhemian methodological point which they underline. This methodology allows that clashes between theory and experiment occur all the time and that they will normally be resolved by assuming the theory true and using the clash as an indication that some auxiliary assumption or observational theory needs to be replaced.[28] In this way the 'protective belt' or auxiliary theories surrounding the 'hard core' theory will be articulated and modified.[29]

There is, however, a problem here, a problem to whose discovery at least the more naive forms of falsificationism would never have led (although it is a problem towards whose solution the sophisticated form of falsificationism outlined by Watkins *above* goes a long way). There are many cases in which the practice of defending theories from refutation by the modification and elaboration of protective hypotheses has been intuitively regarded as unsatisfactory – even as the tell-tale characteristic of pseudo-science. An example which Imre Lakatos used to give in his lectures was that of Marxism. He claimed that the scientific unacceptability of Marxism stems not from its failure to make any testable predictions at all (it made many such predictions, for example, that the working classes would be impoverished absolutely and that once a society has been through its revolution and become socialist it would be free from further revolutions); nor from the fact that these predictions were unsuccessful (even the best scientific theories made unsuccessful predictions). Rather, Lakatos alleged, Marxism's pseudoscientific character is betrayed by its proponents' reaction to the lack of success of its predictions. Marxists explain away this lack of success in various ways rather than regard it as a refutation of Marxism itself.[30] But then, as we have seen, this was also the reaction of Newtonians in analogous circumstances. When is it scientifically satisfactory to blame inconsistencies between fact and theory on auxiliary assumptions and when is it unsatisfactory? What, if anything, distinguishes the 1846 Newtonian who claims that the unexplained perturbations in Uranus's orbit are not genuine refutations of Newton's theory but rather of some auxiliary assumption, from the 1956 Marxist who claims that the events in Hungary in that year are not genuine refutations of Marxist theory but rather of some auxiliary assumptions?

Many have thought that one can only leave it to 'scientific commonsense' to distinguish between acceptable and unacceptable defences of a theory from refutation. But the methodology of scientific research programmes claims to provide a characterisation of this distinction which is generally applicable, objective and explicit. It says that what distinguishes the 1846 Newtonian from the 1956 Marxist is that the latter resolves the clash between his theory and the facts by shifting to a theory which is supported by no more facts than is his previous theory. He may, for example, simply dismiss the Hungarian revolt as not a genuine counter-revolution without giving a more explicit general account of what *would* constitute a genuine revolution – an account which would make his new set of assumptions more, rather than less, testable. Or he may explain away the fact that in the West the working classes have been neither absolutely nor relatively impoverished by invoking a theory of imperialism, *whose only extra empirical support is the non-impoverishment of the Western working classes.* On the other hand, the 1846 Newtonian does not content himself with the claim that the irregularities in Uranus's orbit do not refute Newtonian theory, but rather some (unspecified) auxiliary or observational assumption. Nor even with simply specifying the faulty auxiliary assumption and replacing it with a new assumption. Instead he replaces the faulty assumption with a new assumption of a special kind – one which makes the new total theory capable of receiving genuine support from more facts than the previous total theory. Here, of course, one extra empirical prediction concerned the existence of a hitherto unknown planet. This prediction was subsequently confirmed. Thus the Newtonian's shift was from one set of assumptions to another set which received support from more facts; whereas the Marxist's shift was to a set of assumptions which was incapable of receiving support from more facts than its predecessor. In brief, the difference between Marxism and Newtonianism, is the difference between a *degenerating and a progressive programme.*

One theory is better than another, according to the methodology of research programmes, if it is given genuine support by more facts, whether or not both theories are refuted. If a research programme produces a series of theories such that each theory is better, in this sense, than its predecessor, then the programme is (empirically) progressive. If one research programme produces a theory which is better in this sense than the latest theory produced by its rival then that programme has (for the moment at least) *superseded* it rival.

I should like to highlight two features of this solution of the problem of when one scientific theory is better than another. The first is that it has implicit in it a solution of the so-called Duhem-Quine problem.[31] The problem is: which of the group of theoretical assumptions needed to deduce observational results should one replace (or should regard as falsified) in the event of one of the observational consequences being accepted as false? The solution is: Replace any of the group of the assumptions that you like – the best modification is the one which produces the theoretical system which constitutes most progress over its predecessor. (Which assumption should be regarded as falsified in the event of a clash between theory and observation can only be decided with hindsight: that assumption is falsified which is eventually most successfully replaced.)

The second feature of this definition of progress that I should like to mention is this. It might seem that we can do without all this talk of 'genuine' support from facts and simply characterise one theory as better than a rival if it has excess empirical content over the rival.[32] However, Marxism supplemented by its theory of imperialist exploitation has excess empirical content over its predecessor theory. It predicts that the working classes will not be absolutely impoverished. In other words, it has as excess content precisely the fact which refuted its predecessor and which it was introduced to explain. On the straight-forward excess empirical content criterion, therefore, Marxism would have to be pronounced progressive. And so would almost any series of theories produced by the most blatant *ad hoc* maneouvres. (The only exceptions would be cases of *ad hoc* reduction in content – truly exceptional cases). On the criterion of empirical support developed in my Section 2, on the other hand, such excess 'verifications' do *not* provide genuine empirical support.[33]

We have seen that the methodology of scientific research programmes attempts to correct the Popperian theory of corroboration both over its characterisation of the empirical support relation and over the conditions under which one theory is scientifically superior to another. Involved in both these corrections, however, has been an element whose introduction marks two ways in which the methodology of research programmes *goes beyond* Popper's methodology. These are its recognition of the programmatic aspect of major scientific developments and its importation of this programmatic aspect into methodological appraisals.

4. THE METHODOLOGY OF SCIENTIFIC RESEARCH PROGRAMMES: THE IMPORTANCE OF HEURISTICS

Most clashes between theory and observation are not resolved by giving up the theory under test but rather by modifying one of the auxiliary assumptions needed to deduce the observation result from the theory. This possibility was pointed to by Duhem and the prevalence of its employment in science was pointed out by Kuhn. The methodology of research programmes provides criteria for *evaluating* various possible shifts from one set of assumptions to another. But it also points out that these shifts are not made in a simple trial and error fashion, but are guided by general considerations of an objective and analysable nature.

Indeed, perhaps the most basic difference between the methodology of scientific research programmes and other methodologies (in particular Popper's) is its recognition of the programmatic character of major scientific achievements. It is an historical fact that some important theoretical innovations in science were quickly succeeded by the articulation of a series of scientific theories, related in certain ways to, but *not* implied by, the first theory. So, for example, Einstein's 1905 special theory of relativity was quickly followed by the invention of related theories by Planck (the relativistic law of motion) and by others (for example, the famous $E = mc^2$ law relating mass and energy). These theories bear strong enough resemblances to Einstein's 1905 original to be called 'relativistic theories', but they are not simply deductive consequences of Einstein's original theory. Similarly, Fresnel's invention of the wave theory of the diffraction of light was quickly followed by several other theoretical breakthroughs in optics (mainly by Fresnel himself). These succeeding theories were clearly related to the wave theory of diffraction (as evinced by the fact that they too were called 'wave theories' of various groups of phenomena), yet they were by no means implied by Fresnel's original theory. (For example, Fresnel's wave theory of double refraction is certainly logically independent of his theory of diffraction.)

Facts like these must seem remarkable coincidences to anyone who holds a straightforward trial-and-error, conjectures-and-refutations view of the development of science, for this view can assign no reason why theoretical breakthroughs should come in bursts. The methodology of research programmes, on the other hand, explains these facts, for, according to it, major scientific achievements consist not merely of a set of statements about the world, but *also* of a set of ideas about how to 'fill in', make more precise, draw consequences from,[34] these statements, and also about how to elaborate on them, introduce new assumptions so that they apply to new fields, and how to modify them when difficulties arise. Lakatos called this set of ideas the *positive heuristic* of a programme. But then if major scientific achievements come in the form of programmes, it is easy to see why theoretical innovations often come in bursts – for once a programme exists different scientists may pursue its heuristic and produce new theories. Thus Einstein invented more than a theory in 1905, he invented a programme. By pursuing the heuristic of this programme further theories were produced. These theories were related to, but not logically implied by, Einstein's original *theory*. Similarly what Fresnel invented in the early nineteenth century was not merely a theory but rather a programme. Through pursuing the programme a variety of new theories were produced.

A research programme then consists of a fundamental set of statements about the world and of a positive heuristic. The positive heuristic guides the production of specific theories within the programme.[35] Each specific theory will be 'built around' and will imply the fundamental set of statements about the world (the programme's 'hard core'). The programme (which thus issues in a series of theories) is appraised as 'progressive' if the theories in this series are supported by more facts than their predecessors (in the sense of empirical support elaborated in Section 2). Otherwise the programme is 'degenerating'. This means that the theories produced by the programme either make no genuine excess testable predictions (that is they explain only those facts they were introduced to explain) or such extra predictions as they do make are empirically refuted.

I argued in Section 2 that the heuristic path by which a theory was discovered is relevant to the question of empirical support. Given this, there is a clear, though intuitive, sense in which a programme with a powerful heuristic is 'more likely' to make progress (produce theories which derive support from more facts) than is a programme with a weaker heuristic. (Though of course whether or not it actually does make such progress depends on the objective features of the world.) A powerful heuristic will often point to shortcomings in existing theories in the programme and lay down guidelines for their replacement, *quite independently of any empirical difficulties.* Since new theories produced by such programmes are constructed without the help of empirical considerations, any successful predictions they make will provide them with empirical support in our sense.

The two research programmes I have mainly used as examples both had a clearly definable positive heuristic. Elie Zahar had shown in detail how the positive heuristic of the relativity programme was supplied by the requirements that physical laws be covariant and that the new relativistic laws have the corresponding classical laws as 'limiting cases'.[36] (These requirements were in turn underpinned by certain metaphysical assumptions about the world.) The heuristic of the 19th Century wave optics programme was in large part supplied by the assumptions that the all pervasive ether which carries the light waves is an elastic medium with straightforwardly mechanical properties.[37]

Both these programmes had powerful heuristics. Zahar has shown[38] how the heuristic of the relativity programme did indeed guide the construction of relativistic laws, like Planck's relativistic law of motion and the law that $E = mc^2$. An example of how a powerful heuristic can point to shortcomings in existing theories independently of empirical difficulties is provided by the wave optics programme whose heuristic pronounced inadequate Fresnel's theory of the intensity of reflected beams of polarised light *in spite of* this theory's enormous empirical success.[39]

In some other programmes the heuristic is much weaker, often consisting merely of a series of suggestions for dealing with refutations of existing theories. (The programme may originally have involved much stronger heuristic principles which however did not produce theories which were empirically successful and which were therefore dropped.[40]) One programme with a weak heuristic seems to have been the Ptolemaic programme, whose heuristic amounted to the injunction to 'save the phenomena' (and at the same time the 'hard core' geocentric hypothesis) by a combination of as few uniform and circular motions as possible.[41] This heuristic condemns one to wait for anomalous 'phenomena' to present themselves before proceeding to 'save' them. There seem to be many examples of programmes with weak heuristics in the social sciences. Peter Urbach has argued that one such example is provided by the environmentalist programme in human intelligence – at any rate as it has so far developed. The heuristic of this programme (although originally quite strong) seems to have degenerated to the point where it consists merely of a set of suggestions on how to modify existing theories so that they deal with the anomalies faced by these theories (most of which have been thrown up by the rival hereditarian programme).[42]

5. SCIENTIFIC REVOLUTIONS AND THE PROBLEM OF 'KUHN LOSS'

Having characterised the general features of the methodology of scientific research programmes, I should like briefly to present its account of scientific revolutions, and then to show that it can give some sort of answer to a further problem which besets the Popperian corroboration theory outlined in the preceding chapter.

When do scientific revolutions occur according to the methodology which I have been presenting? They occur when a new programme is introduced which challenges some already accepted programme and which becomes and remains more progressive than its accepted rival. The older programme may actually be degenerating, though this need not be true. Corpuscular optics was degenerating in the early 19th Century when Fresnel revived the wave programme; whereas, if Elie Zahar is right, the classical programme was not degenerating in 1905 when Einstein introduced the relativity programme.[43]

Of course the methodology does not predict that, whenever some new programme comes along which it appraises as more progressive than the old one, all scientists will immediately switch to work on the progressive programme. Nor does the methodology pronounce 'irrational' those scientists who, in such circumstances, stick to the old programme.[44] Such a scientist may, in perfect conformity with this methodology, agree that the new programme is, at that moment, superior, but nevertheless declare his intention to work on the old programme in an attempt to improve it so that it becomes even better than the new programme.[45] Nevertheless we should expect that most scientists will join the most progressive and most promising programme, especially if the early attempts to revitalise the old programme fail.[46]

This account of scientific revolutions allows the methodology of scientific research programmes to go some way towards answering a

criticism which has been brought against falsificationism and to which
that methodology (even in its most sophisticated form outlined in the
preceding chapter) can provide no answer. This criticism is based on
the phenomenon of 'Kuhn loss' of explanatory content.[47]

One of the central arguments produced by the defenders of the
'incommensurability thesis' (in particular Kuhn and Feyerabend) is
that very often the switch which occurs in a scientific revolution from
one theory (or rather research programme) to the next has involved
losses as well as gains in experimentally accepted explanatory
content. In research programme terms: the latest theory produced by
one programme may be supported by facts which do not support the
latest theory produced by its rival and *vice versa*.

The losses involved may be only temporary. Thus to use one of
Feyerabend's favourite examples,[48] while the geocentric theory
(together with Aristotelian physics) could explain why objects do not
fly off the earth, Copernicus's heliocentric hypothesis could not
explain this – at least not until, long after the 'Copernican revolution',
the heliocentric hypothesis was conjoined with a new dynamics. But
if such losses occur at all then they clearly *may* be permanent, that is
they may never be made good by the programme whose adoption
incurred them. For this programme may itself be replaced by a
further one before the explanatory losses are made good.

There seems no doubt that these losses do occur.[49] Indeed that they
do not occur more often may sometimes be merely accidental. (I am
thinking of cases like the following: it is now accepted as a fact that
light exerts a pressure on any screen on which it falls. This is
certainly a direct consequence of the corpuscular theory of light
(assuming, of course, that light-corpuscles possess inertia). On the
other hand, any formalisation of the wave theory of light prior to the
1830s would have predicted no such pressure. The reason why the
switch to the wave theory in the 1820s was not recognised as
involving the loss of explanatory content in this respect was that
experimental techniques were not sufficiently refined to detect this
pressure of light. Several attempts had been made to detect it, but
though at one time success was claimed, the general consensus by the
1820s was that such pressure had not been detected. Indeed this was
taken by some commentators as a crucial experiment in favour of the
wave theory.)[50]

Can the shift from one theory (or programme) to another be

explained as a shift from one theory to a *better* one despite a loss of explanatory content being involved? Not according to the Popperian theory of corroboration outlined by Watkins *above*. No one, so far as I know, has claimed that there are cases of theory shifts in which *only* losses, but no gains, in explanatory content are achieved. This means that these are not cases of shifts from one theory to what Popper would judge to be a *worse* theory. But if a loss of explanatory content occurs then the two theories involved must be incomparable on this Popperian account for each (correctly) 'answers empirical questions' not answered (or answered incorrectly) by the other.[51]

The methodology of research programmes is, on the contrary, much more flexible in this respect and provides several means for discriminating between theories even in these cases. It is true that if the latest theories produced by two research programmes are such that each of them explains facts not explained by the other, then neither programme has 'superseded' the other. But this does not mean that the methodology must remain silent about these cases. Indeed there are at least two things it can say. It may be that one of the theories has been produced by a progressive research programme whilst the other has been produced by a degenerating research programme. And it may be that one of the programmes is heuristically strong, whilst the other is heuristically weak.

When (as will often happen) these two (intra-programme) appraisals go hand in hand by both favouring the same programme, the methodology of research programmes provides a clear rationale for the preference of one programme over the next even if a loss of explanatory content is involved. After all explanatory gains, as well as losses, are involved in these cases, and so failing to prefer the new theory would also incur 'Kuhn loss'. The question then is whether there are any general grounds for regarding one programme as more likely than the other to make good the explanatory losses incurred in adopting it. The methodology of research programmes supplies such grounds, namely the progressive nature and greater heuristic strength of one of the programmes.

A case in which these two intra-programme appraisals do seem to go hand in hand is the early 19th Century revolution in optics. This revolution which occurred round about the late 1820s to early 1830s did involve losses in explanatory power. For example, the existence of dispersion was a straightforward (non '*ad hoc*') consequence of the corpuscular theory. (Admittedly the assumptions involved ran into difficulties when applied to other phenomena, but this is not the issue here – considered simply as an explanation of dispersion they were impeccable.) On the other hand it was not at all clear in the 1830s how dispersion might be explained wave-theoretically. Nevertheless the wave programme was in the 1830s vastly superior to its corpuscular rival and (almost) universally regarded as such. Moreover this superiority can be accounted for by the methodology of research programmes. First, the wave programme had been made enormously progressive by Fresnel. The latest wave theories of light could explain in a non *ad hoc* way many more facts than their predecessors (for example about diffraction and interference of polarised light); whilst the corpuscular programme could at best capture these facts *ad hoc*. That is, the corpuscular programme was degenerating. Secondly, Fresnel had shown that the heuristic of the wave programme could give almost as precise guidelines for research and the development of specific theories as the heuristic of the corpuscular programme. However the heuristic of the wave programme, unlike that of its rival, *was almost completely unexplored.* Thus most scientists began to plump for the wave programme, which eventually began to make good its 'Kuhn loss'. (Cauchy produced wave theoretical explanations of some aspects of dispersion but this phenomenon was not given a full explanation within the wave programme until Gouy explained it, using Fourier analysis, in 1886.[52])

Most cases of switches between theories neither of which explains all that its rival does may thus lose their mystery if we appraise, not the rival theories, but the rival programmes of which they are the latest products. However since this appraisal has two parts, there may be cases in which these two appraisals diverge. (One programme may happen to have been progressive but to have had a weak heuristic, whilst the other although empirically degenerating had a strong heuristic.) Thus this analysis opens up the possibility that the methodology of research programmes may give a general delineation of those cases, if any, in which there is genuine scientific uncertainty between two approaches – the cases of genuine 'incommensurability'. (Clearly one cannot do this using the ideas about incommensurability promulgated by Kuhn and Feyerabend; for there are cases of theories which are pronounced incommensurable by Kuhn and Feyerabend and yet one of which is *clearly* better than the other.) This intriguing possibility awaits historical investigation.

My analysis has at various points involved the notation of one heuristic being stronger than another. Whether a general characterisation can be given of the strength of a heuristic, and hence of a research programme's objective 'promise' is a question which is taken up by Peter Urbach *below*.

Various of my arguments have been designed to show that both in the ways it corrects Popper's methodology and in the ways it goes beyond it, the methodology of research programmes supplies philosophical evaluations which are closer to the intuitive evaluations made in particular cases by scientists working in the more advanced sciences. This point is illustrated in the next chapter by Elie Zahar using a particular episode from the history of science.

London School of Economics

NOTES

[1] I should add that the answers proposed by the methodology of research programmes have also developed over time. The answers I shall give are not those given in Lakatos [1970]. The idea that heuristic considerations have to be imported into theory-appraisal was developed in discussions between Lakatos, myself and Elie Zahar, who was in this respect the prime mover.

[2] Of course the methodology of research programmes is itself 'Popperian' in a wider sense. Many of the corrections and improvements of the explicit methodology which I have, for the sake of definiteness, labelled as Popper's are themselves Popperian in spirit (and Popper scholars could no doubt find sources for many of them in Popper's own work). I regard the methodology of scientific research programmes as the result of a 'creative shift' within Popper's own philosophical research programme.

[3] The two assumptions that all accepted empirical consequences of a theory confirm it and that any two logically independent empirical consequences confirm it equally form the basis of the 'purely logical' position in the age-old 'weight of evidence' debate. See the recent article by Musgrave ([1974]); cp. Lakatos [1968a], p. 387.

[4] Popper actually introduces logical probability considerations and defines the severity of a test as the probability of its outcome given the theory and background knowledge, *minus* the probability of its outcome given background knowledge alone. But these refinements need not concern us.

[5] Popper himself again introduces probability considerations here. Roughly his idea is that if the theory predicts that the outcome of some test will be e, then the more improbable background knowledge makes the occurrence of e, the more severe is the test of theory.

[6] See, *e.g.* Popper [1963], p. 390: "'background knowledge' ... is ... all those things which we accept (tentatively) as unproblematic while we are testing the theory".

There is a slight difficulty here. Popper requires the background knowledge to a theory to be consistent with it. But as he himself points out it is one mark of a very good theory if it corrects (*i.e.* is inconsistent with) previously accepted factual statements. This means that one cannot know in advance of the proposal of a theory what its background knowledge will be! Popper requires that the previously accepted factual statements contradicted by the theory drop out of background knowledge. Thus a theory is no more severely tested by a test whose result it predicts to be different from the result predicted by background knowledge, than it is in a case where background knowledge remains silent about the result. Indeed, on this account, a theory receives *less* credit for successfully contradicting accepted knowledge than it receives for successfully going against a result which accepted knowledge makes 'highly probable'. This is surely contrary to the spirit of the Popperian programme.

On an historical note, this counter-intuitive consequence of Popper's corroboration theory seems to me to have arisen because of the attempt to make background knowledge serve two distinct purposes. It was originally meant to consist of those extra assumptions, both singular and universal, required in the deduction of testable consequences from a scientific theory (see *below*, pp. 52–4). This is indicated in the quotation from Popper *above*. However it was then pressed into service to eliminate trivial confirmations (or corroborations) of theories – a theory should not get credit for simply predicting something that was already part of background knowledge. Indeed, speaking informally, in Popper's definition of the severity of a test whose outcome is e for an hypothesis h given background knowledge b, which definition makes the severity depend on $p(e, h \cdot b)$ *minus* $p(e, b)$, b is playing one role in the first probability function (it is there the set of those extra assumptions we have to make in order to derive e from h), and the second role in the second probability function (there it is the set of already accepted knowledge). In what follows we are essentially investigating how successfully background knowledge performs its second role – that of ruling out trivial confirmations of theories.

[7] Lakatos's [1970] account is essentially the same as this.

[8] Although there are certain general intuitions which favour the atemporal, logical approach – why should we attribute more weight to one of a theory's consequences than to others (beyond the weighting by logical strength)? Perhaps the best way to look at the situation is that the logical, atemporal, confirmation theorists were trying to solve one problem. (Roughly: "To what extent is the truth of a theory guaranteed by the evidence we have available?") While the Popperian approach (and that of the methodology of research programmes) is directed to another problem. (Roughly: "What sort of predictions should a theory make in order for it to have contributed to the growth of knowledge?") The realisation that the two approaches are directed to different problems would perhaps have clarified some aspects of Musgrave's [1974] paper.

[9] See for example his [1971], p. 104.

[10] See *e.g.* Adler, Bazin and Schiffer [1965].

[11] See Whittaker [1910], pp. 135–6.

[12] In fact if we let $N(\lambda)$ be Newton's theory with the parameter λ unspecified, and e

the statement about Mercury's perihelion what such a procedure establishes is the truth of the sentence

$$\exists\lambda(N(\lambda)\to e).$$

I should add that it is, of course, better for a theory for it to be consistent with the facts rather than inconsistent with them. (Indeed some consistency proofs have been important factors in the acceptance of theories. This has been when it had been thought that *no possible form* of some theory could be consistent with some fact. This was true for example of the wave theory of light and rectilinear propagation. Here the consistency proof was provided by Fresnel.) But it is still better for a theory to *predict* a fact. The new characterisation of empirical support which I introduce *below* really amounts to a warning not to confuse consistency proofs with predictions.

[13] See especially Zahar [1973], §3.1.

[14] Elie Zahar in his [1973] argues that this widespread belief is ill-founded, for there were completely independent reasons within Lorentz's programme for giving the specific value to this parameter. Thus the Michelson-Morley result *did* in fact support Lorentz's programme. But of course Zahar would agree that *had* Lorentz's explanation been arrived at in the way described in the text it would not have been supported by the Michelson result.

For the sake of logical clarity I should add that although I speak here of classical theory being provided with a new parameter, in a sense the parameter was already implicit. That is, it was already assumed that there was *no* contraction of rigid rods. If some such assumption had not already been made, this new assumption (unless it introduced inconsistency) could not affect the theory's predictions.

[15] I should make it clear that the methodology of research programmes does not condemn the practice of, for example, reading off the values of parameters from some experimental results. This happens in all the best research programmes. For example, the wave theory arrives at the values of the wavelength λ_i of various kinds of monochromatic light by predicting various interference fringe spacings as functions of λ_i and then reading off the value of λ_i from the observed fringe spacings. The methodology merely states that having *used* these facts to construct their theory, wave theoreticians must look to *other* facts to support the theory. (See my [1975], for the details of the Fresnel case.)

[16] In fact Popper's discussion of conventionalist strategems indicates that he had already spotted the problem in 1934. He meets the problem (more or less) head on in his [1957] paper on 'The Aim of Science'. The problem had often been discovered before. For example, Duhem recognised that it is not difficult to construct 'purely artificial' theoretical systems, but 'we see in the hypotheses on which [such a system] rests, statements skillfully worked out so that they represent the experimental laws already known' (Duhem [1906], p. 28); it is only by avoiding such artificial systems that we can hope to progress toward the 'natural classification'.

[17] This justification can for example be based on Popper's requirement that a theory be given credit only when it has 'stuck out its neck'.

[18] This point is made as a criticism of Lakatos's [1970] criterion of scientific progress by Zahar on p. 102 of his [1973].

[19] Indeed it may have been the seeming unacceptability of these consequences which

prevented those who had spotted the real problem of *ad hoc* explanations from adopting this rather obvious solution of the problem.

[20] Musgrave [1974].

[21] Whereas the Popperian account makes the empirical support relation a three place relation $ES(h, e, b)$, between a hypothesis, some evidence and background knowledge, our new account makes it a three place relation, $ES(h, e, b')$ where b' is only the background knowledge *used in the construction of a theory*.

[22] This is really the basis of Musgrave's claim (see *above*, p. 50) that this approach to empirical support reduces to absurdity.

[23] See pp. 60–1: Whether some fact was used in the construction of a theory is an objective matter – quite separate from any question about whether the theory's inventor knew or 'was aware of' the fact. In the above case of the two scientists who introduce the same theory, if one has to use some fact in order to construct his theory, whilst the second does not, then the second scientist has shown that there are theoretical considerations which *are* supported by this fact (although the first scientist was not aware of it).

Thus in *deciding* whether some fact according to this new account supports a theory one will ask such questions as "Did x's programme give him independent reasons for fixing this parameter in this theory at this value or did its value have to be 'read off' from some observations?" And *not* such questions as "Did x know of this fact or have this fact in mind when he developed this theory?".

[24] *Below*, p. 58ff.

[25] Kuhn [1962]. Similar points were made by Agassi in his attack on what he calls Boyle's rule (see Agassi [1966]) and by Feyerabend (see for example his [1963] and his [1975]).

[26] The fact that a decision is involved here is particularly well emphasised by Popper (see especially his discussion of Fries's trilemma in his [1934] pp. 93–111).

[27] What for example, if the meter-reader was drunk or has bad reflexes?

[28] In the best research programmes the heuristic may give us some indication *which* auxiliary assumption needs to be replaced.

[29] It was a mistake on Lakatos's part to think that a 'protective belt' could get *constructed* in this way. Simply adding extra assumptions to a theoretical system cannot block the derivation of a false observational consequence.

[30] For an example of a 'degenerating research programme' of whose historical accuracy I am more confident, see Chapter 3 of my [1975]. (The example is Biot's development of the corpuscular optics research programme.)

[31] This was already pointed out by Lakatos (see his [1970], pp. 184–8).

[32] This would reduce it to the sophisticated falsificationist account which is essentially that given by Watkins *above*.

[33] This is one important way in which the criterion of progress I have been advocating differs from the one due to Popper; although of course it owes a good deal to the Popper who rejects 'conventionalist strategems' and the like. Further differences are these: (i) (to repeat what I said *above*, p. 52) Popper's corroboration appraisals cannot distinguish between any shifts between refuted theories (the group of Newtonian assumptions amended to include the new planet was still inconsistent with some observational results, *e.g.* about the Moon); (ii) Popper never applied these ideas to the

Duhem-Quine problem, indeed he twice denied that such a problem exists by denying (without argument) that Duhem had shown the inconclusiveness of falsification (see Popper [1934], p. 78, footnote *, and [1963], p. 112); and (iii) that Popper was occasionally confused on these matters is well illustrated by the fact that there are two entries in the subject index of his [1963], (p. 413): 'Marxism-refuted' and 'Marxism-made irrefutable'; these two claims are rather difficult to reconcile unless one has the idea of various *versions* of a Marxist *research programme*, which versions may differ in refutability – but even then the point is not that Marxism has been made completely irrefutable but that there has been no increase in genuine empirical content (and thus in refutability) in the various theory shifts that have been made in response to refutations of previous theories.

[34] Some philosophers have tended to regard the process of drawing consequences from a theory as automatic and unproblematic, but the mathematical machinery a programme provides for drawing out consequences from its theories is an extremely important part of it.

[35] Perhaps a list of *some* of the things a positive heuristic may include will be helpful. The positive heuristic may include mathematics – for example, how theoretical assumptions should be formulated so that consequences may be drawn from them will be guided by the available mathematics; the heuristic may include hints on how to deal with refutations if they arise (*e.g.* 'Add a new epicycle!'); and it may include directions to exploit analogies with previously worked out theories (*e.g.* much of the heuristic power of the corpuscular optics programme was supplied by the assumption that light consists of particles which obey the ordinary laws of particle mechanics, which laws were already highly developed in the late 18th Century).

[36] See Zahar [1973].

[37] See especially my [1975], though some details are to be found in my [1976].

[38] Zahar [1973].

[39] See Whittaker [1910], pp. 132–6, and my [1975]. For another example (Bohr's early quantum programme), see Lakatos [1970], pp. 140–154. I should add that having a powerful heuristic indicates only that the programme is likely to be progressive in the *theoretical* sense – that it will produce theories with extra potential empirical support – over their predecessors. Whether or not some of this extra content is empirically confirmed – so that the programme is also *empirically progressive* – is in the lap of the experimenters.

[40] This seems to be true of the heuristic guidance offered to various classical programmes by the assumption of the existence of the ether. This guidance was very strong at the time of Fresnel but difficulties presented themselves and it had become very weak by the time of Lorentz (see Zahar [1973]; also Schaffner [1972]).

[41] See Lakatos and Zahar [1975].

[42] For this particular example see Urbach [1974]. When I speak of the strength of a heuristic I am referring to its wide applicability, relatively unexhausted state, and ability to operator independently of facts. There is another sense which one might want to speak of a heuristic's strength, namely how near it approaches to being an algorithm. The heuristic of the Ptolemaic programme was strong in this second sense, but weak in mine.

[43] Zahar argues in his [1973] that the classical programme progressed in Lorentz's

hands at least in the empirical sense: it derived *new* support from the result of the Michelson-Morley experiment. (Zahar argues however that Lorentz's programme was not progressive in all senses for heuristically the classical programme had degenerated.)
[44] After all, if it were 'irrational' to work on a degenerating programme we should have to pronounce irrational all those geniuses who took up some old idea which hitherto no one had successfully developed and who turned it into a progressive research programme. (See Section 5 of my [1976].)
[45] A scientist *would* be pronounced 'irrational' (or rather mistaken) by the methodology if he stuck to the old programme denying that the new programme had any merits not shared by the old one, and thus denying that his own programme needed improvement in order to catch up with the new one. It is in such circumstances that we shall begin to suspect the operation of extra-rational motives.
[46] Paul Feyerabend has claimed that unless some time period is specified such that, if a programme consistently degenerates for that period, it is irrational to work on it any further, then the standards supplied by the methodology of research programmes are mere 'verbal ornaments' which hide the fact that a position of 'Anything goes' has been adopted. But anything does *not* go according to the methodology of research programmes – as I pointed out above it is wrong for a scientist to deny that his programme is doing badly if in fact (*i.e.* according to the methodology of research programmes) it is doing badly. If, however, adopting the position of 'Anything goes' simply amounts to denying the validity of the inference from 'Theory or programme *A* is better than *B*' to 'It is rational to work on *A* but not on *B*' then I think we can safely accept Professor Feyerabend's audacious *sounding* claim.
[47] This phenomenon was not discovered for the first time by Kuhn. Adolf Grünbaum pointed out at a recent conference that the phenomenon was noted by Phillip Frank. Paul Feyerabend tells me that he found the phenomenon noted in the work of poet John Donne.
[48] See *e.g.* Feyerabend [1964].
[49] They occur, however, rather less often than Feyerabend would have us believe. In his [1975] he, for example, counts the loss of content about the specific gravity of phlogiston in the Chemical Revolution as an example of incommensurability. But, *of course*, losses in *theoretical* content occur in revolutions, the interesting question is whether losses in *empirical* content occur.
[50] The details of this story are fascinating. Light pressure was accepted as experimentally detected only *after* Stokes had shown that it could also be predicted on the basis of the version of the wave theory then current.
[51] See Watkins *above*.
[52] Even this explanation was far from uncontroversial. For the controversy see Wood [1905], Chapter vi (this was dropped from subsequent editions of Wood's book).

The notes refer to the unified bibliography for Parts I and III on pp. 379–383.

ELIE ZAHAR

'CRUCIAL' EXPERIMENTS: A CASE STUDY

In this section I shall discuss a famous experiment. It was carried out by Kaufmann in 1905 and was intended to be crucial between two rival theories of the electron: a classical theory elaborated by Abraham and the new relativistic theory proposed by Lorentz and Einstein.

Why should the general reader, who may be interested in the *philosophy* of science but not in the technicalities of the history of science, invest effort in reading the present detailed case study of an ingenious but rather elaborate experiment between two rather *complicated* models of the electron? Would not a simpler example have been better for the philosophical problems at hand?

I shall give two different reasons why general methodology may demand detailed, sophisticated case studies.

My first reason is this. I agree with Lakatos's thesis that the history of science acts as an arbiter between different methodologies. It is of course true that methodologies are systems of appraisal consisting of normative propositions; and, as is well-known, there is no direct logical contact between normative propositions and statements of historical fact. However, Lakatos proposed that a methodology M' be preferred to another methodology M if, other things being equal, M' characterises as progressive more of what scientists themselves intuitively consider as good science than does M (or if M' excludes more of what is intuitively regarded as pseudo-science than does M). Thus, one possible way of differentiating between M and M' is to find an historical shift from a theory T to another theory T' where T' was intuitively judged to constitute progress over T and where M', but not M, pronounces T' superior to T. Such a shift would provide historical support for M' as against M.

If M and M' were widely different methodologies, then such a crucial episode should be easy to come by. In our present case however, M is falsificationism and M' is the methodology of scientific

A unified bibliography can be found on pp. 379–383.

G. Radnitzky and G. Andersson (eds.), Progress and Rationality in Science, 71–97.
All Rights Reserved.
Copyright © 1978 by D. Reidel Publishing Company, Dordrecht, Holland.

research programmes (henceforth referred to as MSRP). As John
Worrall showed in the previous section, MSRP is a refinement of
falsificationism or, equivalently, that falsificationism is a coarse-
grained version of MSRP. (This does not prevent MSRP and
falsificationism from contradicting each other in the strict logical
sense.) It is therefore to be expected that MSRP and falsificationism
should give equivalent appraisals in most straightforward situations.
It is even conceivable that MSRP and falsificationism should yield
similar appraisals in *all* actual historical situations: had this been the
case, then MSRP and falsificationism would be analogous to two
empirical theories T and T' such that T' is a theoretical development
of T, but T and T' explain exactly the same known facts; i.e. T and
T' have so far been observationally equivalent. In such a case it might
reasonably be claimed that the excess sophistication of T' over T, or
of MSRP over falsificationism, constitutes *not* progress but *unnecessary* complication. This is why it was important to try to find an
actual historical example over which MSRP and falsificationism disagreed. However, since MSRP is a rather subtle refinement of
falsificationism, it was almost inevitable that the best example I could
find should be rather more complicated than the usual ones.

I now turn to my second reason for regarding a detailed case study
of the present kind as relevant to general methodology. So far I have
spoken of methodologies purely as systems of appraisal which pass
judgement on finished products, i.e. on hypotheses laid on the table.
The question arises as to whether methodologies can play any role in
scientific *discovery*. Many philosophers claim that methodologies do
not play such a role and the construction of methodologies is an
academic exercise irrelevant to scientific praxis. Following the Vienna
Circle, Lakatos made a distinction between the context of 'discovery'
and the context of 'justification'; he also inclined to take the view that
methodologies appraise but give no advice. Such a view however
lands us with the following difficulty. It is agreed on all sides that
science has (globally) progressed at least since the 16th century. It is
an adequacy requirement for any methodology that its general account of scientific progress be in general accord with this historical
progress. Suppose that MSRP fulfils this requirement and that Lakatos is right in denying that methodology plays an effective role in the
actual development of science. How is it then that the scientists have
systematically achieved what they did *not* set out to do; i.e. achieved

progress according to MSRP without being guided by the principles elaborated in MSRP. This suggests a mysterious 'List der Vernunft' whereby scientists are led 'sleepwalking' down the path adumbrated by methodology. My own view is that methodology plays a positive role in scientific progress. I do not of course maintain that scientists possess fully articulated methodologies; but that, in concrete situations, they apply intuitive methodological principles which enable them both to make crucial choices and to construct important hypotheses. And I further maintain that MSRP provides the best available rational reconstruction of the intuitive methodology which some of the best scientists have acted on. It is obvious that this claim calls for detailed case studies in which the genesis of some important theory is examined in order to decide whether and how a certain (intuitive) methodology was operative during the process of discovery. This is the second reason why, in my opinion, the general reader may find the present case study of some interest.

Let us now return to Kaufmann's experiment. My discussion of this experiment illustrates two points of methodological interest: first, the point that deductive logic may play a creative role in the development of the empirical sciences. This claim is denied for example by Descartes and by Feyerabend.[1] (Descartes based his denial of the claim on the triviality of logic; Feyerabend based his denial of the claim on the subservience of logic to the demands of the physical sciences, or rather to the demands of physicists: the scientist freely bends logical rules so as to make them conform to whatever scientific system he wishes to construct.) Against Descartes and Feyerabend I claim that detailed logical analyses – logic-chopping if you like – can further the progress of science in important ways, while logical oversights can seriously impede it.

My second methodological point will be that MSRP, at least in the case of this 'crucial' experiment, reflects rather accurately scientists' intuitive judgements about the significance of the experimental result for the theories in question. In particular it will turn out that Planck's judgement about the significance of Kaufmann's experiment conforms very well with MSRP criteria. Planck subjected that experiment, together with the rival theories under test, to a detailed logical analysis; and the conclusion he thereby reached completely upset the received view about the logical impact of Kaufmann's results.

In 1905 Kaufmann[2] performed an experiment in which he observed the deflection of electrons in an electromagnetic field. The results he obtained were originally taken as having crucially decided in favour of Abraham's classical theory (henceforth referred to as T_A) and against the Lorentz-Einstein theory (henceforth referred to as T_E).

Kaufmann's experiment actually consisted of a sequence of nine sub-experiments, whose respective outcomes may be represented by observation reports of the form $a_1 \& b_1$, $a_2 \& b_2$, ... , $a_9 \& b_9$, where $a_1, ... , a_9$ may be taken as initial conditions and $b_1, ... , b_9$ as the corresponding *observed* outcomes. As to the predicted outcomes: those predicted by T_E will be denoted by b_{E_1}, b_{E_2}, ... , b_{E_9} and those by T_A by b_{A_1}, b_{A_2}, ... , b_{A_9}. So the situation can be schematically represented as follows:

(1) $T_E \rightarrow (a_i \rightarrow b_{E_i})$

and

(2) $T_A \rightarrow (a_i \rightarrow b_{A_i})$

for all $i = 1, 2, ... , 9$.

What were the relations between the outcomes predicted by T_E and T_A and the observed outcomes? It turned out that in all nine cases, the 'values' b_{A_i} predicted by Abraham are closer to the observed 'values' b_i than are the 'values' b_{E_i} predicted by Lorentz-Einstein. This can be pictorially represented as follows:

At first Planck[3] shared the general opinion that Kaufmann's results told against T_E and in favour of T_A; but later he challenged this opinion.[4] He never challenged the truth of the observation reports $a_1 \& b_1$, ... , $a_9 \& b_9$. What he did was to subject the theoretical structure of this whole experimental situation to strict logical analysis.

To begin with, Planck showed that, implicit in the derivation of the predictions b_{E_i} and b_{A_i} from T_E and T_A was an assumption K: K was assumed both by Abraham and by Lorentz-Einstein and was moreover essential for the derivation. Thus the real situation is more accurately represented as follows:

(1') $(T_E \& K) \rightarrow (a_i \rightarrow b_{E_i})$

and

(2') $(T_A \& K) \rightarrow (a_i \rightarrow b_{A_i})$ for all $i = 1, ... , 9$.

Planck also showed that K can be put in the form:

(3) $K \equiv (P \& Q(w) \& (w = w_0))$,

where w is a free parameter, while P and $Q(w)$ are assumptions about the strengths of the magnetic and electric fields respectively. Note that w, which plays an important role in what follows, occurs in $Q(w)$ but not in P.

Planck also showed the following. Both T_E and T_A involve a common principle (which roughly states that the motion of the electron is governed by a Lagrangian function dependent on the velocity and that the velocity of the electron remains smaller than c, the speed of light). Let us call this principle S. Thus;

(4) $T_E \rightarrow S$ and $T_A \rightarrow S$.

Moreover Planck showed that Kaufmann's observation reports logically contradict $S \& K$, i.e.

(5) $((a_1 \& b_1) \& (a_2 \& b_2) \& \dots \& (a_9 \& b_9)) \rightarrow \neg (S \& K)$.

((5) was established by a *reductio ad absurdum*. S includes the principle that the velocity of the electron remains smaller than c. Planck proved that $(a_1 \& b_1) \& \dots \& (a_9 \& b_9) \& S \& K$ implies that the electron reaches velocities greater than c.)

Thus an important interim result of Planck's analysis is that Kaufmann's results are neutral as between T_E and T_A. Those results show that $S \& K$ is false, where S is implied by both theories and K is not logically implied by either theory. Thus, if S were false, T_E and T_A would sink together, whereas if K were false, neither of them need sink with it. We may formally summarise this as follows:

(6) $S \& K$ is false (given Planck's acceptance of Kaufmann's observation results.)

By (4) and (6) it follows that:

(7) $(T_E \& K)$ and $(T_A \& K)$ are both false.

Were there any reasons for imputing the falsity of $S \& K$ to one rather than the other of its conjuncts? There were. To Planck it seemed that *any* theory of the electron, whether classical or not, would have to contain S. By contrast there were physical reasons for suspecting K of being false. (These were roughly that, contrary to K, the electric field ought to be altered by the electrons ionising the gas between the plates of the condenser.) Thus Planck investigated the possibility of modifying K.[5]

Planck proposed two different replacements for K, which we may call K_E and K_A. K_E and K_A are such that both $(T_E \& K_E)$ and $(T_A \& K_A)$ yield Kaufmann's results. Although Planck had independent reasons (already indicated) for doubting K, these did not uniquely determine a replacement for K. To obtain this, he had to use Kaufmann's results. But these yielded alternative solutions, namely K_E and K_A, according to whether T_E or T_A was to be reconciled with those results. Let us now look more closely at what Planck did.

Planck's method in the case of the Lorentz-Einstein theory can be described schematically as follows. (As we shall see, Planck's method in the case of Abraham's theory is the same.) Remember that K is of the form (see equivalence (3) above) $P \& Q(w) \& (w = w_0)$. To simplify the expression I shall henceforth write $R(w)$ for the conjunction $P \& Q(w)$. The deductive test-structure of the Lorentz-Einstein theory as articulated by Planck now reads:

(8) $(T_E \& R(w) \& (w = w_0)) \rightarrow (a_i \rightarrow b_{E_i})$ for all $i = 1, 2, \dots, 9$.

Planck proceeded to regard Kaufmann's results not as tests of T_E but (in the first place at least) as means of determining the value of the free parameter w in the auxiliary hypothesis $R(w)$ needed to deduce Kaufmann's results from T_E. In fact he found that, for each $i \in \{1, 2, 3, \dots, 9\}$, the equation $w = w_0$ can be altered so as to obtain $w = w_{E_i}$, where w_{E_i} is a numerical value uniquely determined by $T_E \& R(w) \& a_i \& b_i$. That is

(9) $(T_E \& R(w) \& a_i \& b_i) \rightarrow (w = w_{E_i})$.

It turns out that we also have:

(10) $(T_E \& R(w) \& (w = w_{E_i})) \rightarrow (a_i \rightarrow b_i)$.

In other words, each of Kaufmann's nine results could be explained on the basis of T_E by making a suitable assumption about the parameter w.

If we let

(11) $K_{E_i} \equiv (R(w) \& (w = w_{E_i}))$

then (10) becomes:

(12) $(T_E \& K_{E_i}) \rightarrow (a_i \rightarrow b_i)$.

which holds for all $i = 1, 2, \ldots, 9$.

Now the important point is that since the K's are all statements about the electric and magnetic fields in a certain experimental situation, and since the same apparatus is used in all of Kaufmann's nine sub-experiments, we ought always to have the same auxiliary hypothesis. In other words we should have:

(13) $K_{E_1} \equiv K_{E_2} \equiv \cdots \equiv K_{E_9}$; i.e.
         $w_{E_1} \equiv w_{E_2} \equiv \cdots \equiv w_{E_9}$.

Planck showed that it does indeed turn out that the w_{E_i}'s are all nearly equal. If we allow ourselves the idealising assumption that the *near* equality was due to experimental error and that really the w_{E_i}'s are all equal, then Planck's procedure (as thus idealised) is as follows.

One singles out one of Kaufmann's experimental results, say $a_m \& b_m$, and determines the auxiliary hypothesis K_{E_m} such that:

$$K_{E_m} \equiv (R(w) \& (w = w_{E_m})), \quad \text{where} \quad (T_E \& R(w) \& a_m \& b_m) \rightarrow (w = w_{E_m}).$$

one then finds that:

(14) $(T_E \& K_{E_m}) \rightarrow (a_i \rightarrow b_i)$ for all $i = 1, 2, \ldots, 9$.

Writing K_E for K_{E_m}, then, if we accept this idealisation, it follows from MSRP and my criterion of non *ad hocness*[6] that $T_E \& K_E$ provides a non *ad hoc* explanation of eight out of Kaufmann's nine experimental results. In fact Planck thought that all nine results would thus be explained. His view seems to have been that if $w_{E_1} = w_{E_2} = \cdots = w_{E_9}$, then any one among the observation statements $a_1 \& b_1, \ldots, a_9 \& b_9$ can be used in (9) in order to determine w_E and hence K_E. None of the statements $a_1 \& b_1, \ldots, a_9 \& b_9$ plays a privileged role with respect to the remaining ones. Planck therefore intuitively concluded that, if one of these statements is explained in a non *ad hoc* way, then all of them are. However, this intuitive conclusion is incorrect for it fails to distinguish between a possible case in which K_E would be constructed totally independently of Kaufmann's results and the actual case where at least one result is needed.[7]

So much for Planck's view of the explanation of Kaufmann's

results by T_E. What was the situation with regard to T_A? In fact Planck showed that the relation of T_A to Kaufmann's result is almost exactly parallel to that of T_E: a series of auxiliary assumptions K_{A_i} can be constructed in such a way that:

(15) $(T_A \& K_{A_i}) \rightarrow (a_i \rightarrow b_i)$.

The procedure is completely parallel to the one used in the Lorentz-Einstein case.[8]

Planck again found that the $w_{A_1}, \ldots, w_{A_9}$, which determine $K_{A_1}, \ldots, K_{A_9}$ are nearly equal. Hence, under the same assumptions as above, T_A explains eight out of the nine experimental results in a non *ad hoc* way.

However the parallel between T_A and T_E is not perfect. It turned out that the numbers $w_{E_1}, \ldots, w_{E_9}$ are more nearly equal to one another than are $w_{A_1}, \ldots, w_{A_9}$. Thus T_E has a definite (but inconclusive) advantage over T_A.

How do these historical considerations about T_E, T_A and Kaufmann's experiment support the philosophical claims outlined at the beginning of this section?

The claim that 'logic-chopping' may be important in physics is borne out in at least three ways.

First, Planck derived, by means of a purely logical analysis, the negative result that Kaufmann's experiments were not crucial between Abraham and Lorentz-Einstein: what the experiments actually refuted was the conjunction of the common component S of both theories plus the non-trivial auxiliary hypothesis K.

Second, with the help of the logical implication (9), which has no physical content, Planck was able to modify K into K_E so that $(T_E \& K_E)$ yields Kaufmann's results.[9] He was likewise able to modify K into K_A so that $(T_A \& K_A)$ also yields Kaufmann's results.

Lastly, and very importantly, consider what effect Kaufmann's experiment had on the scientific development of someone who did not have the benefit of Planck's logical analysis, namely Lorentz. In his *Theory of Electrons* Lorentz claimed that, because of Kaufmann's experiment, he could not accept the full covariance of Maxwell's equations.[10] Had Lorentz been aware of the correct logical situation as just presented, *i.e.* had he realised that Kaufmann's experiment is neutral between Abraham's and Einstein's theories, then he might have been converted to Relativity more quickly – especially when we remember that Planck's logical analysis shows that the Lorentz-Einstein theory fits the facts more closely than does Abraham's classical hypothesis. That is: had Lorentz realised what logical connections link the experiment to the theories, then, first he would have accepted Poincaré's corrections to Lorentz's electrodynamics which made them indistinguishable from Einstein's. Secondly, there is good reason to suppose that, once he had accepted the covariance of Maxwell's equations, Lorentz would have gone the whole hog and joined the Relativistic camp. In an earlier paper[11] I argued that the heuristic power of a programme can be judged independently of its empirical success or failure. The covariance of Maxwell's equations, which implies that the electromagnetic ether is in principle undetectable, meant that the ether as it stood in 1905 had lost all heuristic power.

It is this complete loss of heuristic power which would have impelled Lorentz to join the Relativity camp. The only (apparent) obstacle was Kaufmann's result which seemed decisively to go against the fully covariant T_E.

My conjecture that Lorentz was prepared to switch to Relativity, and that Kaufmann's result was a main obstacle to this, is further supported by the following facts. In the Notes to his *Theory of Electrons*[12] Lorentz cited Bucherer's 1908 experiment,[13] as one decisive reason for his switch to Relativity. But Bucherer's experiment is in fact a variant of Kaufmann's with this difference: it takes account of Planck's logical analysis. We have seen that the auxiliary hypothesis K (a conjunction of the form $P \& Q(w) \& (w = w_0)$) where $Q(w)$ is an assumption about the intensity of the electric field) contains the free parameter w. It is by suitably adjusting w to fit the experimental outcome that *both* T_E and T_A can be squared with Kaufmann's results. Every value of w reflects an assumption $Q(w)$ about the strength of the electric field. One way of ruling out this possibility of post hoc adjustment (which saves both theories) is to devise an experiment whose outcome is independent of any specific assumption about the electric field. This is precisely what Bucherer did, and his results told unequivocally in favour of the Lorentz-Einstein theory, as against Abraham's theory.[14]

As regards Lorentz's decision to switch to Relativity, this would not have been directly motivated by Bucherer's result, since *that result* provides equal support for Lorentz's and for Einstein's *pro-*

grammes. Furthermore, in a paper on gravitation[15] published in 1914 Lorentz indicated that he had long accepted the Relativity Principle as a heuristic tool. Thus Lorentz's acceptance of Relativity antedates Einstein's explanation of Mercury's perihelion in 1915, *i.e.* it antedates the *empirical* supersession of his old programme by Einstein's. It seems plausible therefore that Lorentz was converted to Relativity by the realisation that, while covariance was opening up new possibilities, the ether had become heuristically sterile.[16] Thus, had Lorentz known right from the beginning that Kaufmann's experiment was not crucial, he would most probably have accepted the covariance of Maxwell's equations and joined the Relativity Programme at its inception in 1905.

In order to vindicate my second philosophical claim made at the beginning of this section, I shall examine how falsificationism compares with MSRP as regards the historical episode I have described.

Let us see how falsificationism and MSRP might answer the question of whether or not Kaufmann's results support either of the competing theories.

(1) Had the original belief that Kaufmann's experiment decided in favour of Abraham and against Lorentz-Einstein, been correct, then falsificationism and MSRP would have yielded parallel appraisals. According to falsificationism, Kaufmann's experiment would have confirmed Abraham and refuted Lorentz-Einstein. According to MSRP, Kaufmann's experiment would have constituted an anomaly for Lorentz-Einstein and a novel fact successfully predicted by Abraham.

(2) Had Kaufmann's experimental results been predicted in advance on the basis of both theories, then again no difference between MSRP and falsificationism would have emerged. According to both methodologies both theories would have been confirmed by Kaufmann's results.

(3) In the actual case, the two methodologies yield different appraisals. The question with which I shall conclude is: which appraisal fits in best with the intuitions of working scientists – or, more specifically in this case, with Planck's intuitions?

According to falsificationism, Planck's modifications of K (into, respectively, K_E and K_A) were *ad hoc*, perhaps even 'conventionalist stratagems': they saved both T_E and T_A from Kaufmann's results.

Those results provide no corroboration for either T_E or T_A: they were not the results of tests on these two theories but were already part of background knowledge. Nor were those results adequately *explained* by theories which had been adjusted to fit them.

The appraisal clashes with Planck's. Admittedly, Planck did not speak of confirmation or corroboration in this connection, but he did say that, if $w_{E_1} = w_{E_2} = \cdots = w_{E_9}$, then 'all the deflections actually observed by Mr Kaufmann *were completely explained*' by T_E.[17]

The MSRP appraisal comes out much closer to Planck's than does the falsificationist appraisal. According to MSRP, Planck's claim that T_E (or T_A) explains all of Kaufmann's *nine* results was a slight exaggeration (it only explains any eight of them); but it was only a slight exaggeration.

TECHNICAL APPENDIX

In this Appendix I propose to give a brief account of Kaufmann's experiment. I do not intend to give a detailed description of the experiment, for such a description would be irrelevant from the standpoints both of theoretical physics and of the philosophy of science. The first section of this paper contains, as we have seen, the uninterpreted symbols 'S', 'K', 'a_i', 'b_i' etc. The purpose of the second section is on the one hand to state the hypotheses and observation reports which 'S', 'K', 'a_i', 'b_i',... stand for, and on the

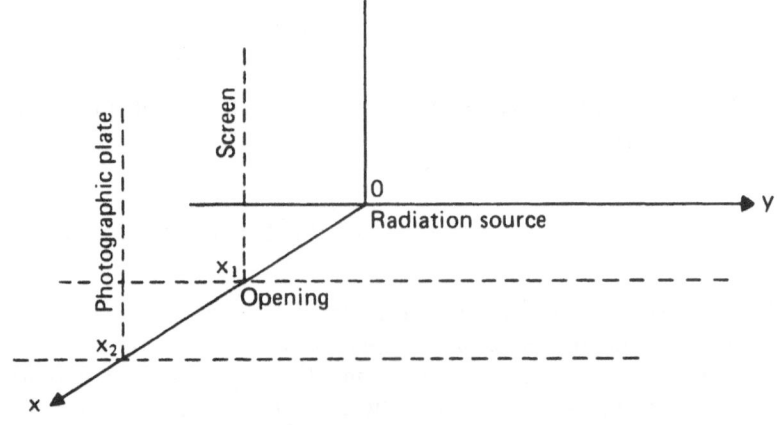

Fig. 1.

other hand to establish those logical relations between S, K, a_i, b_i, ... on which the philosophical arguments presented in the first section hinge.

Kaufmann's experiment consists in letting electrons move in an electromagnetic field and in then observing the deflection of these electrons by the field. The electric field $\mathbf{E}$ is generated by a condenser whose plates are parallel to the xz-plane of the chosen frame of reference. Thus $\mathbf{E}$ is of the form: $\mathbf{E} = (0, E, 0)$.

A magnet, whose poles are parallel to the plates of the condenser, creates a constant magnetic field $\mathbf{H}$ which, being parallel to $\mathbf{E}$, is of the form: $\mathbf{H} = (0, H, 0)$.

At the origin there is a grain of radium which acts as an electron source. The plane $x = x_1$ is occupied by a screen which has an opening at the point $(x_1, 0, 0)$. $\ulcorner x_1 \urcorner$ stands for the numerical value $\ulcorner 1.994 \urcorner$. At $x = x_2$ there is a photographic plate on which the electrons impinge. $\ulcorner x_2 \urcorner$ stands for the numerical value $\ulcorner 3.963 \urcorner$. The plates of the condenser extend from $x = 0$ to $x = x_1$ and are parallel to the xz-plane. Thus the equations of the plates are of the form $y = \pm d$, where d is some numerical value. The poles of the magnet can be regarded as two infinite planes parallel to $X\,0\,Z$.

Some electrons leave the origin, pass through the opening at $(x_1, 0, 0)$ and then impinge on the plate at $x = x_2$. The outcome of Kaufmann's experiment can be described by saying that nine electrons hit the plate where the measured coordinates of the electrons are $(x_2, \bar{y}_1, \bar{z}_1)$, $(x_2, \bar{y}_2, \bar{z}_2)$, ... , $(x_2, \bar{y}_9, \bar{z}_9)$. '$\bar{y}_1$', '$\bar{z}_1$', '$\bar{y}_2$', '$\bar{z}_2$', ... , $\bar{y}_9$, $\bar{z}_9$ represent certain numerical values; e.g. $\bar{y}_3 = 0.0506$ and $\bar{z}_3 = 0.2423$. For each $i \in \{1, 2, ... , 9\}$, the observation reports a_i and b_i are as follows:

(16) $a_i \equiv$ (the z-coordinate of the ith electron as it hits the plate is $\bar{z}_i$),

and

$b_i \equiv$ (the y-coordinate of the ith electron as it hits the plate is $\bar{y}_i$).[18]

Let us now turn to the interpretation of 'K', 'S', 'T_E' and 'T_A'. By (3) above, K is of the form $P\&Q(w)\&(w = w_0)$ where w is a free parameter, w_0 is a numerical value, and P and $Q(w)$ are propositions about the strengths of the magnetic and electric fields respectively. We have seen that the modulus H of the magnetic field $\mathbf{H}$ is constant.

The calculated value of H turns out to be 142.8. Thus:

(17) $P \equiv (H = 142.8 = \text{constant})$.

Turning to $Q(w)$, let us first observe that, since the plates of the condenser do not extend beyond the screen, $E = 0$ for all $x \in [x_1, x_2]$. ⌐'$[x_1, x_2]$' denotes the closed interval from x_1 to x_2⌐. Between the origin and the screen the electric field **E** is for the most constant and then linearly drops to zero in the neighbourhoods of $x = 0$ and $x = x_1$. Let us write $E = 10^8 \, wE_1$, where w is a free parameter and E_1 is the function of x represented by the following graph:

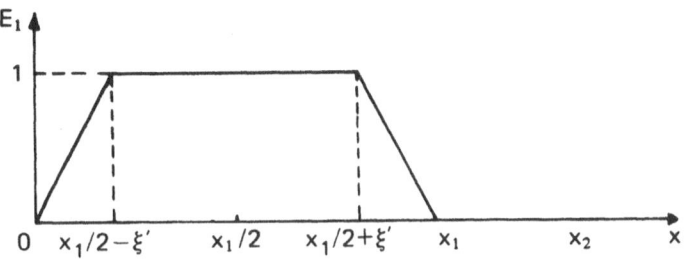

Fig. 2. $x_1 = 1.994$, $x_2 = 3.963$, $\xi' = 0.593$.

We see that E_1 is normalised (i.e. $E_1 = 1$) in the region where E is homogeneous. If we change the independent variable from x to $\xi = x - x_1/2$, E_1 becomes a function $E_1(\xi)$ of ξ such that:

(18)
$$\begin{cases} E_1(\xi) = 0 \text{ for } \xi \in [x_1/2, x_2 - x_1/2], \text{ and} \\ E_1(\xi) = 1 \text{ for } \xi \in [0, \xi'], \text{ and} \\ E_1(\xi) = \alpha - \lambda\xi \text{ for } \xi \in [\xi', x_1/2], \text{ and} \\ E_1(-\xi) = -E_1(\xi) \text{ for } \xi \in [0, x_1/2], \text{ where} \end{cases}$$

(19)
$$\begin{cases} x_1 = 1.994, x_2 = 3.963, \xi' = 0.593, \\ \alpha = 2.468 \text{ and } \lambda = 2.475. \end{cases}$$

Kaufmann took w to be equal to 2500/0.1242. Thus, if we write w_0 for the numerical value 2500/0.1242, the auxiliary hypothesis K can be expressed as follows:

(20) $K \equiv P \& Q(w) \& (w = w_0)$
         $\equiv (H = 142.8) \& (E = 10^8 \, wE_1) \& (w = 2500/0.1242)$, where E_1 is the function of ξ determined by (18).

Let us now turn to the interpretation of T_E and T_A. Since T_E and T_A are high-level all-embracing theories, I shall not spell them out in detail. I shall rather specify those consequences of T_E and T_A that are relevant to Kaufmann's results. Let us start by considering the component S common to T_E and T_A which is such that $S\&K$ is falsified by Kaufmann's experiment.[19] S states that the speed q of the electron remains smaller than c and that there exists a function L of q such that:

$$\frac{d}{dt}\left(\frac{\partial L}{\partial \dot{x}}, \frac{\partial L}{\partial \dot{y}}, \frac{\partial L}{\partial \dot{z}}\right) = \text{Lorentz force acting on the electron.}$$

By a suitable choice of units the Lorentz force can be expressed as: $e(\mathbf{E} + \mathbf{q} \wedge \mathbf{H})$, where $e = $ (charge of the electron)$/c$, $\mathbf{q} = $ velocity of the electron $= (\dot{x}, \dot{y}, \dot{z})$, and $q = $ speed of the electron $= [\mathbf{q}] = (\dot{x}^2 + \dot{y}^2 + \dot{z}^2)^{1/2}$.

Therefore:

$$\frac{d}{dt}\left(\frac{\partial L}{\partial \dot{x}}, \frac{\partial L}{\partial \dot{y}}, \frac{\partial L}{\partial \dot{z}}\right) = e(\mathbf{E} + \mathbf{q} \wedge \mathbf{H}),$$
$$= e((0, E, 0) + (\dot{x}, \dot{y}, \dot{z}) \wedge (0, H, 0)),$$
$$= (-eH\dot{z}, eE, eH\dot{x}).$$

This vector equation splits into the following 3 scalar equations:

(21) $\dfrac{d}{dt}\left(\dfrac{\partial L}{\partial \dot{x}}\right) = -eH\dot{z},$

(22) $\dfrac{d}{dt}\left(\dfrac{\partial L}{\partial \dot{y}}\right) = eE,$

(23) $\dfrac{d}{dt}\left(\dfrac{\partial L}{\partial \dot{z}}\right) = eH\dot{x}.$

The component S can therefore be spelt out as follows:

(24) $S \equiv (q \leqslant c)\&(21)\&(22)\&(23),$
$$\equiv \left((\dot{x}^2 + \dot{y}^2 + \dot{z}^2 \leqslant c^2)\&\left(\frac{d}{dt}\left(\frac{\partial L}{\partial \dot{x}}\right) = -eH\dot{z}\right)\right.$$
$$\left.\&\left(\frac{d}{dt}\left(\frac{\partial L}{\partial \dot{y}}\right) = eE\right)\&\left(\frac{d}{dt}\left(\frac{\partial L}{\partial \dot{z}}\right) = eH\dot{x}\right)\right).$$

Note that L, as it occurs in S, is an unspecified function of q. L acts somewhat like a free parameter. T_E and T_A differ in that they ascribe different 'values' to the function L. According to T_E:

$$L = -m_0c^2((1 - q^2/c^2)^{1/2} - 1) \text{ and according to } T_A:$$

$$L = -\tfrac{3}{4}m_0c^2\left(\frac{c^2 - q^2}{2cq}\log\left(\frac{c + q}{c - q}\right) - 1\right),$$

where m_0 = mass of the electron at rest.

For the purpose of assessing the methodological significance of Kaufmann's results, 'T_E' and 'T_A' can be regarded as representing the following two propositions:

(25) $T_E \equiv S\&(L = -m_0c^2((1 - q^2/c^2)^{1/2} - 1))$ and

(26) $T_A \equiv S\&\left[L = -\tfrac{3}{4}m_0c^2\left(\frac{c^2 - q^2}{2qc}\log\left(\frac{c + q}{c - q}\right) - 1\right)\right].$

We are now ready to give an interpretation of b_{E_i} and b_{A_i}. Let $(x_2, \bar{y}, \bar{z})$ be the coordinates of an arbitrary electron as it hits the plate. By (16), a_i is the proposition that $\bar{z}_i$ is the z-coordinate of the ith electron; b_i is the proposition that $\bar{y}_i$ is the value of $\bar{y}$ when $\bar{z} = \bar{z}_i$. Our notation already suggests that b_{E_i} will be a proposition about the value of $\bar{y}$ which, according to T_E, ought to correspond to $\bar{z} = \bar{z}_i$. But how does T_E predict such a value of $\bar{y}$? Let me give a simplified answer which will be more fully clarified below.[20] The conjunction $T_E\&K$ determines an equation of the form $\bar{y} = f_E(\bar{z})$ where the right-hand side is an expression involving $\bar{z}$. Substituting $\bar{z}_i$ for $\bar{z}$ and computing $f_E(\bar{z}_i)$ we obtain a numerical value $\bar{y}_{E_i}$ such that $\bar{y}_E = f_E(\bar{z}_i)$. The proposition b_{E_i} is defined by:

(27) $b_{E_i} \equiv$ (the y-coordinate of the ith electron as it hits the plate is $\bar{y}_{E_i}$).

Thus

(28) $\begin{cases} (T_E\&K\&a_i) \to b_{E_i}. \text{ That is} \\ (T_E\&K) \to (a_i \to b_{E_i}). \end{cases}$

Similarly, $T_A\&K$ determines an equation[21] of the form $\bar{y} = f_A(\bar{z})$. Substituting $\bar{z}_i$ for $\bar{z}$, we obtain $\bar{y}_{A_i} = f_A(\bar{z}_i)$, where $\bar{y}_{A_i}$ is some numerical value. Defining:

(29) $b_{A_i} \equiv$ (the y-coordinate of the ith electron as it hits the plate is $\bar{y}_{A_i}$),

we have:

(30) $(T_A \& K) \to (a_i \to b_{A_i})$.

It was found that $\bar{y}_i$ is different both from $\bar{y}_{E_i}$ and from $\bar{y}_{A_i}$, but that $|\bar{y}_i - \bar{y}_{A_i}| < |\bar{y}_i - \bar{y}_{E_i}|$ for all $i = 1, 2, ... , 9$. This is why Kaufmann's experiment was taken to refute Lorentz-Einstein while confirming Abraham.

This brings us to the end of the interpretative part of this appendix I shall now tackle the problem of establishing the logical relations on which my philosophical claims rest. This last part of the paper is divided into the following subsections.

(i) *Important consequences of* $S \& P \& Q(w)$

We consider a single electron which leaves the origin, passes through the opening at $(x_1, 0, 0)$ and then hits the plate at $(x_2, \bar{y}, \bar{z})$. Let

(31) $p \underset{\text{def}}{=} \dfrac{dL}{dq}$.

Remembering that $q = \text{speed} = (\dot{x}^2 + \dot{y}^2 + \dot{z}^2)^{1/2}$, we have

$$\frac{\partial L}{\partial \dot{x}} = \frac{dL}{dq}\frac{\partial q}{\partial \dot{x}} = p\tfrac{1}{2}(\dot{x}^2 + \dot{y}^2 + \dot{z}^2)^{-1/2}2\dot{x} = p\frac{\dot{x}}{q}.$$

Thus Equation (21) becomes:

(32) $\dfrac{d}{dt}\left(p\,\dfrac{\dot{x}}{q}\right) = -eH\dot{z}$.

Similarly, Equations (22) and (23) can be rewritten as:

(33) $\dfrac{d}{dt}\left(p\,\dfrac{\dot{y}}{q}\right) = eE$

and

(34) $\dfrac{d}{dt}\left(p\,\dfrac{\dot{z}}{q}\right) = eH\dot{x}$

respectively.

We recall that H is constant, so that (32) and (34) can be immediately integrated to yield:

(35) $p\,\dfrac{\dot{x}}{q} = -eHz + \mu$

and

(36) $p \dfrac{\dot{z}}{q} = eHx + \nu,$

where μ and ν are constant. Dividing (35) by (36):

$$\frac{\dot{x}}{\dot{z}} = \frac{-z + a}{x + b}, \text{ where } a \text{ and } b \text{ are also constant.}$$

Hence: $\dot{x}(x + b) = \dot{z}(-z + a)$. Integrating with respect to t:

$\frac{1}{2}(x + b)^2 = -\frac{1}{2}(z - a)^2 + k$, i.e.

$(x + b)^2 + (z - a)^2 = 2k.$

⌐k = constant⌐. This is the equation of a circle Γ in the xz-plane. Since the electron passes through the points $(0, 0, 0)$, $(x_1, 0, 0)$ and $(x_2, \bar{y}, \bar{z})$, Γ must go through the three points $(0, 0)$, $(x_1, 0)$ and $(x_2, \bar{z})$ in the xz-plane. Of course these three points uniquely determine Γ, i.e. Γ is uniquely determined by x_1, x_2 and $\bar{z}$. Let us give a parametric representation of Γ, taking as our parameter the angle ϕ which the tangent to Γ at a variable point P makes with the x-axis (see Figure 3).

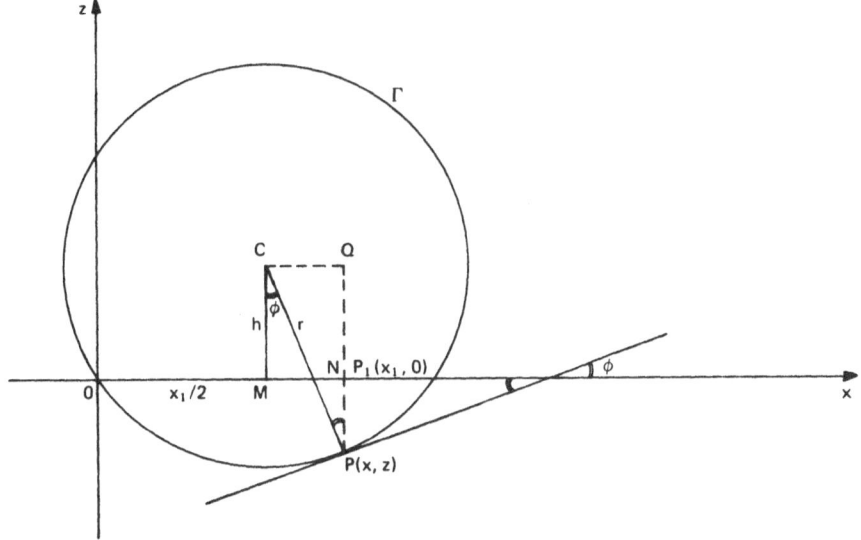

Fig. 3.

Let r = radius of the circle Γ and $h = MC$, where M is the midpoint of OP_1 and C is the centre of Γ. The coordinates x and z of a variable point P of Γ can be expressed in terms of ϕ as follows:

(37) $\qquad \begin{cases} x = OM + MN = x_1/2 + r \sin \phi, \\ z = NP = -(PQ - NQ) = -(PQ - MC) = -r \cos \phi + h. \end{cases}$

We now propose to express r and h in terms of x_1, x_2 and $\bar{z}$. This is possible because x_1, x_2 and $\bar{z}$ uniquely determine Γ. Let ϕ_1 and ϕ_2 be the values assumed by ϕ at the points $P_1(x_1, 0)$ and $P_2(x_2, \bar{z})$ respectively (see Figure 4).

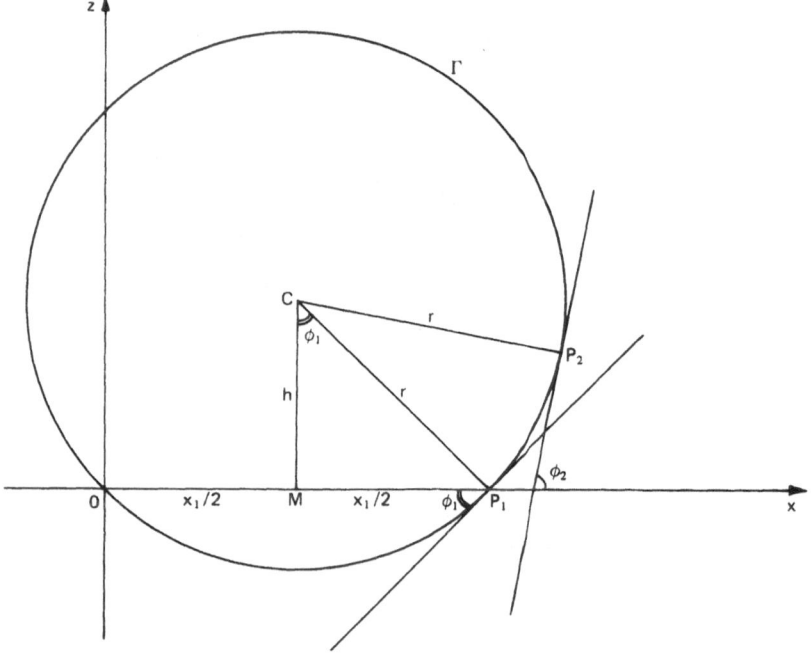

Fig. 4.

In the right-angled triangle $MCP_1 : \tan \phi_1 = (x_1/2)/h$. Hence $h = (x_1/2)/\tan \phi_1$, i.e. $h = (x_1/2) \cot \phi_1$.

Substituting in (37):

(38) $\qquad \begin{cases} x = r \sin \phi + x_1/2, \\ z = -r \cos \phi + (x_1/2) \cot \phi_1. \end{cases}$

Similarly, in the triangle MCP $(x_1/2)/r = \sin \phi_1$, i.e.

(39) $r = x_1/2 \sin \phi_1$.

It remains for us to express ϕ_1 in terms of x_1, x_2 and $\bar{z}$. We recall that ϕ_2 is the value of ϕ at $P(x_2, \bar{z})$.

Substituting in (38):

$$x_2 = r \sin \phi_2 + x_1/2 \quad \text{and} \quad \bar{z} = -r \cos \phi_2 + (x_1/2) \cot \phi_1.$$

Substituting for r from (39) into $x_2 = r \sin \phi_2 + x_1/2$, we obtain

(40) $\sin \phi_2 = (2x_2 - x_1) \sin \phi_1/x_1$.

Equation (40) expresses ϕ_2 in terms of ϕ_1. Let us now turn to

$$\bar{z} = -r \cos \phi_2 + (x_1/2) \cot \phi_1.$$

Thus:

$$r \cos \phi_2 = x_1/2 \cot \phi_1 - \bar{z}.$$

Squaring:

$$R^2 \cos^2 \phi_2 = (x_1/2 \cot \phi_1 - \bar{z})^2;$$

i.e.

$$r^2(1 - \sin^2 \phi_2) = (x_1/2 \cot \phi_1 - \bar{z})^2.$$

By (39) and (40).

$$\frac{x_1{}^2}{4 \sin^2 \phi_1} \left(1 - \frac{(2x_2 - x_1)^2 \sin^2 \phi_1}{x_1{}^2}\right) = (x_1/2 \cot \phi_1 - \bar{z})^2.$$

Expanding both sides of this last equation and noting that

$$1 + \cot^2 \phi_1 = \frac{1}{\sin^2 \phi_1},$$

we obtain:

(41) $\tan \phi_1 = \dfrac{x_1 \bar{z}}{(x_2 - x_1)x_2 + \bar{z}^2}.$

We are now home and dry. (41) gives us ϕ_1 in terms of x_1, x_2 and $\bar{z}$. Substituting in (39) and (40), we obtain r and ϕ_2 in terms of x_1, x_2 and $\bar{z}$. Equations (38) express x and z in terms of ϕ, ϕ_1, x_1 and r, and hence indirectly in terms of ϕ, x_1, x_2 and $\bar{z}$.

We shall now determine ϕ as a function of t. Differenting equations (38):

(42) $\qquad \dot{x} = r \cos \phi \dot{\phi}$ and $\dot{z} = r \sin \phi \dot{\phi}$.

Substituting from (42) and (38) into (35):

$$\frac{p}{q} (r \cos \phi \dot{\phi}) = eHr \cos \phi + k,$$

where k is some constant. Substituting $\Pi/2$ for ϕ in this last equation: $0 = 0 + k$. Thus $k = 0$. Therefore

$$\frac{p}{q} (r \cos \phi \dot{\phi}) = eHr \cos \phi + k = eHr \cos \phi.$$

Dividing by $r \cos \phi$:

(43) $\qquad \frac{p}{q} \dot{\phi} = eH$; that is $\dot{\phi} = eHq/p$.

We recall that $q^2 = \dot{x}^2 + \dot{y}^2 + \dot{z}^2 = r^2 \cos^2 \phi \dot{\phi}^2 + \dot{y}^2 + r^2 \sin^2 \phi \dot{\phi}^2$ (by (42)) $= r^2 \dot{\phi}^2 + \dot{y}^2$.
Since $\dot{y}^2$ is negligibly small in comparison with $\dot{x}^2 + \dot{z}^2$,

(44) $\qquad q^2 = r^2 \dot{\phi}^2$; that is $q = r\dot{\phi}$ (approximately).

Substituting in (43):

(45) $\qquad p = eHr = eHx_1/2 \sin \phi_1 \ulcorner$by (39)$\urcorner$.

p is therefore constant. We recall that $p \underset{\text{def}}{=} dL/dq$ where L is a function of q. Solving $p = dL/dq$ for q in terms of p, we conclude that q is a function of p. Like p, q is therefore constant. Since both p and q are constant, it follows from (43) that

(46) $\qquad \dot{\phi}$ = constant. Integrating:

(47) $\qquad \phi = at + b$, where a and b are both constant.

Let us now consider Equation (33), i.e.

$$\frac{d}{dt} \left(\frac{p}{q} \dot{y} \right) = eE.$$

Since both p and q are constant, this equation can be written as:

(48) $\qquad eE = \dfrac{p}{q} \dfrac{d\dot{y}}{dt} = \dfrac{p}{q} \dfrac{d^2 y}{dt^2}.$

But

$$\frac{d}{dt} = \frac{d\phi}{dt} \frac{d}{d\phi} = \dot{\phi} \frac{d}{d\phi}.$$

Noting that $\dot{\phi}$ is constant:

$$\frac{d^2}{dt} = \dot{\phi} \frac{d}{d\phi} \left(\dot{\phi} \frac{d}{d\phi} \right) = \dot{\phi}^2 \frac{d^2}{d\phi^2}.$$

Substituting in (48)

(49) $\qquad \dfrac{p\dot{\phi}^2}{q} \dfrac{d^2 y}{d\phi^2} = eE.$

By (43) and (45)

(50) $\qquad \dfrac{d^2 y}{d\phi^2} = \dfrac{r}{qH} E.$

It remains for us to integrate (50) with respect to ϕ. We recall that $E = 10^8\, wE_1(\xi)$ where $\xi = x - x_1/2$. (See Figure 2 and Equations (18).) Substituting in (50):

(51) $\qquad \dfrac{d^2 y}{d\phi^2} = 10^8\, \dfrac{rw}{qH} E_1(\xi).$

We have to express ξ in terms of ϕ. By (38):

(52) $\qquad \xi = x - x_1/2 = r \sin \phi + x_1/2 - x_1/2 = r \sin \phi.$

Hence $r \sin(-\phi) = - r \sin \phi = - \xi$; so $-\xi$ corresponds to $-\phi$.

(53) $\qquad$ Put: $E_1(\xi) = \bar{E}_1(\phi).$

Since $E_1(-\xi) = - E_1(\xi)$, we have: $\bar{E}_1(-\phi) = E_1(-\xi) = - E_1(\xi) = - \bar{E}_1(\phi)$. In order to integrate (51), we have to express $E_1(\phi)$ explicitly in terms of ϕ. Let ϕ' be the value of ϕ which corresponds to $\xi' = 0.593$ (See (19)). By (52)

(54) $\qquad \xi' = r \sin \phi'$ i.e. $0.593 = r \sin \phi'.$

Since r is a known function of x_1, x_2 and $\bar{z}$, this last equation determines ϕ' in terms of x_1, x_2 and $\bar{z}$. By (18):

(55)
$$\bar{E}_1(\phi) = E_1(\xi) = 1 \quad \text{for} \quad \xi \in [0, \xi'],$$
$$\text{i.e. for} \quad \phi \in [0, \phi']; \text{ and}$$
$$\bar{E}_1(\phi) = E_1(\xi) = \alpha - \lambda\xi = \alpha - \lambda r \sin \phi$$
$$\text{for} \quad \xi \in [\xi', x_1/2], \text{ i.e. for} \quad \phi \in [\phi', \phi_1]; \text{ and}$$
$$\bar{E}_1(-\phi) = E_1(-\xi) = -E_1(\xi) = -\bar{E}_1(\phi)$$
$$\text{for} \quad \xi \in [0, x_1/2], \text{ i.e. for} \quad \phi \in [0, \phi_1].$$

Integrating (51) from 0 to ϕ_1:

$$\left(\frac{dy}{d\phi}\right)_{\phi_1} = 10^8 \frac{rw}{qH} \int_0^{\phi_1} \bar{E}_1(\phi)\, d\phi$$

$$= 10^8 \frac{rw}{qH} \left(\int_0^{\phi'} \bar{E}_1(\phi)\, d\phi + \int_{\phi'}^{\phi_1} \bar{E}_1(\phi)\, d\phi \right)$$

$$= 10^8 \frac{rw}{qH} \left(\int_0^{\phi'} 1\, d\phi + \int_{\phi'}^{\phi_1} (\alpha - \lambda r \sin \phi)\, d\phi \right),$$

Hence

(56) $$\left(\frac{dy}{d\phi}\right)_{\phi_1} = 10^8 \frac{rw}{qH} (\phi' + \alpha(\phi_1 - \phi') + \lambda r(\cos \phi_1 - \cos \phi')).$$

By the first equation in (18), $E_1(\xi) = 0$ for $\xi \in [x_1/2, x_2 - x_1/2]$; i.e. $\bar{E}_1(\phi) = 0$ for $\phi \in [\phi_1, \phi_1]$. By (51)

$$\frac{d^2y}{d\phi^2} = 0 \text{ for all } \phi \in [\phi_1, \phi_2]. \text{ Integrating between } \phi \text{ and } \phi_1:$$

$$\frac{dy}{d\phi} - \left(\frac{dy}{d\phi}\right)_{\phi_1} = 0, \text{ i.e. } \frac{dy}{d\phi} = \left(\frac{dy}{d\phi}\right)_{\phi_1}$$
$$= \text{constant, for } \phi \in [\phi_1, \phi_2].$$

Integrating between ϕ_1 and ϕ_2:

(57) $$(y)_{\phi_2} - (y)_{\phi_1} = (\phi_2 - \phi_1)\left(\frac{dy}{d\phi}\right)_{\phi_1}.$$

But $(y)_{\phi_1} = 0$ because the electron passes through the point $(x_1, 0, 0)$; and $(y)_{\phi_2} = \bar{y}$ because the electron hits the plate at $(x_2, \bar{y}, \bar{z})$. Substituting in (57):

(58) $\bar{y} = (\phi_2 - \phi_1)\left(\dfrac{dy}{d\phi}\right)_{\phi_1}$. By (56) and (58)

(59) $\bar{y} = 10^8 \dfrac{rw}{qH}(\phi_2 - \phi_1)(\phi' + \alpha(\phi_1 - \phi') + \lambda r(\cos \phi_1 - \cos \phi'))$.

Note that ϕ_1, ϕ_2, ϕ' and r are known functions of x_1, x_2 and $\bar{z}$, and that x_1, x_2, α and λ are numerically given quantities. Thus (59) can be written as:

(60) $\bar{y} = F(w, q, \bar{z})$, where F is a known function of the three variables w, q and $\bar{z}$.

Equation (60) plays a central role in Planck's assessment of Kaufmann's results. This is because (60) enables us, given $\bar{z}$, to calculate each of the three quantities $\bar{y}$, w, q in terms of the remaining two.

(ii) *Experimental Refutation of S&K*

We recall that $K \equiv P \& Q(w) \& (w = w_0)$, where w_0 stands for the numerical value $2500/0.1242$.[22] By (60):

(61) $\bar{y} = F(w_0, q, \bar{z}) = F(2500/0.1242, q, \bar{z})$.

Consider the observation report $a_1 \& b_1$ which tells us that $\bar{y} = \bar{y}_1$ when $\bar{z} = \bar{z}_1$.[23] Substituting $\bar{z}_i$ for $\bar{z}$ and $\bar{y}_i$ for $\bar{y}$ in (61), we obtain:

(62) $\bar{y}_i = F(w_0, q_i, \bar{z}_i)$,

where q_i denotes the speed of the ith electron as it hits the photographic plate. (62) enables us to calculate q_i, since $\bar{y}_i$, w_0 and $\bar{z}_i$ are all numerically given. Planck found that, for $\bar{y}_i = 0.0247$ and $\bar{z}_i = 0.1354$.
$q_i/c = 1.034$ (approximately), where c is the velocity of light. Hence $q_i > 0$, which contradicts an implication of S, namely that the speed of an electron remains smaller than c. Thus $S\&K$ is experimentally refuted by Kaufmann's results. Since S is entailed both by T_E and by T_A, it follows that $T_E\&K$ and $T_A\&K$ are also refuted.

(iii) *Important consequences of $T_E\&P\&Q(w)$ and of $T_A\&P\&Q(w)$*

We recall that $T_E \to S$ and $T_A \to S$ because S is a component common to T_E and T_A.[24] Hence:

$$T_E\&P\&Q(w) \to S\&P\&Q(w)$$

and

$$T_A\&P\&Q(w) \to S\&P\&Q(w).$$

We can therefore make use of all the results established in (i).

By (25), $L = -mc^2((1 - q^2/c^2)^2 - 1)$. Noting that $p \underset{\text{Def}}{=} dL/dq$ it follows that T_E implies:

$$p = \frac{d}{dq}(-m_0 c^2((1 - q^2/c^2)^{1/2} - 1)) = m_0 q/\sqrt{1 - q^2/c^2}.$$

In conjunction with (45), this last equation yields:

$$\left(\frac{e}{m_0}\right) \frac{Hx_1}{2 \sin \phi_1} = \frac{q}{\sqrt{1 - q^2/c^2}}.$$

This enables us to calculate the (constant) speed of the electron. Let us denote this calculated value of q by q_E, where the subscript E indicates the dependence of the calculation on the theory T_E. Thus $T_E \& P \& Q(w)$ implies:

(63)$_E$ $$\left(\frac{e}{m_0}\right) \frac{Hx_1}{2 \sin \phi_1} = \frac{q_E}{\sqrt{1 - q_E^2/c^2}}.$$

Substituting q_E for q in (60):

(64)$_E$ $\bar{y} = F(w, q_E, \bar{z})$.

Similarly, $T_A \& P \& Q(w)$ implies that

$$L = -\tfrac{3}{4} m_0 c^2 \left(\left(\frac{c^2 - q^2}{2qc}\right) \log\left(\frac{c + q}{c - q}\right) - 1\right);$$

hence

$$p = \frac{dL}{dq} = \frac{3}{4}\left(\frac{m_0 c^2}{q}\right)\left(\left(\frac{c^2 + q^2}{2qc}\right) \log\left(\frac{c + q}{c - q}\right) - 1\right).$$

In conjunction with (45), this last equation yields:

(63)$_A$ $$\frac{e}{m_0} \frac{Hx_1}{2 \sin \phi_1} = \frac{3c^2}{4q_A}\left(\left(\frac{1 + q_A^2/c^2}{2q_A/c}\right) \log\left(\frac{c + q_A}{c - q_A}\right) - 1\right),$$

where q_A is the speed of the electron according to the theory $T_A \& P \& Q(w)$. Substituting q_A for q in (60)

(64)$_A$ $\bar{y} = F(w, q_A, \bar{z})$.

Note that q_E and q_A are calculated numerical quantities and that $q_E \neq q_A$. Thus T_E and T_A predict different values for the speed of the electron.

(iv) *'Confirmation' of T_A&K and Refutation of T_E&K*

Let us remember that $K \equiv P\&Q(w)\&(w = w_0)$ where $w_0 = 2500/0.1242$. Substituting w_0 for w in $(64)_E$, we see that $T_E\&K$ implies: $\bar{y} = F(w_0, q_E, \bar{z})$. Since both w_0 and q_E denote known numerical values, the equation can be written as:

$(65)_E$ $\bar{y} = f_E(\bar{z})$, where $f_E(\bar{z}) \underset{\text{Def}}{=} F(w_0, q_E, \bar{z})$.

For each value of $\bar{z}$, $(65)_E$ determines the value of $\bar{y}$ which, according to $T_E\&K$, corresponds to $\bar{z}$. Substituting the observed value $\bar{z}_i$ for $\bar{z}$, let us calculate the value $\bar{y}_{E_i}$ predicted by $T_E\&K$ for $\bar{y}$. Thus:

$(66)_E$ $\bar{y}_{E_i} = f_E(\bar{z}_i)(= F(w_0, q_E, \bar{z}_i))$.

Similarly $T_A\&K$ entails:

$(65)_A$ $\bar{y} = f_A(\bar{z})$ where $f_A(\bar{z}) \underset{\text{Def}}{=} F(w_0, q_A, \bar{z})$;

and

$(66)_A$ $\bar{y}_{A_i} = f_A(\bar{z}_i)(= F(w_0, q_A, \bar{z}_i))$.

As we have already remarked,[25] it turns out that:

$$|\bar{y}_i - \bar{y}_{A_i}| < |\bar{y}_i - \bar{y}_{E_i}| \quad \text{for all} \quad i = 1, \dots, 9.^{[26]}$$

This inequality led people to the erroneous conclusion that Kaufmann's experiment confirmed Abraham while refuting Lorentz-Einstein.

(v) *Adjustment of parameters in $T_E\&P\&Q(w)$ and in $T_A\&P\&Q(w)$*

We recall that $T_E\&P\&Q(w)$ implies:

$(64)_E$ $\bar{y} = F(w, q_E, \bar{z})$.

By (16), $a_i\&b_i$ tells us that $\bar{y} = \bar{y}_i$ when $\bar{z} = \bar{z}_i$. If we adjoin $a_i\&b_i$ to $T_E\&P\&Q(w)$, we can substitute $\bar{y}_i$ for $\bar{y}$ and $\bar{z}_i$ for $\bar{z}$ in $(64)_E$. Thus:

$(67)_E$ $\bar{y}_i = F(w, q_E, \bar{z}_i)$.

Solving (67) for w, we obtain

$(68)_E$ $w = w_{E_i}$,

where w_{E_i} is some numerical value. Hence:

$(69)_E$ $(T_E\&P\&Q(w)\&a_i\&b_i) \to (w = w_{E_i})$.

Note that w_{E_i} is a solution of $(67)_E$. Thus, substituting w_{E_i} for w in $(67)_E$, we obtain the identity:

$(70)_E$ $\bar{y}_i = F(w_{E_i}, q_E, \bar{z}_i)$.

From (70), it follows that $\bar{y} = \bar{y}_i$ when $\bar{z} = \bar{z}_i$, $q = q_E$ and $w = w_{E_i}$. But $T_E \& P \& Q(w)$ entails $q = q_E$.[27] Hence:

$(71)_E$ $\begin{cases} (T_E \& P \& Q(w) \& (w = w_{E_i}) \& a_i) \to b_i, \text{ i.e.} \\ (T_E \& P \& Q(w) \& (w = w_{E_i})) \to (a_i \to b_i).[28] \end{cases}$

Similarly it can be shown that:

$(69)_A$ $(T_A \& P \& Q(w) \& a_i \& b_i) \to (w = w_{A_i})$;

and

$(71)_A$ $(T_A \& P \& Q(w) \& (w = w_{A_i})) \to (a_i \to b_i)$,

where w_{A_i} is a solution of

$(67)_A$ $\bar{y}_i = F(w, q_A, \bar{z}_i)$.

That is:

$(70)_A$ $\bar{y}_i = F(w_{A_i}, q_A, \bar{z}_i)$

is an identity.

Planck found that $w_{E_1}, \dots, w_{E_9}$,[29] are more nearly equal among themselves than are $w_{A_1}, \dots, w_{A_9}$.

This tips the balance slightly in favour of T_E.

London School of Economics

NOTES

[1] See Feyerabend [1972].
[2] See Kaufmann [1905].
[3] See Planck [1906a].
[4] See Planck [1906b].
[5] See Planck [1907].
[6] This is explained in detail above in Section 2.
[7] Thus we have two conflicting intuitions: the intuition that, if *any one* of Kaufmann's results is satisfactorily explained, then *all* of them are; and the intuition that Planck's achievement would have been greater, had he not worked backwards from at least one experimental result to the determination of the auxiliary hypothesis K_E. I would argue

(see also Section 2) that the conflict ought to be resolved in favour of the second intuition; this anyway does not involve much modification of the first intuition.

[8] In fact, this procedure is as follows: T_A is Abraham's hypothesis. K_{A_i} is of the form $R(w)\&(w = w_{A_i})$, where w_{A_i} is a numerical value uniquely determined by $T_A\&R(w)\&a_i\&b_i$. That is: $(T_A\&P(w)\&a_i\&b_i) \rightarrow (w = w_{A_i})$. See equivalence (9).

[9] Though this involved an idealisation, as indicated above.

[10] See Lorentz [1906], p. 212.

[11] See Zahar [1973].

[12] See Lorentz [1906], p. 339, Note 86.

[13] See Bucherer [1908].

[14] See Bucherer [1908], p. 525.

[15] See Lorentz [1914].

[16] I argue this more fully in my [1973].

[17] My italics and translation. Planck's actual words were: "Wäre [w] bei einer Theorie für alle Ablenkungen gleich gross, so könnte man sagen, dass für diese Theorie alle von Herrn Kaufmann gemessenen Ablenkungen vollständig erklärt werden ...". See Planck [1907], p. 213.

[18] In fact both $\bar{z}_i$ and $\bar{y}_i$ are obtained through an averaging process over a (finite) range of measured entities. Thus $\bar{z}_i$ and $\bar{y}_i$ are not, strictly speaking, the 'observed' coordinates of a single electron.

[19] Cf. p. 84.

[20] See e.g. $(65)_E$.

[21] See e.g. $(65)_A$.

[22] See (20).

[23] See (16).

[24] See (25) and (26).

[25] See p. 83.

[26] e.g. $\bar{y}_3 = 0.0506$, $\bar{y}_{A_3} = 0.0526$, $\bar{y} = 0.0555$. Cf. Planck [1906b], p. 129.

[27] See (iii).

[28] For the meanings of a_i and b_i see (16).

[29] It turns out that: $w_{E_1} > w_{E_2} > \cdots > w_{E_9}$ and $w_{A_1} > w_{A_2} > \cdots > w_{A_9}$ and e.g.: $w_{E_1} = 17\,970$, $w_{E_9} = 17\,590$ while $w_{A_1} = 18\,840$, $w_{A_9} = 18\,040$.

The notes refer to the unified bibliography for Parts I and III on pp. 379–383.

PETER URBACH

THE OBJECTIVE PROMISE OF A RESEARCH
PROGRAMME

1. INTRODUCTION

Philosophers of science have put forward various methodologies for
appraising the scientific merits of theories. These methodologies have
usually been backward-looking in the sense that they treat theories as
finished products and appraise their scientific value mainly in terms of
their empirical performance to date.

When appraising theories which are still being developed and on
which scientists are still carrying out their enquiries, the question
arises whether there is a *forward-looking* methodology which would
enable us to appraise the future performance of a line of research.
Such a question is likely to be raised, for instance, by a scientist faced
with the problem of which of various lines of research he should
invest his time and energy in or by a research sponsor who is
interested in supporting that research project which has the greatest
chance of success.

The orthodox view among modern philosophers of science is that
any attempt to appraise the future performance of scientific research
is bound to be futile. In this section I will argue that the orthodox
view represents a considerable over-reaction to the old Cartesian
philosophy and that, with regard to future scientific performance, the
methodology of scientific research programmes provides an approach
which tempers the pessimism of the orthodox position while avoiding
the pitfalls of a naively over-optimistic conception of the heuristic
power of some 'method' of science.

2. THE ORTHODOX VIEW ON SCIENTIFIC CREATIVITY

Philosophers of science who have addressed themselves to the question of future scientific performance have come to conclusions which
fall into two main categories. On the one side, classical philosophers
such as Descartes held that the philosophy of science vouchsafed a
very definite and reliable method for the advancement of science.

A unified bibliography can be found on pp. 379–383.

99

G. Radnitzky and G. Andersson (eds.), Progress and Rationality in Science, 99–113.
All Rights Reserved.
Copyright © 1978 by D. Reidel Publishing Company, Dordrecht, Holland.

Given such a method, the problem of appraising future performance is trivially solved: provided that the method is applied carefully and systematically, its fruits are bound to be good. On the other
side, most modern philosophers of science, for example Whewell and
Popper, have claimed that the scientific future is essentially unknow-
able: all we can do is wait and see what theories are proposed, and
then appraise them.

The Cartesian scientific ideal, which admitted into science only
'demonstrable knowledge', excluding mere 'pretty and probable conjectures', was accompanied by a *heuristic ideal*. According to this
heuristic ideal, science was to be pursued according to strict
mechanical procedures whose reliability was guaranteed.

Although Descartes granted that desultory and undirected investigations may occasionally lead to true discoveries, he regarded
any success achieved in this way as entirely fortuitous and as likely to
occur so infrequently that "it were far better never to think of
investigating truth at all than to do so without a method".[1] For this
reason Descartes set down his "rules for the direction of the mind":

"certain simple rules such that if a man observe them accurately, he shall never assume
what is false as true, and will never spend his mental efforts to no purpose, but will
always gradually increase his knowledge and so arrive at a true understanding of all
that does not surpass his powers".[2]

This idea was still alive two centures later. Sir John Herschel
thought (following a vulgarised version of Baconian philosophy) that
science is a more or less mechanical procedure requiring for its
reliable application merely a clear and unprejudiced use of the senses
and a knowledge of elementary logic:

"Experience, once recognised as the fountain of all our knowledge of nature, it follows
that, in the study of nature and its laws, we ought [to] ... content ourselves with
observing, as a plain matter of fact, what *is*. But before experience itself can be used
with advantage, there is one preliminary step to make, which depends wholly on
ourselves; it is the absolute dismissal and clearing the mind of all prejudice, from
whatever source arising, and the determination to stand or fall by the result of a direct
appeal to facts in the first instance, and of *a direct logical* deduction from them
afterwards."[3]

Now if any of these 'methods' actually constituted a true method of
scientific discovery, the problem of deciding which line of research
will be the most profitable to pursue would be rendered much more
tractable. Given that the applicants for research grants are conscien-

tious workers in possession of the right method and willing to apply it, research councils may allocate funds simply in the light of the cost of the various projects and of the social and political utility of the knowledge which will most certainly be produced.

The Cartesian idea of a discovery-making method is, of course, totally discredited nowadays. Certain modern philosophers of science may have tried to retain different parts of the old *aim* of science, but every one of them repudiates the idea of an infallible heuristic. Indeed most of them look askance at the idea of even a fallible heuristic for scientific theorising.[4]

Popper shares this widespread opinion. Indeed he had a special reason for claiming that there can be no general method for advancing from a given scientific theory to a better one: the more advanced theory will, typically, be *inconsistent* with its predecessor(s). His chief illustration for this was Newton's theory which not only transcends any finite set of observation statements in precision and in its reference to all space-time regions but is also inconsistent with Kepler's and Galileo's laws. The often voiced opinion that Newton logically derived his laws from those of Kepler and Galileo must, in view of these considerations, be rejected.

Popper went on to argue that these considerations provide "a very strong argument against induction"[5]; they "refute the Baconian myth that we must begin with observations in order to derive our theories from them". Popper adds, claiming Kant as the originator of the idea, that behind the historiographical refutation of induction lies "a logical fact; that there [are] ... logical reasons why this kind of thing did not occur in the history of science: that it is logically impossible to derive theories from observations".[6]

But if theories are not derived from our previous knowledge and if we reject the Kantian view that they are not derived from the laws of human understanding, where do new scientific theories come from? Popper's answer to this question is that new theories are:

"the *free* creation of our own minds, the result of an almost poetic intuition".[7]

Similar views concerning the origins of scientific theories were expressed by Popper in his earliest writings:

"The initial stage, the act of conceiving or inventing a theory seems to me *neither to call for logical analysis nor to be susceptible of it*. The question how it happens that a new idea occurs to a man – whether it is a musical theme, a dramatic conflict or a scientific theory – may be of great interest to empirical psychology, but it is irrelevant to the logical analysis of scientific knowledge."

"If it is the processes involved in the stimulation and release of an inspiration which are to be reconstructed then I should refuse to take it as the task of the logic of knowledge. Such processes are the concern of empirical psychology but hardly of logic."

"... there is no such thing as a logical method of having new ideas, or a logical reconstruction of this process."

"... every discovery contains an 'irrational element' or a 'creative intuition' in Bergson's sense."[8]

Hempel[9] has endorsed these views of Popper's and a similar account of scientific discovery has been put forward by, for example, Whewell and Reichenbach. Thus, Whewell:

"The conceptions [i.e. theories] by which facts are bound together are suggested by the sagacity of discoverers. *This sagacity cannot be taught.* It commonly succeeds by guessing; and this success seems to consist in framing several tentative hypotheses and selecting the right one. *But a supply of appropriate hypotheses cannot be constructed by rule, nor without inventive talent.*"[10]

Similarly Reichenbach claims that:

"The act of discovery escapes logical analysis; there are no logical rules in terms of which a 'discovery machine' could be constructed which would take over the creative function of the genius."[11]

For Popper, Whewell and others, the origins of new scientific theories are essentially mysterious. It clearly follows from this thesis that there is no possibility of predicting future theoretical developments in science. As Popper puts it, after demonstrating that Galileo's and Kepler's theories are inconsistent with Newton's:

"It is important to note that from Galileo's or Kepler's theories we do not obtain *even the slightest hint* of how these theories would have to be adjusted – what false premises would have to be adapted, or what conditions stipulated – should we try to proceed from these theories to another and more generally valid one such as Newton's."[12]

Popper adds that the logical gap between successive scientific theories can only be bridged by 'ingenuity', the outcome of which can never be scientifically predicted.[13]

On this view of scientific discovery, it may well be possible to arrive at rational decisions concerning sponsorship for the *empirical testing* of already formulated scientific hypotheses, but the elaborate scrutiny of a line of *theoretical* research to decide whether it deserves sponsorship would be an empty ritual: one might just as well settle the issue by the toss of a coin.

In view of what I shall claim later in this section, it is, perhaps, rather ironic that Lakatos followed Popper in discounting any possibility of appraising the future performance of research programmes. Modern methodologies, including the methodology of scientific research programmes consist he claimed, "*merely* of a set of rules for the appraisal of ready, articulated theories".[14] And since degenerating programmes 'can always stage a comeback', while progressive programmes can begin to degenerate at any time, the normative rules of the methodology of scientific research programmes can never advise scientists which programme they should pursue.[15]

However, Lakatos also expressed '*doubts*' about whether the modern shift in methodology from the old Cartesian heuristic ideal had not 'gone a bit too far' and he added that "following Polya, I have held that there might well be a *limbo* for a 'genuine' heuristic which is rational and non-psychologistic".[16] It is precisely this aspect of the methodology of scientific research programmes, about which Lakatos himself expressed doubts, that I shall exploit in Section 5 in identifying the objective promise of a research programme.

3. SOME OBJECTIONS TO THE ORTHODOX VIEW

The orthodox thesis that successive scientific theories arise as mysterious, free acts of creative intuition and that consequently their construction could not have been guided by any rational method, seems to me an over-reaction to Descartes' philosophy. In this part, I shall indicate some of the difficulties which the orthodox view encounters and in the remaining parts of this paper I will argue that the methodology of scientific research programmes overcomes many of these difficulties and, through its theory of heuristics provides a more sanguine approach to appraising the promise of a research programme.

One consideration which detracts somewhat from the plausibility of the orthodox view is that many leading scientists do not seem to agree with it.

Consider, for example, Planck's and Ostwald's views on the relative promise of the kinetic theory and its main rival as they stood at the end of the nineteenth century. Both of these scientists claimed to have objective reasons for regarding the future of the kinetic theory as bleak.

"If it were possible to carry out the kinetic approach, it would constitute the most perfect account of the phenomenon of heat. *Further development seems impossible however for obstacles, at present insurmountable, stand in the way of future progress.* These are due not only to the highly complicated mathematical treatment, but principally to essential difficulties ... in the mechanical interpretation of the fundamental principles of thermodynamics."[17]

"It became clear to me how much more *fertile* energetics was than the kinetic atomistic theory which had become almost completely *sterile*. It had got into a thicket of mathematical difficulties which *robbed it of mobility* and almost completely prevented its adherents from going new experimental ways."[18]

The question whether there are rational grounds for supporting one line of pure research rather than another was much discussed around the Second World War. The editor of *Nature*, for example, recoiled in horror against the then much canvassed view that one cannot make any rational appraisal of the promise of a line of research.

"What is axiomatic is that we cannot escape the responsibility for prosecuting scientific effort where, according to our best judgement, it is likely to be most fruitful. That judgement may prove to be mistaken, but if it is honestly made, based on a critical survey of the facts of the whole situation so far as they can be determined we are at least not open to the charge of prostituting science or wasting scientific resources. We have made an attempt at something *more than a guess at the direction of further progress* and the mistake itself can prove a stepping stone for further advance."[19]

A second consideration with a negative bearing on the orthodox view has an undermining rather than a refuting character.

As we saw, the orthodox view as formulated by Popper relies heavily on the fact that in a major scientific advance from T to T' – for instance, from Kepler's laws to Newton's law of gravitation – it typically happens that T' is logically inconsistent with T. This does of course mean that T' could not have been logically derived from T. However, it does not mean that T' could not have been derived from T in accordance with certain extra-logical heuristic rules. I shall argue in Sections 4 and 5 that new scientific theories have, as a matter of historical fact, often been obtained from older ones by the application of certain rules; and even (*pace* both Popper and Polanyi) by the *deliberate* application of *articulated* rules.[20]

There is a third consideration which tells more directly against Popper's theory of scientific creativity. This consideration is connected with the occurrence in science of independent, 'simultaneous', discoveries. On John Watkins's interpretation of Popper's position there is "no general reason to expect ... simultaneous discoveries".[21] But this is surely an understatement. On Popperian theory, there is every reason *not to expect* simultaneous discoveries and the existence of any must be adjudged a sheer coincidence. After all, even if two independent workers are presented with identical problem situations there is an infinite number of possible theoretical solutions from which to choose. And if selection from among these is not guided by any publicly accessible rules, the likelihood that more than one scientist will make a particular choice is negligible. Nevertheless, two or more scientists working in complete independence, often in different countries, frequently arrive at identical solutions to common problems.

The discovery of the conservation of energy, of non-Euclidean geometries, of the planet Neptune, of Mendelian genetics and of the infinitesimal calculus are well-known examples of independent, simultaneous discoveries by two or more scientists. Robert Merton and his colleagues have investigated many hundred simultaneous discoveries. These sociologists have also highlighted the traditional fear amongst even the most eminent scientists that their achievements would be forestalled by others working on the same problems and they have described the various institutions which the scientific community has developed in order to protect its members from having their ideas scooped by competitors.[22] If we were to accept Popper's account of the process of discovery, we would have to regard the almost universal fear among scientists of having their discoveries preempted as totally irrational and the elaborate precautions they often adopt to prevent others falsely claiming priority to a discovery as pathological.[23] It is extremely difficult to see how the existence of so many simultaneous discoveries could be explained on Popperian lines. For the inductivist, on the other hand, the existence of simultaneous discoveries presents no problem at all. Since problem situations and reliable methods for solving problems are public property, all that is required for the emergence of a simultaneous discovery is a few hard-working scientists with similar theoretical interests. On this view, it is no more remarkable that scientists arrive at identical solutions independently than that schoolchildren versed in the same algorithms frequently produce similar solutions to algebraic problems. But the inductivist account explains too much. By making

scientific discovery so difficult, Popper cannot explain why there are
so many simultaneous ·discoveries; but inductivists, by making
science so easy, have the converse problem of explaining why there
are so few.

I shall show in the next part that the methodology of scientific
research programmes is often able to predict simultaneous discoveries and that this fact highlights a difference between this
methodology and most modern philosophies of science which is
relevant to our present purpose, namely that it is not silent about
future scientific developments.

4. THE ROLE OF HEURISTICS IN SCIENCE

The methodology of scientific research programmes is a philosophy
of science developed by Imre Lakatos out of a thoroughgoing critique
of falsificationism. This methodology takes the arguments against
inductivism seriously and it thus incorporates no infallible rules for
increasing our knowledge; yet unlike its Popperian predecessor, it
does not look on the method by which new scientific ideas are created
as an entirely closed book. Lakatos's methodology denies that most
scientific theories spring suddenly into existence, *ex nihilo*, in a
completely mysterious inspirational flash of genius; and in particular,
through its account of heuristics, it describes the elements of a
(fallible) method of discovery within a research programme.

From our present point of view, a key difference between
falsificationism and the methodology of scientific research programmes is that the latter takes on-going research-programmes as the
typical and outstanding achievements of science, whereas the former
takes given theories, which it treats as fixed. Now, a research-
programme has a future dimension. As well as its past achievements it
displays a potentiality for generating a succession of new theories in
accordance with its heuristic, which may have been articulated from
the start.

The fact that a research-programme sets out much of its research
plan and its strategy for resolving anomalies in advance, and that this
is public property, enables the methodology of research-programmes
to deal with simultaneous discoveries (which, as we saw, constitute a
puzzle for the Popperian view). Since the problems also have a public
character, it is not surprising if scientists working independently of

one another, perhaps in different parts of the world, but working within the same research-programme, come up with well-nigh identical solutions to their common problems. For example, the Newtonian programme incorporated the (publicly accessible) rule: "postulate the existence of a previously unobserved gravitating body whenever a planetary orbit deviates from its predicted value". When the planet Uranus was discovered to have an anomalous motion, at least two scientists working independently put forward the same hypothesis which led to the discovery of the planet Neptune.

It is only in regard to simultaneous discoveries made within different programmes that the methodology of scientific research programmes is at a loss for an explanation.

5. HEURISTIC POWER AND OBJECTIVE PROMISE

As we have just seen, its heuristic is an important part of a research-programme. Recognition of this fact introduces a novel element into the appraisal of science. In addition to the kinds of appraisal which the methodology of research programmes largely takes over from Popperian falsificationism (content-increasing adjustments to anomalies, successful predictions of novel facts, etc.), we can also assess research-programmes in terms of their *heuristic power*. In this part I shall indicate how the power of heuristics can be compared and I shall argue that *the power of a heuristic provides a suitable measure for the objective promise of the research programme in which it is embedded.*

Lakatos never offered a clear, general characterisation for the power of a heuristic although he frequently employed this intuitive notion, and wrote occasionally of heuristics 'running out of steam' and of research programmes being near their 'natural saturation point'.[24]

The characterisation I offer for the power of a heuristic is only a first approximation which will no doubt be filled out and improved after further philosophical and historiographical research.

The power of heuristics may, I suggest, be compared along at least three dimensions.

First, heuristics may vary in the *precision* with which they guide the construction of successive theories. Other things being equal, the most powerful heuristic will determine precisely the succeeding theory in the programme; a weaker heuristic will merely set out a greater or smaller number of hints for the construction of theories.

Secondly programmes may vary in their *resourcefulness* in the face of empirical difficulties. New versions of a research programme may be generated on the occasion of an anomaly by invoking some standard procedure for resolving such anomalies and the more powerful a programme, the more anomaly – resolving techniques it possesses. It needs to be stressed here that these standard procedures only contribute to the power of a programme if they can resolve anomalies in a content-increasing manner.

Thirdly, powerful programmes often display a considerable *autonomy* in which new theories are developed independently of the outcome of any empirical test by applying the heuristic to fields other than those already covered by existing theories.

The independent testability of succeeding theories in a programme is a necessary condition for its having any power whatever. In this regard, the status of background knowledge and of contemporary mathematical theories are particularly important in comparing programmes for the power of their heuristics. For example, a research programme which makes its predictions in the form of equations which are insoluble within any current mathematical theory is heuristically weaker, other things being equal, then one which incorporates an adequate mathematical theory. Similarly, the availability of corroborated background knowledge may augment the power of a heuristic by supplying independent means for determining initial conditions.

An initially powerful programme may gradually become heuristically weakened by being carried through, whether this is effected successfully or unsuccessfully. Thus I wish to separate the heuristic power of a programme from its progressiveness. It may also happen that two programmes differing greatly in power at their inceptions become weakened at different rates so that, after a time, the originally more powerful programme becomes the weaker. (Of course, the occurrence of such an inversion of the relative power of two programmes does not mean that one or other of them has achieved the greater empirical success.) A programme may, after all the instructions of the heuristic have been followed, finally reach a 'point of saturation'. At this point, scientists engaged on that programme will have nothing further to do.

An example of a programme which was very powerful in the first sense (*precision*) is the Ptolemaic programme. Here, successive theories were almost completely determined by the heuristic and virtually no latitude was allowed to the inventiveness of scientists. It is important to note that this algorithmic character of a heuristic does not automatically render the theories it produces *ad hoc* and, in fact, the Ptolemaic programme made several predictions. For example, it predicted a certain sequence of phases of Venus. However, the Ptolemaic programme was weak in the third sense, in that it had virtually no *autonomy*. Had the Sun and the five planets turned out to move in perfect circles around the Earth, this programme would have been immediately extinguished and it was only because this ideal (and weaker versions of this ideal) were not empirically realised that activity on the programme was sustained. Moreover, although the Ptolemaic programme could add any number of further epicycles to absorb anomalies, this fact hardly added to the programme's *resourcefulness* since successive versions of the programme brought about by such modifications were bound, given the observational techniques of the time, to be progressively less susceptible to independent testing.

The relativity programme was much weaker than the Ptolemaic programme *in the first sense* since it merely laid down certain *desiderata* which successive theories should satisfy and a good deal of leeway was given to creative ingenuity. On the other hand, this programme had considerable autonomy and was able to develop new, testable theories without any reference to the failures of its previous theories.[25]

The Cartesian vortex programme was much weaker than the relativity programme when compared on the first criterion. Inspired by metaphysical objections to the existence of a vacuum, the vortex theory asserted that celestial and gravitational phenomena are caused by a whirling celestial fluid surrounding the stars. But before the programme could get off the ground and make some testable predictions it required a theory of fluid mechanics.[26] This it did not have and consequently the main prescriptive element of the programme exhorted scientists to construct such a theory; but this exhortation was accompanied by no more precise instruction than to try to construct a liquid analogy of a celestial vortex, on a laboratory scale. Even this vague hint on how to get the vortex programme going lost most of its force after Huygens showed that if a liquid vortex induced in a laboratory cylinder were to provide a satisfactory analogue for the celestial vortex, then the action of this vortex could only explain gravity if the celestial matter were denser then all known matter.[27]

Let me now turn to my claim that a satisfactory measure of the objective promise of a research programme can be found in the power of the programme's heuristic.

Ideally we prefer a programme which we know in advance leads to well corroborated theories. But since the collapse of Cartesian philosophy has rendered this ideal unsatisfiable, we may now *merely hope* that Nature will oblige our theories. My claim that we can appraise a programme's objective promise through its heuristic power will not appeal to any guarantee of future empirical success nor to any assignment of probabilities to future empirical success. This would be impossible. I claim, however, *that a programme with a powerful heuristic is likely to produce a longer sequence of testable scientific theories, that is, greater theoretical progress than a weaker rival.* This claim can be justified with reference to the three criteria of relative heuristic power which I outlined earlier.

In the first place, programmes may differ, as I mentioned above, in the *precision* with which they direct the construction of new, testable theories. Thus, other things being equal, the more powerful programme places a smaller 'creative gap' between successive theories in a research programme, so that in such programmes the element of ingenuity and luck is reduced. Clearly, the more important the role of ingenuity and luck in the construction of new testable versions of a research programme, the less likely is such a construction to be effected and hence the less likely is the programme to progress. This is the first ground for claiming that a programme with a powerful heuristic may have greater objective promise than a weaker one.

I have already mentioned that programmes may differ in their autonomy, i.e. in the number of new areas of enquiry to which the heuristic may be extended independently of the empirical success of previous versions. If two programmes differ in their degree of autonomy from empirical tests and if their heuristics give hints of similar precision for the independent construction of new, testable theories, then the more powerful programme has, I claim, the greater objective promise.

Although programmes cannot be distinguished from the point of view of their likelihood of achieving empirical success, they may differ in a way which becomes significant when they face a disobliging Nature. As I indicated earlier, if other things are equal, the more powerful heuristic has the greater number of techniques for the non-*ad hoc* resolution of anomalies, and hence has the greater resourcefulness in the face of refutation. Thus, if experimental tests are as adverse for a powerful programme as for a weaker one, the former will suffer a slower demise and will survive even after the latter has been extinguished. The surviving programme will thus retain its ability to generate new, testable scientific theories, some of which may be empirically confirmed. On the other hand, its extinguished rival is no longer able to resolve its difficulties.

I suggest that greater resourcefulness in the face of anomalies provides a further rationale for regarding the programme with the more powerful heuristic as the more promising. While this greater resourcefulness does not imply that the more powerful heuristic is likely to make more *empirical* progress than a weaker rival, it does imply that, if attended by a disobliging Nature, it will survive longer and retain its ability to progress theoretically when its weaker rival has lost this power.[28]

6. CONCLUSION

According to the Popperian theory of discovery, the history of science consists of a series of logically and psychologically disconnected theories and nothing at all can be predicted about the future theoretical developments in science.

I have argued in this paper that, *unlike other philosophies of science, the methodology of scientific research programmes enables one to make some assessment of the potential for future development of a research programme.* I regard this as one of the most important and original aspects of this methodology. Admittedly, what one can say about a research programme's future falls short of predicting how it will fare in empirical tests and those who regard any view of objective promise as acceptable only if it incorporates an assurance of or assigns some probability to future *empirical* success will certainly be disappointed by the account I have suggested. Although it falls short of the ideal interpretation of promise, I have argued that the power of a heuristic has some features which make it a suitable candidate for a weaker concept of promise. In particular, I claim that a programme with a powerful heuristic is likely to lead to greater *theoretical* progress than a weaker rival.

The interpretation of objective promise which I offer seems to explain the appraisals by scientists such as Planck and Ostwald of certain research programmes.[29] It seems likely that scientists have actually used an appraisal of heuristic power in deciding which lines of research have the greatest promise and that most historians of science have ignored this important dimension to their subject. This is admittedly a historical conjecture which requires considerable further research before it can be satisfactorily tested.[30]

London School of Economics

NOTES

[1] Descartes, [1628], p. 9.

[2] *Ibid.*

[3] Herschel [1831], p. 59; the second set of italics are mine.

[4] Polanyi is the modern philosopher whose views of heuristics in science seem to lie closest to the Cartesian view. For example he claims: "There is no doubt ... of the scientist's capacity to assess in outline the course of an enquiry that will lead to a result which at the time he makes the assessment, is essentially indeterminate". It is on account of this capacity, Polanyi argues, that in science "it is rare to come across years of futile efforts wasted". (Polanyi [1972], p. 44.)

[5] Popper [1972], p. 198.

[6] Popper [1963], p. 189.

[7] Popper [1963], p. 192.

[8] Popper [1934], section 2; my italics. These statements of Popper's actually contain several theses and I cite them only to bring out Popper's view that the invention of a new idea cannot be logically reconstructed, that its origins are essentially mysterious. I will argue against this thesis in the next part. However, I do not dissent from the other major thesis expressed in these quotations, namely that the origins of a theory are irrelevant to an appraisal of its objective merits.

[9] Hempel [1945], p. 4.

[10] Whewell [1858], p. 59; my italics.

[11] Reichenbach [1951], p. 231. Medawar, in the same vein, claims that 'the generative act in scientific enquiry, "having an idea", ... represents the imaginative or logically unscripted episode in scientific thinking' (Medawar [1969], p. 55).

[12] Popper [1972], p. 201; my italics.

[13] *Ibid.*, p. 202.

[14] Lakatos [1970], p. 92.

[15] Lakatos [1971a], p. 101 and Lakatos [1971b], p. 174.

[16] Lakatos [1971c], p. 77.

[17] Planck [1897].

[18] Ostwald [1927], in retrospect. See Clark [1976] for a detailed examination of the kinetic and thermodynamics research programme.

[19] *Nature*, August 7, 1943; my italics. I am grateful to Roy Wolfe for drawing my attention to this quotation.

[20] Duhem ([1906], p. 221) poked gentle fun at the 'layman' who naively looks on a scientific hypothesis as the product of a sudden creative leap and who believes that: "this fairy who ... he calls by the name of science has touched with his [sic] magic wand the forehead of a man of genius and that the theory immediately appeared alive and complete like Pallas Athena emerging fully armed from the forehead of Zeus. He thinks it was enough for Newton to see an apple fall in the orchard in order that the effects of falling bodies, the motions of the earth, the moon and the planets and their satellites, the trips of comets, the ebb and flow of the ocean, should all come to be suddenly summarised and classified in that one proposition: Any two bodies attract each other proportionally to the product of their masses and inversely to the square of their mutual distance."

[21] Watkins [1974], p. 406.

[22] See Merton [1957], [1961] and [1963].

[23] In the past, scientists locked up their discoveries in coded summaries and learned societies provided special facilities for recording the date of a discovery. More recently, rapid-publishing journals such as *Nature* have provided a similar service.

[24] See e.g. Lakatos [1970], p. 155. It seems that it was Noretta Koertge's criticisms in her [1971] which first persuaded Lakatos that "in appraising research programmes one has to take into account the different powers ... of their positive heuristics". (Lakatos [1971b], p. 177.)

[25] See Zahar [1973] for a detailed account of the relativity research programme.

[26] Descartes' laws of impact were not only underivable and known to be underivable from Cartesian metaphysics, they were also irrelevant to the vortex theory since they only applied under the imaginary conditions of bodies colliding in a vacuum. They could not be transferred to a continuous fluid medium.

[27] See Mach [1912], p. 199.

[28] Stephen Toulmin discusses a similar notion to my objective promise, which he calls 'ripeness' and which he too regards as an 'intrinsic feature' of a 'field of enquiry'. However, he gives no general criteria by which ripeness is appraised nor any description of the unit of appraisal. However Toulmin's excellent examples suggest that he regards what I called 'autonomy' as the major feature of a ripe 'field of enquiry'. (See Toulmin [1972], pp. 211–2.)

[29] See *above*, p. 104.

[30] Zahar provides one significant piece of evidence for this conjecture by showing that, in their early transfer of allegiance from Lorentz's to Einstein's programme, scientists' major consideration was the greater heuristic power of the relativity programme. (See Zahar [1973].)

The notes refer to the unified bibliography for Parts I and III on pp. 379–383.

PART II

REFLECTIONS ON THE LSE POSITION

ADOLF GRÜNBAUM

POPPER VS INDUCTIVISM*

1. INTRODUCTION

To my knowledge, the large literature on the rivalry between the Popperian and inductivist methodologies has not come to grips with some of the crucial issues on which the appraisal of their conflicting theses turns. And I wish to issue a new challenge to the cardinal arguments on which Popper rests his indictment of inductivism.

The present paper is drawn from a larger Lakatos Memorial essay in which I have also included (i) a critique of Popper's historiography of inductivism, (ii) a demonstration of a major inconsistency in his 1972 account of corroboration, and (iii) a critical examination of his objections to Duhem's views on falsifiability.[1] The larger essay, in turn, is one of a series of critical studies in which I have examined the major pillars of Popper's philosophy of science.[1a]

In the present paper, I shall select some salient portions from the aforecited Lakatos memorial essay FT which are especially germane to the topic of the Workshop.

2. INDUCTIVIST SUPPORTABILITY AND FALSIFIABILITY OF HYPOTHESES AS CRITERIA OF DEMARCATION BETWEEN SCIENCE AND NON-SCIENCE

Did Popper misportray inductivism when arguing for his falsificationist criterion of demarcation as an alternative to it?

He invokes his own historiography of inductivism to conclude the following: In 1919 the account of testability furnished by all forms of inductivism was such that "there clearly was a need for a different criterion of demarcation".[2] Popper then enunciates falsifiability as the linchpin of the *scientific entertainability* of a hypothesis. Thus, falsifiability by potential negative instances is claimed to play a distinguished role to the exclusion of inductive supportability or probabilistic confirmability by positive instances. Says he: "Testability is falsifiability" (C&R, p. 36). I shall speak of Popper's 'demarcation asymmetry' in order to allude to the asymmetry in the roles

117

G. Radnitzky and G. Andersson (eds.), Progress and Rationality in Science, 117–142.
All Rights Reserved.
Copyright © 1978 by D. Reidel Publishing Company, Dordrecht, Holland.

which he assigns to the falsifiability and the supportability of a hypothesis vis-à-vis its scientific status.

Philosophers whom Popper disapprovingly calls 'inductivists' try to use supportive instances of hypotheses or theories to make either absolute or relative *credibility assessments* of them. An example of an absolute credibility judgment would be to say that a given hypothesis is more likely to be true than false. A relative credibility judgment might take the form of saying that a theory is more credible than a certain rival. *Some* inductivists propose to 'probabilify' hypotheses on given evidence by holding that, in principle, hypotheses can be assigned numerical degrees of confirmation which are mathematical probabilities in the sense of satisfying the mathematical calculus of probability. But other inductivists deny that either absolute or relative degrees of credibility of hypotheses *must* be construed as mathematical probabilities.[3]

As I have documented in FT, Popper's historiography of inductivism is unfortunately oblivious of a paramount fact. It is the fact that there are very important differences among inductivist epistemologies concerning the requirements which must be met by an observational finding, if this finding is to count as a *bona fide supportive* instance of a hypothesis. And, as we shall see, just these crucial differences render quite untenable Popper's declaration that "The fundamental doctrine which underlies all theories of induction is *the doctrine of the primacy of repetitions*. ... According to this doctrine, repeated instances furnish a kind of *justification* for the acceptance of a universal law".[4] Indeed, the following conclusion, emerges in FT from my critique there of Popper's historiography of inductivism: The inductive use of supportive instances to 'credibilify' hypotheses in one way or another does *not* automatically commit inductivism as such to grant *credible scientific status* or even to lend at least *some* credence to a hypothesis H *merely* because there are numerous observationally true consequences of H, and no known negative instances. For the mere quest for credibilifying a hypothesis by supporting evidence does NOT require at all that *every* positive instance of a hypothesis be automatically held to be supportive of the hypothesis to some *non*-zero degree or other! *A fortiori*, the program of inductive credibilification *as such* in no way requires that all positive instances count as *equally* supportive to some *non*-zero degree.

In the first chapter of Popper's C&R, he adduces astrology, Freudian psychoanalysis, the Marxist theory of history and Adlerian psychology in an endeavor to show the following: (1) all forms of inductivism are logically committed to grant scientific status to these four theories in the face of the numerous positive instances marshalled by them, and (2) whereas the requirements of inductive supportability are hopelessly incompetent to derogate the theories in question as pseudo-scientific, an alternative demarcation criterion based on falsifiability succeeds in excluding them from the body of science.[5]

Long before Popper's injunction that "Confirmations should count only if they are the result of *risky predictions*" (C&R, p. 36), Francis Bacon made a vital contribution toward distinguishing *merely positive* from *supportive* instances of a theory by emphasizing that some kinds of positive instances can differ radically from others in *evidential value* and treating the evidentially significant ones under the heading of '*Prerogative Instances*' (*Novum Organum* II, Section 21ff).[6] I shall say in more modern parlance that an instance is a 'positive' one with respect to a *non*-statistical theory T, if its occurrence or being the case can be deduced from T in conjunction with suitable initial conditions. But an instance is supportive of T, if it is positive *and* has the probative significance of conferring a stronger truth presumption on T than T has without that instance. As will be shown in our impending discussion of the very important case of causal hypotheses, post-Baconian inductivists appealed to the distinction between merely positive and supportive instances of a theory: They did so in order to guard against unsound causal inferences such as *post hoc ergo propter hoc* and against other ravages of the unbridled use of the hypothetico-deductive method in inductive inference.

Not only Bacon's tables of comparative instances but also J. S. Mill's 'inductive methods' for appraising *causal* hypotheses had long ago led inductivists to demand *controlled experiments* or so-called *controls* as an indispensable check on whether positive instances do have the probative significance of being supportive instances! But in fairness to Popper, it should be pointed out that there has been one important school of inductivists among both philosophers and eminent scientists who championed the doctrine that *any* positive instance of a hypothesis also necessarily qualifies as a *supportive* instance of the hypothesis. This doctrine that any positive case of a hypothesis is automatically also supportive to *some* degree is sometimes called 'the instantiation condition' (Nicod's criterion).[7] In an *unbridled* use of the instantiation condition *if there be such*, any positive instance of a hypothesis will *increase* the credibility of the hypothesis *as much as any other*. And hence in *such* an invocation of the instantiation condition, one considers the mere *number* rather than the relative weight of positive instances when assessing the credibility of a hypothesis. On the latter *stronger* version of the instantiation condition, degrees of credibility could not be construed as mathematical probabilities, since a sufficiently large finite number of positive instances would then yield a 'probability' *greater* than unity, in contravention of the probability calculus. More generally, the stronger version is clearly the less interesting one. Indeed, it is close to a *caricature* of Nicod's criterion!

But it was none other than the confirmed inductivist J. S. Mill who emphatically rejected the instantiation condition after Bacon had rejected induction by simple enumeration of positive instances as puerile. Thus, when discussing the case of a *spurious* plurality of causes in his *Logic*, Mill *denies* the probative value of *mere* repetitions of positive instances just as much as Popper does (LSD, p. 269), and indeed for much the same reasons.[8] Mill stresses the need to find positive instances of a specified causal hypothesis such that these instances *also* refute one or more *rival* hypotheses as to the cause. And he writes:

Most people hold their conclusions with a degree of assurance proportioned to the mere *mass* of the experience on which they appear to rest; not considering that by the addition of instances to instances, all of the same kind, that is, differing from one another only in points already recognized as immaterial, nothing whatever is added to the evidence of the conclusion. A single instance eliminating some antecedent which existed in all the other cases [i.e., an instance which refutes a *rival* hypothesis as to the cause] is of more value than the greatest multitude of instances which are reckoned by their number alone.[9]

Clearly, there is at least one historically influential inductivist conception of testability which makes three assertions as follows: (i) it calls attention to the difference between the genus of positive instances of a hypothesis H and the species S of those special positive instances which serve to eliminate one or more specified rivals of H, (ii) it rejects the instantiation condition by denying that every member of the genus of positive instances automatically qualifies as supportive, and (iii) it regards the species S of positive instances as supportive in the sense of conferring some degree of credibility on *H*, even if the amount of available supportive evidence is *insufficient* to make it more likely that *H* is true rather than false. Note that among these three assertions, only the third endorses inductive inference.

Yet Popper is so preoccupied with dissociating himself from the third of these assertions (LSD, p. 419) that he ignores the first two. Thus, he is prompted to misportray *all* post-Baconian theories of induction as assigning *supportive* probative significance to all positive instances *alike*. In this way, he is *misled* into charging that all such versions of inductivism are unable to indict psychoanalysis, for example, as unscientific in the face of its many positive instances. He gives a concise statement of this misportrayal by making the following previously cited statement (LSD, p. 420, italics in original): "The fundamental doctrine which underlies all theories of induction is *the doctrine of the primacy of repetitions*. ... According to this doctrine, repeated instances furnish a kind of *justification* for the acceptance of a universal law." But this claim is untenable in the face of the espousal of assertions (i) and (ii) by the eliminative inductivism of Bacon and Mill. And, as Section 3 below will show, Popper's portrayal of all forms of inductivism is likewise refuted by the fact that Bayesian probabilistic inductivism disavows the instantiation condition. Popper's preoccupation with opposing assertion (iii), to the exclusion of appreciating assertions (i) and (ii), is made evident by his appraisal of the eliminative inductivism of Bacon, Whewell and Mill. Speaking of the latter on the immediately preceding page of LSD (p. 419), he says:

... the sole purpose of the elimination advocated by all these inductivists was to establish *as firmly as possible the surviving theory* which, they thought, must be the *true* one (or perhaps only a *highly probable one*, in so far as we may not have fully succeeded in eliminating every theory except the true one).

As against this, I do not think that we can ever seriously reduce, by elimination, the number of the competing theories, since this number remains always infinite.

When we discuss Popper's theory of corroboration in Section 3, the last sentence of this quotation from him will return to haunt his own theory. Indeed, Bacon's conception of eliminative induction, defective though it was, will then be seen to be understandably plausible while Popper's purportedly deductivist conception of corroboration will be seen to be even devoid of plausibility.

For brevity, I shall use the term 'instantionist inductivism' to refer to the very special version of inductivism which does espouse the instantiation condition. Then I can say that Popper's characterization of all theories of induction as being instantionist also runs counter to the fact that Mill demanded a check by the Joint Method of Agreement and Difference on the *probative* significance of the positive instances which had been accumulated by the merely *heuristic* Method of Agreement! Hence we reach the following important conclusion: At least some post-Baconian inductivists *deny* that every positive instance of a hypothesis H is automatically supportive of H to some non-zero degree or other. And *a fortiori* these inductivists – as distinct from the putative exponents of *unbridled instantionist* inductivism! – deny that all positive instances count as *equally* supportive. In view of the importance of this conclusion, let us restate it by saying that all *anti*-instantionist versions of inductivism espouse the following cardinal epistemic principle: *The ability of a theory T to 'explain' and/or predict certain phenomena deductively* (with the aid of suitable initial conditions) *is generally only a necessary and NOT a sufficient condition for qualifying these phenomena to count as T-supporting instances over and above being merely positive instances of T!*

Note here that within the genus of positive instances, the term 'confirming instance' is disastrously ambiguous as between a supportive and a non-supportive species of positive instances. By the same token, logical mischief has been wrought by the weasel word 'verification', and by the equivocal verb 'verify'. Popper's misportrayal of inductivism as universally instantionist is abetted by the ambiguity of the terms 'confirmation' and 'verification', when he says misleadingly: "It is easy to obtain confirmations, or verifications, for nearly every theory – if we look for confirmations" (C&R, p. 36).

Like Popper, Watkins misportrays inductivism as invariably *instantionist* when Watkins tells us in his position paper what an 'inductivist theory of confirmation' is. As Watkins has it, it would be self-contradictory for *any* kind of inductivist to *deny* the instantiation condition, which it plainly is not. I am not, of course, denying that people are free to give any stipulative definition of the mere *word* 'inductivist'. But I submit that, in this case, the function of this definition in the Popperian literature has been misleading and *persuasive* rather than *stipulative*. In the same vein, I note that Worrall introduces his account of Popper's theory of corroboration (in his Section 2.2) by pointing out that Popper *denied* the instantiation condition. And Worrall's presentation appears to suggest that there was something *innovative* in that denial of Popper's, or at least that Popper pioneered by demanding that only the successful outcomes of so-called 'severe' tests count as supportive.

Worrall's account seems to overlook that Bacon not only denied the instantiation condition but also deemed positive instances of a theory T to be genuinely supportive (or 'prerogative') only because these same instances were also negative ones for T's rivals. Thus, in *these* respects, Popper was hardly innovative vis-à-vis Bacon. Of course, I do not deny that at a given historical juncture, it may be very useful to call attention to an insight that was achieved much earlier by others.

Let me merely mention, but not rehearse, the criticism in the literature that Popper's deductive falsifiability of a hypothesis is much too strong a requirement for scientific entertainability, as illustrated by the mere example of the hypothesis 'All men are mortal'. For the latter says that for each man, *there is some time or other* at which he dies, i.e., that no man lives forever. To *falsify* this deductively by a test statement, we must produce at least one man who NEVER dies. For no matter how long any Methusalah lives, he may still die later on. Hence no basic observation statement can deductively falsify the old saw that we are all mortal. Other examples are furnished by assertions of non-existence concerning *perpetual* motion machines of the first and second kinds, as in the first two laws of thermodynamics.

Surely I am not being captious if I conclude from my analysis that Popper's demarcation asymmetry is either unsound or too strong: It is unsound if it is claimed to have demarcational capabilities with respect to psychoanalysis and astrology, for example, which are not also possessed by the requirements of anti-instantionist inductivism, or it is too strong as just explained. As shown by my examples in FT, the mere fact that anti-instantionist inductivists try to use supportive instances to 'probabilify' or credibilify hypotheses does NOT commit them to granting credible scientific status to a hypothesis *solely* on the strength of existing positive instances, however numerous. Indeed as shown by thousands of cases of people who are improved *after* psychiatric treatment, Mill's inductivism can discount even such a multitude of positive instances as *non*-supportive of the hypothesis of therapeutic effectiveness. But I now need to forestall a possible misunderstanding of the moral I draw from my comparison of Popper's demarcation asymmetry with the conception of scientificality advocated by an inductivist who rejects the instantiation condition.

As already noted, Popper is primarily concerned with scientific *entertainability*, whereas the inductivist is intent upon the scientific *credibility* of a hypothesis. For Popper, the mere falsifiability of a hypothesis suffices for according scientific status to it, but the inductivist may be prepared to grant no more than *potential* scientific status to it in virtue of its inductive support*ability* in principle. And Popper's *corroborated* scientific hypothesis is the *counterpart* of the inductivist's *actually* scientific hypothesis. Hence, if one wished, one could treat Popper's falsifiability and the inductivist's genuine support*ability* as counterpart criteria in their respective endeavors to effect a demarcation between non-science, on the one hand, and those theories which are at least scientifically *entertainable*, on the other.

Yet it is very important to note that Popper *himself* (C&R, ch. I) did claim *greater restrictiveness* even for his falsifi*ability* criterion of demarcation than for the actual inductive support*edness* required for inductive credibility. And he adduced psychoanalytic theory as an illustration of the allegedly greater stringency of his falsifi*ability* criterion.

Can it be held that the inductivist's criterion avoids any defect corresponding to the weakness of Popper's falsifiability requirement, which fails by being too strong? I think not, because the inductivist's genuine supportability likewise fails to the extent that there is no satisfactory general theory of evidential support which states unambiguous necessary and sufficient conditions that are successfully applicable to *every* concrete case. We must be mindful in this connection of the fact that science covers a whole gamut of kinds of claims which pose problems of empirical validation: Comprehensive theories like neo-Darwinism or general relativity, causal hypotheses like 'Asbestos is carcinogenic', statistical hypotheses, and even simple generalizations like 'All ravens are black'. Consider just the last of these examples, which lends itself to the statement of Hempel's paradox of confirmation. If the inductivist avoids that paradox by rejecting the instantiation condition, then it is still incumbent upon him to tell us under what conditions a positive instance of 'All ravens are black' *does* count as supportive. Yet no satisfactory general answer is available, as far as I know. Nor is it clear that inductivism can furnish a generally satisfactory method for assigning non-zero posterior mathematical probabilities to even the empirically most successful universal law-hypotheses of physics, for example. Hence I think that the inductivists have no more succeeded than Popper in stating *general* criteria for effecting a *neat* demarcation of science from non-science.

Thus, the upshot of my comparison of inductivist conceptions of scientificality with Popper's is *not* the claim that there is a viable inductivist counterpart to Popper's defective demarcation criterion for scientific entertainability. Instead the moral I draw is the following: Popper was seriously mistaken in claiming that IN THE AB-SENCE OF NEGATIVE INSTANCES, all forms of inductivism are necessarily committed to the (probabilified) scientific credibility of a theory, merely because that theory can adduce numerous positive instances.

And, as we shall soon see, the inductivist can justly complain that Popper's pure deductivism has no non-trivial answer to the question: "What does it mean to say that a successful risky prediction COUNTS in favor of the theory that made it?"

3. POPPER'S THEORY OF CORROBORATION VERSUS BAYESIAN PROBABILISTIC INDUCTIVISM

3.1. *Popper on the Probability of Universal Laws*

One of Popper's major reasons for alleging the 'impossibility of an inductive probability' is as follows: In a universe containing infinitely many distinguishable things or spatio-temporal regions, '*the probability of any (non-tautological) universal law will be zero*' (LSD, Appendix vii, p. 363; italics in original). Commenting on the meaning of the term 'probability' in this sentence, he says (LSD, p. 364):

By 'probability', I mean here either the *absolute* logical probability of the universal law, or its probability *relative to some evidence*; that is to say, relative to a singular statement, or to a finite conjunction of singular statements. Thus if a is our law, and b any empirical evidence, I assert that

(1) $p(a) = 0$

and also that

(2) $p(a, b) = 0$

These formulae will be discussed in the present appendix. The two formulae (1) and (2) are equivalent.

And for the case in which the universal statement a is interpreted as entailing an infinite conjunction of singular statements, Popper then goes on to give several arguments for concluding that $p(a) = 0$. He regards one of these arguments to be 'incontestable' (LSD, p. 366).[10]

The same conclusion is reiterated and amplified in C&R, where Popper writes:

... in view of the high content of universal laws, it is neither surprising to find that their probability is zero, nor that those philosophers who believe that science must aim at high probabilities cannot do justice to facts such as these: that the formulation (and testing) of *universal laws* is considered their most important aim by most scientists (C&R, p. 286).[11]

Furthermore, he contrasts the inevitably vanishing inductive probabilities of universal law statements with the *non*-vanishing degrees of corroboration of which such statements are capable on appropriate evidence in his theory of corroboration, saying:

But it can be shown by purely mathematical means that *degree of corroboration can never be equated with mathematical probability*. It can even be shown that all theories, including the best, have the same probability, namely zero. But the degree to which they are corroborated (which, in theory at least, can be found out with the help of the calculus of probability) may approach very closely to unity, i.e. its maximum, though the probability of the theory is zero (C&R, pp. 192–193).

But Popper's conclusion that, for any universal statement a, $p(a) = 0$ can now be shown to be incompatible with his *quantitative* theory of the content and verisimilitude of a hypothesis. For when we turn to the latter theory (C&R, pp. 390–397, 234, 218), we find the following:

(1) Both in C&R and in his *Objective Knowledge*,[12] Popper defines the *measure* ct(a) of the content of a theory a as ct(a) $= 1 - p(a)$, where '$p(a)$' is 'the logical probability' of a. He further characterizes the latter probability very meagerly (OK, p. 51) as 'the logical probability that it [a] is true (accidentally, as it were)'.

(2) When speaking of the logically *incompatible* theories of Einstein and Newton – respectively denoted by 'E' and 'N' – Popper tells us without any explanation that 'the content measures ct(N) and ct(E)' bear out the intuition that, as between these two theories, 'Einstein's has the greater content' (OK, p. 53). But if ct(E) > ct(N), then $p(N) > p(E)$, and hence $p(N) > 0$ even though N is a conjunction containing universal statements and should therefore have *zero* probability according to Popper's cited claim. Let us now see

how this inconsistency carries over into his theory of quantitative verisimilitude.

(3) Popper states his view as to what outcome of attempted falsifications of E would epistemically warrant the conjecture that the quantitative falsity-content of E does not exceed the corresponding falsity-content of N (OK, p. 53). And he then deduces (OK, p. 53) that E has greater quantitative verisimilitude than N from the conjunction of the following three assertions: (i) the aforementioned claim that $ct(E) > ct(N)$, (ii) the stated conjecture as to the comparative falsity-contents of E and N, and (iii) the contention that "the stronger theory, the theory with the greater content, will also be the one with the greater verisimilitude *unless its falsity content is also greater*" (OK, p. 53). In any case, his cardinal example of greater verisimilitude makes the claim that $ct(E) > ct(N)$.

But since N and E are each replete with universal statements, consistency with his claims in LSD (Appendix vii) requires Popper to say that both $p(N) = 0$ and $p(E) = 0$. And since $ct(a) = 1 - p(a)$, this has the consequence that $ct(E)$ and $ct(N)$ have the *same* value unity. Yet, as we just noted, he told us in OK (p. 53) that $ct(E) > ct(N)$. Furthermore, Popper asserts (C&R, pp. 218 and 256) that (1) 'with increasing content, probability decreases, and *vice versa*', and that (2) if one (universal) theory B *unilaterally entails* another (universal) theory A, then $ct(B) > ct(A)$ (OK, pp. 51 and 53). These assertions (1) and (2) require that if B unilaterally entails A, then $p(A) > p(B)$, and hence that at least one of the two probabilities $p(A)$ and $p(B)$ be *non*-zero. But for universal non-tautological theories A and B, the latter requirement contradicts the cited LSD claim that *both* of them must be zero. Moreover, even for the case in which B *unilaterally* entails A, the probability calculus requires only that $p(A) \geqslant p(B)$ rather than that $p(A) > p(B)$, as Popper does here. I have discussed the ramifications of this latter fact elsewhere (cf. Section 3 C(ii) of the *second* of my articles cited in fn. 1a).

Thus, Popper hoists himself on his own petard: If he is to furnish an indictment of inductive probabilities by his LSD claim that $p(a) = 0$ and $p(a, b) = 0$ for any non-tautological universal a and any empirical evidence b, then his theory of quantitative content and quantitative verisimilitude is untenable; conversely, if the latter is affirmed, the claim $p(a) = 0$ is no longer available as a premiss for a *reductio ad absurdum* argument against inductive probabilities.

3.2. *Popper on Probability vis-à-vis the Aims of Science*

One of the citations in the preceding subsection I contains Popper's complaint that the aim to achieve high probabilities as in Bayesian inductive inference *fails* to vindicate the quest for universal laws: As Lakatos has summarized it (PDI, p. 259): "... *if* inductive logic was possible, *then* the virtue of a theory was its improbability rather than its probability, given the evidence". Concerning this complaint, Ronald Giere has aptly made a personal remark to the following effect: The probabilistic inductivist does *not* say that the aim of science is or ought to be to play it safe. Instead, this inductivist tells us (i) what risks we take, *if* we choose *not* to play it safe, and (ii) that *if* we wish to play it safe, then we should act on the *less* daring of two theories whenever possible. On this understanding of probabilistic inductivism, Popper's complaint no longer applies.

3.3. *Comparison Between Popper's Methodology and Bayesian Inductivism*

It has been or ought to have been well-known since the nineteenth century that the prior probabilities in Bayesian inference pose formidable if not insoluble difficulties, at least if they are to be construed *non*-subjectively. Hence, if I now proceed nonetheless to assess the capabilities of Bayesian inductivism to implement Popper's methodological prescriptions, I do *not* do so in the spirit of d'Alembert, who is said to have declared: "Allez en avant, et la foi vous viendra." Rather I do so, because I hope to show that this assessment is instructive, although it is moot whether the status of Bayes's prior probabilities will be satisfactorily clarified in the future. Can Popper do any better?

A. *The Epistemic Capabilities of Popper's Methodology*

The following two epistemic contentions of Popper's are pertinent here:

(i) "Confirmations should count only if they are the result of *risky predictions*; that is to say, if, unenlightened by the theory in question, we should have expected an event which was incompatible with the theory – an event which would have refuted the theory" (C&R, p. 36).

(ii) "When trying to appraise the degree of corroboration of a theory we may reason somewhat as follows. Its degree of corroboration will increase with the number of its corroborating instances. Here we usually accord to the first corroborating instances far greater importance than to later ones: once a theory is well corroborated, further instances raise its degree of corroboration only very little. This rule however does not hold good if these new instances are very different from the earlier ones, that is if they corroborate the theory in a *new field of application*. In this case, they may increase the degree of corroboration very considerably" (LSD, p. 269).

Popper clarifies the latter of these two demands by writing elsewhere (C&R, p. 240) as follows:

(iii) "A serious empirical test always consists in the attempt to find a refutation, a counter example. In the search for a counter example, we have to use our background knowledge; for we always try to refute first the *most risky* predictions, the *most unlikely* ... consequences' (as Peirce already saw [footnote omitted here]); which means that we always look in the *most probable kinds* of places for the *most probable* kinds of counter examples – most probable in the sense that we should expect to find them in the light of our background knowledge. Now if a theory stands up to many such tests, then, owing to the incorporation of the results of our tests into our background knowledge, there may be, after a time, no places left where (in the light of our new background knowledge) counter examples can with a high probability be expected to occur. But this means that the degree of severity of our test declines. This is also the reason why an often repeated test will no longer be considered as significant or as severe: there is something like a law of diminishing returns from repeated tests (as opposed to tests which, in the light of our background knowledge, are of a *new kind*, and which therefore may still be felt to be significant). These are facts which are inherent in the knowledge-situation; and they have often been described – especially by John Maynard Keynes and by Ernest Nagel – as difficult to explain by an inductivist theory of science. But for us it is all very easy. And we can even explain, by a similar analysis of the knowledge-situation, why the empirical character of a very successful theory always grows stale, after a time" (C&R, p. 240).

Before commenting on the capabilities of the Bayesian inductivist to implement these methodological prescriptions of Popper's, let me point out why I think that Popper himself cannot justify these demands within his own deductivistic framework.

(i) As for Popper's demand (i) that positive instances should count only if they are the results of successful *risky* predictions, let us ask: What does the word *'count'* mean in his deductivist framework, when Popper declares that "Confirmations should count only if they are the result of *risky predictions*"? Count toward or for what? Qua *pure* deductivist, can Popper possibly maintain without serious inconsistency, as he does, that *successful* results of initially risky predictions should 'count' in favor of the theory making the risky prediction in the sense that in these "crucial cases ... we should expect the theory to fail if it is not true" (C&R, p. 112)? The latter statement is part of his characterization of 'crucial cases' in which he says:

A theory is tested not merely by applying it, or by trying it out, but by applying it to very special cases – cases for which it yields results different from those we should have expected without that theory, or in the light of other theories. In other words we try to select for our tests those crucial cases in which we should expect the theory to fail if it is not true (C&R, p. 112).

Note Popper's claim that – on the basis of background theories – we should expect the [new] theory to fail [a 'crucial' test] *if it is not TRUE*. As against this assertion, I say that what is warranted *deductivistically* is the following:

(a) The older theories (*cum* initial conditions) predict an outcome contrary to the result *C* predicted by the new theory *H*, or the older theories are at least deductively non-committal with respect to *C*, thereby making the prediction of the new theory 'risky' in *that special sense*. Popper is here concerned with the more highly risky kind of prediction *C*, which is *contrary* to each of the older theories rather than logically independent of them. For this reason as well as in order to simplify the discussion, I shall sometimes concern myself with just the more highly risky kind of *C*.

(b) If the new theory *H* does *not* fail the crucial test but passes it, clearly nothing follows deductively about *its* truth. What does follow from the observation statement *C* is the truth of the 'infinitely' weaker disjunction *D* of ALL and only those hypotheses which individually entail *C*. For *C* itself is one of the disjuncts in the infinite disjunction *D*. And thus *C* entails *D* in addition to being entailed by *D*.[13]

Note that for any true disjunct in *D*, the latter has an infinitude of *false* disjuncts which are pairwise *incompatible* with that true disjunct in *D* and, of course, also incompatible with the background knowledge. Yet – contrary to the background knowledge – each of this infinitude of false theories predicts C no less than H does, and indeed H may be false as well. Hence of what avail is it to Popper, *qua deductivist*, that by predicting C, H is one of an infinitude of theories in D incompatible with those *particular* theories with which scientists had been working by way of historical accident, and that scientists happened to have thought of the particular disjunct H in D?

According to Popper's definition of the term 'severe test', the experiment E which yielded the riskily predicted C does qualify as a 'severe test' of H. But surely the fact that H makes a prediction C which is incompatible with the prior theories constituting the so-called 'background knowledge' B does *not* justify the following contention of Popper's: A *deductivist* is entitled to expect the experiment E to yield a result *contrary* to C, *unless H is true*.

Indeed, Popper's reasoning here is of a piece with the reasoning of the Bayesian proponent of *inductive* probabilities which has been stated by Salmon as follows:

A hypothesis risks falsification by yielding a prediction that is very improbable unless that hypothesis is true. It makes a daring prediction, for it is not likely to come out right unless we have hit upon the correct hypothesis. Confirming instances are not likely to be forthcoming by sheer chance.[14]

It would seem, therefore, that when Popper assessed the epistemic significance of severe tests in the quoted passage, he was unmindful of his own admonition against Baconian eliminative inductivism. As we recall from our citation of LSD in Section 2, this admonition reads as follows:

... the sole purpose of the elimination advocated by all these inductivists was to *establish as firmly as possible the surviving theory* which, they thought, must be the *true* one (or perhaps only a *highly probable one*, in so far as we may not have fully succeeded in eliminating every theory except the true one).

As against this, I do not think that we can ever seriously reduce, by elimination, the number of the competing theories, since this number remains always infinite (LSE, p. 419).

Popper's allegedly deductivist valuation of the corroborative epistemic significance of successful severe tests can now be seen to be unsound in the light of Popper's own premises, whereas Bacon's contrary (though false!) premises do justify Bacon's eliminative inductivism. For Popper is avowedly aware that the number of competing theories in the disjunction D is indeed *infinite*. And yet he claims that an outcome C successfully predicted by some one disjunct H in D 'should count' as corroborative of H, just because C happens to be risky with respect to the background knowledge B. On the other hand, Bacon made the quite mistaken assumption that scientists could devise a *finite* yet *exhaustive* list of all logically possible alternative theories relevant to a given phenomenon. But having made that unsound assumption, Bacon quite reasonably felt justified in maintaining that scientists could then irrevocably establish the truth of *one* of these specified theories, if they succeed in refuting all of its rivals in the purportedly exhaustive *finite* disjunctive class of relevant theories.

Thus, it is all the more incongruous that the avowed deductivist Popper accords epistemic significance to successful predictions just because they are risky, amid chiding Bacon for having overlooked that we can never seriously reduce by elimination the *infinite* number of competing theories. Popper's deductivistically gratuitous valuation of severe tests is echoed by Watkins in his Position Paper, where he says the following: "we want highly testable theories because we hope that, if such theories are severely tested they have a high chance of being weeded out if they are in fact false". Given the infinitude of relevant competing theories as stressed by Popper, I do not understand the basis on which a deductivist can hope for such a high chance, and *hence I cannot see any purely deductivistic rationale for advocating severe tests as contrasted with non-severe tests*.

What then would be a *deductivistically* sound construal of Popper's injunction that "Confirmations should count only if they are the result of *risky predictions*"? It would seem that deductively speaking, C can be held to *count* for H only in the following Pickwickian sense: C deductively refutes or *counts* AGAINST those *particular* rivals of H which belong to B while leaving fully intact an infinitude of other rivals in the disjunction D.

So much for the first of Popper's demands, stated under (i) above.

(ii) Is it 'all very easy' for him in keeping with his major anti-inductivist tenets to espouse the *reasons* offered by him for *not* repeating the same corroborating experiment *ad nauseam* and for claiming that "there is something like a law of diminishing returns from repeated tests" (C&R, p. 240)? Alan Musgrave has argued cogently that it is not. And Musgrave makes the following telling points:

(a) Let p qualify as a *very risky* kind of prediction with respect to the initial background knowledge B_0, and let p be tested successfully, say, ten times, thereby yielding ten corroborating instances for the new theory T entailing p. Musgrave says:

... after each performance of the test, the particular results are incorporated into our background knowledge, changing it successively from B_0 to B_1, to B_2, and so on, up to B_{10}. Popper's claim seems to be that the probability of the prediction in the next instance *gradually rises* in the light of our successively augmented background knowledges. Hence the severity of the successive repetitions of the same test *gradually falls*, and so does the degree of corroboration afforded by them to T.

But this clearly involves a straightforward *inductive argument*. It involves the inductive argument that each positive instance of the universal test-implication p increases its probability. (Alternatively, if p is construed as a singular prediction, it involves the inductive argument that the accumulation of past instances renders the next instance of the same kind more and more probable.) It would seem, therefore, that if we eschew inductive arguments, then the incorporation of past results of a severe test into background knowledge can do nothing to reduce the severity of future performances of that test. It would seem, in other words, that if Popper's theory of corroboration really is a non-inductivist one, then it cannot provide us with diminishing returns from repeated tests.[15]

(b) Now suppose that Popper were to attempt to *escape* the charge of inductivism as follows: On grounds *whose specification remains urgent*, he takes the aforementioned ten corroborations of p to be 'sufficiently many' repetitions of the experiment to accept p. Commenting on this putative approach, Musgrave writes:

Suppose that we test p ten times, that we always get the same result, and that ten repetitions is 'sufficiently many' for us to accept the universal test-implication p and reject the falsifying hypothesis. Suppose that we incorporate into background knowledge, not the *particular* results of our ten experiments, but the *universal* statement p. Now we need no inductive argument to make the severity of future tests of p decline. Indeed, it will decline sharply to *zero*, because p follows trivially from our augmented background knowledge as well as from the theory T.

This account is perfectly consistent with Popper's claim to have a non-inductivist theory of corroboration. But it can hardly be said to provide us with *diminishing* returns from repeated tests, if this means *steadily diminishing* returns. Instead, we have a 'one-step function': before 'sufficiently many' repetitions of the same test have been performed, each one has the same severity; after 'sufficiently many' repetitions, the severity of all future ones is zero. Perhaps this is what Popper had in mind all along: for he only claimed that he could provide *something like* a law of diminishing returns, and one step down is, I suppose, something like a gradual slide.[16]

(c) But the latter 'saltation'-version of the law of diminishing returns leaves unsolved for Popper the problem of '*how* often is sufficiently often' when repeated performances of a corroborating test are to be held adequate for incorporating the *universal* hypothesis *p* into the background knowledge.

The arguments given so far against Popper's ability to accommodate his own demands (i) and (ii) might be no more than *tu quoque* arguments, unless the Bayesian inductivist can do better in regard to justifying the gradual diminution of the epistemic returns from repeated trials. I shall now try to show in what sense the latter can indeed do better. We already saw earlier (Section 2) that the kind of inductivism which espouses the *unbridled* invocation of the instantiation condition can indeed *not* do better. For in *this* particular version of inductivism, all positive instances alike, including those resulting from repetitions of the same kind of test, will raise the credibility of the relevant hypothesis by the *same* amount. In this *latter* sense, there is warrant for the concessions by Keynes and Nagel which are mentioned by Popper.

Hence let us now look at the capabilities of the Bayesian framework.

B. *Bayesian Inference: The Rejection of the Instantiation Condition, and the Law of Diminishing Epistemic Returns from Repeated Tests*

In comparing Bayesian inference with Popper's epistemology, it behooves us to note first that the Bayesian schema rejects even the weaker version of the instantiation condition. I have given the specifics of that rejection in FT. After furnishing here a summary of the latter, we shall see how the Bayesian obtains diminishing epistemic returns from repetitions of the same kind of experiment.

We are about to consider a form of inductivism which regards degrees of credibility to be mathematical probabilities and thus holds Bayes' theorem to be applicable to the probability of hypotheses. But I disregard here whether a subjectivistic or an objectivistic construal is given to these probabilities and will use the term 'Bayesian inductivism' in a sense neutral to the difference between them.

Let *B* assert an initial condition relevant to the universal statement or deterministic (causal) law-hypothesis *H*, and let *B* also assert some other background knowledge. Let *C* be the phenomenon which is predicted deductively by *B* and *H*. Assume that *C* occurs.

Then in the Reichenbach notation, where $P(X, Y)$ is the probability *from* X *to* Y, Bayes' theorem (division theorem) can be written for this case as

$$P(B\&C, H) = \frac{P(B, H)}{P(B, C)} \times P(B\&H, C),$$

where $P(B\&H, C) = 1$.

Now assume that $P(B, C) \neq 0$: Unless this assumption is made, the right-hand-side ('rhs') of the equation and hence the posterior probability would be undefined, and it would be incoherent to have gone ahead with a test for C. And be mindful of the fact that, in our case of a universal (non-statistical) hypothesis H, $P(B\&H, C) = 1$. Then the necessary and sufficient condition that the occurrence of the positive instance C will yield a *posterior* probability which is *equal* to H's *prior* probability is as follows: Either $P(B, H) = 0$ or $P(B, C) = 1$. When the latter disjunctive condition is not fulfilled, the ratio of the posterior and prior probabilities of the universal hypothesis H will always exceed 1 even if only slightly, so that the occurrence of C will assure that the posterior probability of H is greater than its prior probability.

In FT, I deal with two matters as follows:

(1) I consider whether the case in which $P(B, H) = 0$ or $P(B, C) = 1$ are only marginal and hence trivial. For if they were thus both trivial, then in the case of a universal H, the weaker version of the instantiation condition would be *violated* in *only trivial* cases. And I conclude there that a verdict of 'non-trivial' is arguable.

(2) Furthermore, I consider the status of the instantiation condition for *non*-universal hypotheses H, i.e., for hypotheses such that $P(B\&H, C) < 1$. And I note that for such hypotheses, the weaker instantiation condition is violated if $P(B\&H, C) = P(B, C)$, i.e., when the evidence C is equally probable whether or not H is true.[17]

Let us now turn to cases in which the weaker form of the instantiation condition *is* satisfied *and* in which $P(B\&H, C) = 1$. Then we are concerned with supportive rather than non-supportive positive instances, and we ask: To what extent, if any, can the Bayesian inductivist comply with Popper's two stated demands? As we recall, these were that positive instances should count only if they are furnished by successful risky predictions and that repeated trials yield diminishing epistemic returns. Assume that neither of the two prior probabilities in the fraction on the rhs of our statement of Bayes' theorem above vanishes. Then we can write

$$\frac{P(B\&C, H)}{P(B, H)} = \frac{1}{P(B, C)}.$$

We are now concerned with the case of a successful *risky* prediction in which the probability of C on the background knowledge and given initial condition *without H* is *very* low – say 10^{-6} – so that $P(B, C)$ is *near* zero. Then the rhs is *very* large, say 1 000 000. But in that case the *ratio* of the posterior probability of H to its prior probability is likewise huge, viz. 1 000 000.

But this means that according to this theorem of the probability calculus, the following is the case: If a hypothesis H predicts a phenomenon C whose occurrence is very unlikely without that hypothesis on the basis of the background knowledge alone, then the occurrence of C confers strong support on H in the sense that the *factor* by which C *increases* the probability of H is enormous, say a *factor* of 1 000 000.

It is important to note here that a huge *factor* of probability increase need not be tantamount to a large *amount* of increase in the probability of H. To see this, note first from an expanded equivalent of the term $P(B, C)$ in an alternate form of Bayes' theorem, that in our putative case of $P(B\&H, C) = 1$, the value of $P(B, H)$ will typically be even smaller than $P(B, C)$. (This is also intuitive because hypotheses *other than H* whose prior probability on B is likewise very low will also entail the prediction C.) Hence for a sufficiently small value of $P(B, H)$ – say 10^{-8} – even a *million-fold* increase in the probability of H will yield an *amount* of increase of even less than 1/100 in the probability of H.[18] For the *difference* between the prior and posterior probabilities of H will be only 10^{-2}–10^{-8}.

This *caveat* concerning the distinction between the *factor* by which the probability of H changes and the *amount* of such change should now be borne in mind when considering the Bayesian's ability to implement Popper's demand for diminishing epistemic returns from repeated trials. We now ask: What is the Bayesian *supportive* significance of *repeated* occurrences of the positive kind of event instantiated by C (hereafter 'C-like' event)? Can the Bayesian succeed where unbridled instantianist inductivism failed, and show why the same confirming experiment should *not* be repeated *ad nauseam*?

In the Bayesian inductivist framework, each occurrence of a C-like event can be held to change the *content* of the background knowledge B such that the prior probability $P(B, C)$ of the next C-like event will *increase*, so that its reciprocal will *decrease*. Therefore, the *ratio* of the posterior and prior probabilities of H will *decrease* with such repeated positive occurrences. Hence the successive *factors* by which repeated occurrences of C-like events will increase the probability of H will get ever smaller. But this decrease in the successive *factors* of probability-increase does *not* itself suffice at all to assure a *monotonic* diminution in the *amounts* of probability-increase yielded by repeated occurrences after the first occurrence C! For suppose as before that prior to the first such occurrence $P(B_0, H) = 10^{-8}$, so that just before the *second* such occurrence the revised prior probability $P(B_1, H)$ becomes 10^{-2} after the above putative million-fold increase. Suppose that before the second occurrence, $P(B_1, C)$ were, say, 1/50. Then the second occurrence would increase the probability of H by a factor of only fifty as contrasted with a million. But the *amount* of such probability-increase effected by the first occurrence would be less than 1/100, whereas the amount of increase yielded by the second occurrence would have the much greater value 49/100. It is true that the posterior probability can become at most 1, so that the successive amounts of probability-increase can neither keep growing on and on nor even indefinitely remain above some non-zero lower bound. But the latter restriction by itself is a far cry from *assuring* a *monotonic* diminution in the successive amounts of probability-increase. Such a diminution might be assured, however, if the decrease in the successive values of the *ratios* of the posterior and prior probabilities of the hypothesis were sufficiently drastic each time.

Clark Glymour[19] has argued that there are cases in scientific practice in which the successful outcome of a 'severe' test nonetheless justifiably does *not* redound to the credibility of the hypothesis in the scientific community. For example, measurements of the gravitational red shift predicted by the general theory of relativity ('GTR') would seem to qualify as a 'severe' test of the latter's field equations in the sense that such a red shift is not predicted by the earlier Newtonian rival theory of gravitation. But Glymour adduces Eddington's analysis of this case to exhibit a justification for the view of physicists that the gravitational red shift is a very weak test of the GTR, while other phenomena are rather better tests, although the latter do not qualify as equally 'severe'.

The foregoing comparison between Popper's methodology and Bayesian inductivism in regard to the capability of *implementing* Popper's avowed 'law of diminishing returns from repeated tests' shows the following: Vis-à-vis Bayesian inductivism, Popper is hardly entitled to claim that 'it is all very easy' for him to show that his methodology has a superior capability of justifying that 'law'.

One important point remains to be considered in our comparative appraisal of Bayesian inductivism: Does Bayesianism sanction the credibilification of a new hypothesis by a sufficient number of positive instances which are the results of NON-risky predictions? For example, would a sufficient number of cases of people afflicted by colds who drink coffee daily for, say, two weeks and recover not confer a *posterior* probability greater than $\frac{1}{2}$ on the new hypothesis that such coffee consumption *cures* colds? And how, if at all, does the Bayesian conception of inductive support enjoin scientists to employ a *control group* in the case of this causal hypothesis K with a view to testing the rival hypothesis that coffee consumption is causally irrelevant to the remission of colds?

As for the first question, note that in the case of a test of H by a *non*-risky prediction, $P(B, C)$ is even initially quite *high*, i.e., $P(B, C) > \frac{1}{2}$, as in our example of the recoveries from colds. And as we accumulate positive instances C_i ($i = 1, 2, 3, ...$), the content of B changes after each new positive instance C_i such that $P(B, C)$ increases further. Hence the latter's reciprocal will decrease. But that reciprocal is the previously discussed *factor* $1/P(B, C)$ of probability-increase of the hypothesis under test. In our example of the hypothesis K concerning the therapeutic benefits of coffee drinking, the *factor* of its probability-increase will therefore itself decrease. The *general* question is whether the probability increases in the hypothesis which ensue from the accumulation of indefinitely many positive instances will ultimately raise its probability to nearly 1 or whether these increases will be such as to yield a posterior probability which is *only asymptotic to very much less than* $\frac{1}{2}$.

Unfortunately, apart from some rather restricted conditions which Allan Gibbard has investigated (unpublished), it would seem that more typically the Bayesian formalism does indeed permit the following: In the case of those *successful* but *non*-risky predictions which *are* supportive, a sufficient number of them will credibilify a hypothesis as being more likely to be true than false.

And it would seem that one must likewise *supplement* the formalism by the injunction to employ experimental controls in testing causal hypotheses like K: Such an injunction could be based on Hilpinnen's version of the principle of *total* evidence which enjoins us to *seek out* as much potentially relevant evidence as possible.[20]

In any case, the Bayesian probabilistic form of inductivism, no less than the Baconian version, is able to *deny* that a mere mass of positive instances of a hypothesis must automatically be held to confer a substantial measure of credibility on the hypothesis. And in Popper's theory of corroboration, the corroboration of a theory T by an observable fact F which was successfully predicted by T is *parasitic* on the logical relations of F to T's *known* rivals R_1, $R_2, ... , R_n$ as follows: For each known rival R, the accepted fact F *either* deductively falsifies R *or* F is *novel* with respect to R in the sense of being logically independent of R. In other words, a successfully predicted positive instance F of T corroborates T, if for each R, the fact F is either contrary to R or not predicted by R but without being contrary to R.

On the basis of our comparative scrutiny above of Popper's methodology vis-à-vis two major versions of inductivism, it will be apparent why I conclude the following: Popper is not justified if he claims to have advanced over all forms of post-Baconian inductivism "in explaining satisfactorily what is ... a 'supporting instance' of a law" (OK, p. 21; see also item (5) on p. 20). In saying this, I am also being mindful of my arguments above that Popper has given us no *purely deductivistic* rationale for advocating *severe* tests as being at all conducive to the discovery of true theories.

In conclusion, let me merely mention here that in FT, I have argued as follows: In his OK (pp. 18–20), Popper *tacitly* makes an important *inductivist* commitment when assigning *epistemic* significance to his so-called truth-preferability of corroborated theories over refuted ones.

In the light of the several groups of considerations presented in this paper, how can Popperians justify adhering to Popper's portrayal and indictment of inductivism while maintaining that he has given us a viable epistemological alternative on genuinely deductivist foundations?

University of Pittsburgh

NOTES

* This contribution by Professor Grünbaum is the paper which he presented at Kronberg, with one important difference to be described in a moment. Since that time he has been working indefatigably on topics dealt with in the present paper, and much of this more recent work has already been published by him in 1976 in the following tetralogy of papers:

'Is Falsifiability the Touchstone of Scientific Rationality? Karl Popper Versus Inductivism', in *Essays in Memory of Imre Lakatos, Boston Studies in the Philosophy of Science,* Vol. 39, edited by R. S. Cohen *et al.*, pp. 213–252, D. Reidel Publ. Co., Dordrecht, Holland.

'Can a Theory Answer More Questions Than One of Its Rivals?', *The British Journal for the Philosophy of Science* 27, 1–23.

Is the Method of Bold Conjectures and Attempted Refutations *Justifiably* the Method of Science?', *The British Journal for the Philosophy of Science* 27, 105–136.

'*Ad Hoc* Auxiliary Hypotheses and Falsificationism', *The British Journal for the Philosophy of Science* 27, 329–362.

The *first* of these publications overlaps rather considerably with the present paper and Professor Grünbaum would have preferred to have provided a new piece instead of it for this volume. However, Professor Watkins's Reply largely concentrates on Grünbaum's present paper, which therefore needed to be included if this volume was to be self-contained. Hence Professor Grünbaum acquiesced in this inclusion at the behest of the others concerned, although he would have greatly preferred to update the present paper in various ways. Yet there is a historical interest in reprinting it in an unrevised form, since the Kronberg Conference witnessed the first direct confrontation between this leading critic of the Popperian philosophy and one of its best known defenders.

The one significant change in his paper is as follows. It originally contained a brief excursion into Freudian psychoanalysis as a test case for Popper's methodology. This, together with some other passing references to Freud, has now been cut out. But the deleted material is available in considerably expanded form in the following very recent papers by Professor Grünbaum:

'How Scientific is Psychoanalysis?', in *Science and Psychotherapy*, edited by R. Stern, L. Horowitz and J. Lynes, pp. 219–254, Haven Press, New York 1977.

'Is Freudian Psychoanalytic Theory Pseudo-Scientific by Karl Popper's Criterion of Demarcation?', *American Philosophical Quarterly* 16 (April 1979).

(EDS.)

[1] This much larger essay is entitled 'Is Falsifiability the Touchstone of Scientific Rationality? Karl Popper versus Inductivism'. It has appeared in *Essays in Memory of Imre Lakatos*, R. S. Cohen and M. W. Wartofsky (eds.), *Boston Studies in the Philosophy of Science*, Vol. 39, D. Reidel Publ. Co., Dordrecht, Holland, Boston, U.S.A., 1976, pp. 213–252. And it will be cited below under the acronym 'FT'.

I wish to thank the editors and publisher of the Lakatos Memorial Volume for their kind permission to use material from my essay FT here.

[1a] Other studies in that series of critiques are A. Grünbaum, 'Can a Theory Answer more Questions Than One Of Its Rivals?' *British Journal for the Philosophy of Science* 27, 1–23 (1976); 'Is the Method of Bold Conjectures and Attempted Refutations Justifiably the Method of Science?' *British Journal for the Philosophy of Science* 27, 105–136 (1976); and '*Ad Hoc* Auxiliary Hypotheses and Falsificationism', *British Journal for the Philosophy of Science* 27, 329–362 (1976). Two such additional studies, published *after* 1976, are listed in the concluding paragraph of the *Editors'* note.

[2] Popper, K. R., *Conjectures and Refutations*, New York and London, Basic Books, 1962, p. 256. Hereafter this work will be cited as 'C&R' within the text.

[3] Cf. Russell, B., *Human Knowledge*, New York, Simon & Schuster, 1948, p. 381.

[4] Popper, K. R., *The Logic of Scientific Discovery*, London, Hutchinson, 1959, p. 420 (italics in original). Hereafter this work will be cited within the text as 'LSD'.

[5] According to Lakatos [Lakatos, I., 'The Role of Crucial Experiments in Science', *Studies in History and Philosophy of Science* 4, 315 (1974) and 'Popper on Demarcation and Induction' (hereafter 'PDI'), *The Philosophy of Karl Popper*, The Library of Living Philosophers, P. A. Schilpp (ed.), LaSalle: Open Court, 1974, Book I, pp. 245–246], Popper *tailored* his demarcation criterion to the requirement of *not* according scientific status to the aforementioned four theories.

[6] Burtt, E. A. (ed.), *The English Philosophers From Bacon to Mill*, New York, Random House (Modern Library Series), 1939.

[7] Cf. R. Giere's illuminating discussion ['An Orthodox Statistical Resolution of the Paradox of Confirmation', *Philosophy of Science* 37, 354–362 (1970)] of the resolution of Hempel's paradox of confirmation in statistical theory by means of *rejecting* a certain version of the instantiation condition. But note the *caveat* in note 17 below concerning the difference between our concept of 'positive instance' and the corresponding concept relevant to Giere's analysis.

[8] In Section 3 below, we shall consider to what extent, if any, Popper's own methodology entitles him to claim that *mere* repetitions of an initially corroborating type of instance increases the corroboration of a hypothesis only very little, if at all.

[9] Mill, J. S., *A System of Logic*, 8th ed., New York, Harper & Bros., 1887, p. 313.

[10] Rebuttals to Popper's arguments for $p(a) = 0$ are given in Colin Howson's paper 'Must the Logical Probability of Laws be Zero?' *British Journal for the Philosophy of Science* 24, 153–160 (1973). And on pp. 161–162, he offers a counterexample to Popper's $p(a) = 0$. See also J. Hintikka, 'Carnap and Essler Versus Inductive Generalization', *Erkenntnis* 9, 235–244 (1975).

[11] I am indebted to Noretta Koertge for some of these references.

[12] Popper, K. R., *Objective Knowledge*, Oxford, Oxford University Press, 1973, p. 51. Hereafter this work will be cited as 'OK'.

[13] That D is indeed equivalent to C becomes intuitive by reference to Tarskian logical contents as follows: The logical content of a disjunction is the *intersection* of the respective contents of the disjuncts. And the intersection of the contents of the infinitude of disjuncts in D will be just the content of C. To illustrate, take the *entire* infinitude of pairwise different curves, all of which go through or contain each of a finite set S of points in the xy plane. This totality T of curves or point sets will have exactly the set S as its set-theoretical intersection. For no point outside of S can belong

142   ADOLF GRÜNBAUM

to ALL of the curves in T. The points in S play the role of 'data points' in analogy to the observation statement C. And the curves in T play the role of the hypotheses in D each one of which entails C.

[14] Salmon, W. C., *The Foundations of Scientific Inference*, Pittsburgh, University of Pittsburgh Press, 1966, p. 119.

[15] Quotations from Musgrave here are from his article "Popper and 'Diminishing Returns from Repeated Tests'", which appears in the *Australasian Journal of Philosophy* **53**, 250–251 (1975).

[16] Ibid., p. 251.

[17] Good, I. J., [*British Journal for the Philosophy of Science* **17**, 322 (1966–67)] has given a perhaps far-fetched example having the following features: C states that a randomly selected bird is a black raven, H states that 'all ravens are black', and $P(B\&H, C) \ll 1$ (i.e., $10^2/10^6$). Good takes a black raven to be a 'case' (positive instance) of H *in a sense DIFFERENT from our above sense*, since his B does *not* assume the initial condition that the randomly selected bird is a raven! And despite the universality of H, the special feature of Good's example then is that not only is $P(B\&H, C) \ll 1$, but also $P(B\&H, C) < P(B, C)$, so that the ratio on the rhs is *less than* 1. But this means that the 'case' C of the hypothesis yields a *posterior* probability of H which is *smaller* than its prior probability: A perhaps somewhat far-fetched but even more resounding repudiation of the instantiation condition than the case of *equal* prior and posterior probabilities of H.

I am indebted to Wesley Salmon, William Harper and Laurens Laudan for helpful discussions of one or another facet of Bayesian inference.

[18] The need to distinguish the *factor* of probability increase from the *amount* was overlooked in this context in Salmon's *The Foundations of Scientific Inference, op. cit.*, pp. 118–120.

[19] Glymour, C., 'Relevant Evidence', *The Journal of Philosophy* **LXXII**, 403–426 (1975).

[20] Hilpinnen, R., 'On the Information Provided by Observations', J. Hintikka and P. Suppes (eds.), *Information and Inference*, D. Reidel Publ. Co., Dordrecht, Holland, 1970, Section II, esp. pp. 100–101. I am indebted to Teddy Seidenfeld not only for this reference but also for very clarifying comments on Allan Gibbard's results, which I mentioned above.

PAUL FEYERABEND

IN DEFENCE OF ARISTOTLE: COMMENTS ON THE CONDITION OF CONTENT INCREASE

TABLE OF CONTENTS

G. Radnitzky and G. Andersson (eds.), Progress and Rationality in Science, 143–180.
All Rights Reserved.
Copyright © 1978 by D. Reidel Publishing Company, Dordrecht, Holland.

centuries' fails for many reasons. The actions of the 'scientific elite' are neither rational
nor uniform and the rules which Lakatos produces have no force. Besides, they are
based on science, at least in intention; and are therefore parties in rather than judges of the
debate.

(vii) Most of the objections which critical rationalists raise against unloved views are
formal objections: the views do not agree with certain standards. But the criticism can
be reversed. If the views are correct, then the standards are unrealistic. For example, if
Aristotelianism is correct, then the search for 'depth' is as illusory as we now regard the
search for hell. Critical rationalists give no reasons why standards can be used against
views but views not against standards – and they have never considered the matter. This is
another reason why their objections against Aristotelian procedures cannot be taken
seriously.

(viii) The normative criticism which critical rationalists occasionally use against
opponents is restricted to a narrow domain and is mostly circular. Much better is Mill's
suggestion to judge a theory of rationality not only by its intellectual consequences, but
by the consequences it has in the lives of those who adopt it. And these consequences must
be seen in the widest possible way. For example, we must not forget the deterioration of
theology and religion that followed the rise of modern science. Intellectuals are not likely
to shed many tears over such matters, but other people are starting to take them seriously
and may in the end decide to criticize their intellectuals by no longer permitting them to
rule knowledge in their own way at taxpayers' expense. No doubt this *financial* criticism
of ideas will be more effective than their intellectual criticism, and it should be used.

(ix) Summary.

(i) *To discuss the idea of content increase it does not suffice to
consider its 'intrinsic' merits for this just amounts to comparing the
idea with itself. Nor is it sufficient to point out that it has advantages
over alternative theories of confirmation that have been developed in
recent years. Such theories never aided research and never led to any
results outside confirmation logic. A proper comparison must find a
world view that encouraged research, led to results but does not
conform to the condition of content increase. Aristotelian commonsense is such a world view. It created physics, astronomy, biology,
theology, the history of ideas, an art theory, there is a theory of
mathematics, there are special results such as a theory of the
continuum and all these subjects conform to a stable set of concepts
that is never changed, though it may be temporarily covered by error.
The condition of content increase is not applied to the domain of these
concepts.*

Content increase, say critical rationalists, occurs *de facto* during
scientific change, and it is also eminently *rational*. Let us first
examine the question of rationality.

This is best done by presenting a world view (i.e. a cosmology with

an epistemology and a methodology) that does not contain content
increase, by examining its merits and possible arguments against it. The
world view I have chosen is 'Aristotelian commonsense'. I call it
commonsense because it contains elements that can be found in many
different cultures and are connected with the mastery of nature and of
society.[1] I call it 'Aristotelian' because Aristotle examined properties
and invented a methodology and a theory of knowledge suitable for it.
It is obvious that a brief presentation such as this has to be severely
streamlined. It is hoped that the streamlining will not cut it off from
the historically identifiable advantages I want to quote in its favour.

In Aristotle universals *arise* from sense experience and principles
are *tested* by comparing them with the results of observation.[2] 'It is
the business of experience to find the principles which belong to each
subject. In astronomy, for example, it was astronomical experience
that provided the principles of the science, for it was only when the
phenomena were adequately grasped that the proofs in astronomy
were discovered. And the same is true of any science whatsoever'
(*An. Pr.* 46a17ff). Accordingly, 'the loss of any one of the senses entailed the loss of the corresponding portion of knowledge' (*An. Post.*
81a38). Principles not consistent with observation 'are wrongly
assumed ... [for] ... principles require to be judged by their results,
and particularly from their final issue. And in the case of knowledge
that issue is the perceptual phenomenon that is reliable when it
occurs' (*de coelo* 306a7). It is not advisable to 'force ... observations
and try to accommodate them to one's theories and opinions ...
looking for confirmation to theory rather than to the facts of observation' (*de coelo* 293a27 – sounds very much like Newton!). Nor is it
advisable to 'transcend' sense perception and to disregard it on the
ground that 'one ought to follow the argument' (*de gen. et corr.*
325a13 – against Parmenides). The best course is "to follow the
method already mentioned, to begin with phenomena ... and, when
this is done, to proceed to state the causes of those phenomena and to
deal with their development".

These methodological requirements are combined with a theory of
perception that makes them plausible and lends them force. "The
sensitive and cognitive faculties of the soul" says Aristotle, (*de anima*
431b26ff "are potentially these objects, viz. the sensible and the
knowable). These faculties, then, must be identical either with the
objects themselves, or with their forms. Now they are not identical

with the objects; for the stone does not exist in the soul, but only that
form of the stone." "The sentient subject, as we have said, is potentially such as the object of sense is actually. Thus during the process
of being acted upon it is unlike, but at the end of the process it has
become like that object, and shares its equality" (418a2ff). *"That
which sees does in a sense possess colour;*[3] for each sense organ is
receptive of the perceived object, but without its matter. This is why,
even when the objects of perception are gone, sensations and mental
images are still present in the sense organ" (425b23ff). In the act of
perception the very forms of nature are present in the mind and not
merely images of them. Going against perception therefore means
going against nature itself. Following perception means giving a true
account of nature.

The theory of perception in turn receives support from experience
and from a theory of motion that comprises all types of change, in
living beings as well as in inanimate matter and that agrees with
evidence of the most convincing kind.[4] It is in agreement both with
experience and with general physics.

On the other hand, it is not asserted that every single act of
perception gives a true account of the world. The Aristotelian theory
describes what happens during perception *in the normal case*. But the
normal case may be distorted and even entirely concealed disturbances. "Error ... seems to be more natural for living beings, and
the soul spends more time on it" (*de anima* 427b3ff). The disturbances must be studied and removed so that knowledge can be
obtained.

The process by which universals are 'established' in the soul
depends on particulars as well as on 'low level universals' already
imprinted in it (*An. Post.* 100a3ff, esp. 100b2) which means that an
idiosyncratic history of perception leads to idiosyncratic perceptions
later on. Also the senses, being acquainted with our everyday surroundings are liable to give misleading reports of objects outside this
domain as is proved by the appearance of the Sun and the Moon: on
Earth large but distant objects in familiar surroundings such as
mountains are seen as being large, but far away. The Moon and the
Sun, however, 'appear to measure one foot across even to men who
are in health, and know [their] real measurements' (*de somn.* 458b28;
cf. *de an.* 428b4ff). The discrepancy is due to the imagination which is
"some kind of movement ... caused by actual sensation" (*de an.*

428b12ff). The movement "resides in us and resembles sensations" (429a5), but "it may be false ... especially when the sensible object" appears in unusual conditions, such as large distance (428b30f) and removed from supervision by the 'controlling sense' (*de somn.* 460b17). A combination of unusual conditions and absence of control thus leads to illusions; for example, patterns on the wall are sometimes seen as animals (460b12). Cf. also *Met.* 1010b14 on the perception of objects that are 'foreign', or 'strange' to the sense perceiving them as well as *de part. animal.* 644b25 where it is said that the objects of astronomy, though 'excellent, beyond compare, and divine' are less accessible to knowledge. The evidence that "might throw light on them and the problems concerning them is but scantily furnished by observation" and so error is likely to arise.

Reading such passages we see that Aristotle was aware of the difficulties of astronomical observation,[5] he knew that the senses, used under exceptional circumstances can give exceptional and erroneous reports, he knew how to explain these reports and so he would not at all have been surprised by the problem of the first telescopic observations. Compared with him the first observers approached the matter with a great and quite naive confidence. Ignorant of the psychological problems of telescopic vision, unfamiliar with the physical laws governing light in a telescope they went ahead and changed our world view. This was seen very clearly by Ronchi and by some of his followers.[6]

Apart from unusual conditions, error may also be due to the reaction of the senses themselves (*de somn.* 460b24), it may come from processes such as imagination which are triggered by sensations (*de anima* 428a10), from 'mistakes in the operation of nature' comparable to 'monstrosities' in biology (*Phys.* 199a38), it may occur because the senses have been overstrained – "the excitation is too strong, ... the ratio of the adjustment (between sense and surroundings) is destroyed" (*de an.* 425b25), or when emotion, illness, too great distance or other unusual conditions interfere with the proper working of the senses (*de somn.* 460b11 – the examples used here and their explanations show that Aristotle could have given a perfectly acceptable account of the strange phenomena reported by the first telescopic observers). There are subliminal stimuli (*de divin. per somn.* 463a8) which produce large scale actions of the organism affected (463a29) and there are imperceptible events (*Meteor.* 355b20).

Objects not appropriate to the sense by which they are perceived are more likely to lead to error than the proper objects of the senses concerned (colour, in the case of sight: *Met.* 1010b14, *de anima* 428b18), but even in the latter case there can be 'mistakes in the operation of nature', as we have seen. Misguided by such events we may become inclined to believe a false theory 'as based on experience' and may have to reject it *"because one can see no reasonable cause why it should be so"* (*de divin. per somn.* 462b14): Aristotle is prepared to "squar[e] a recalcitrant fact with an empirical hypothesis".[7] All this refutes Randall's assertion that Aristotle "had no sense of the possibility of correcting by more accurate means of observation".[8] It also shows that the empiricism of Aristotle was more sophisticated than either his critics or even some of his followers seemed to realise.

The difference between Aristotelian empiricism and the empiricism *implicit* in modern science (as opposed to the empiricism that turns up in the more philosophical pronouncements of scientists) thus does not lie in the fact then the former overlooks observational error while the latter is aware of it. *The difference lies in the role which error is permitted to play.* In Aristotle error beclouds and distorts particular perceptions *while leaving the general features of perceptual knowledge untouched.* However great the error, these general features can always be restored and it is from *them* that we receive information about the world we live in. Aristotelian philosophy corresponds to commonsense. Commonsense admits error, it has found means of dealing with it, some forms of science included, but it will never concede that it is false throughout. Error is a *local phenomenon*, it does not distort our *entire outlook.* Modern science, on the other hand (and the Platonic and Democritian philosophies it absorbed) postulated just such *global distortions.* When it arose in the 16th and 17th centuries, "it call[ed] in question a whole system, not just a particular detail; moreover, it was an attack not only on the physicists, but on almost all sciences, and all received opinions ..." (*Physics* 253a31ff – speaking about Parmenides). The moving force behind the attack was an accidental combination of special discoveries in astronomy, physics, geography, meteorology, optics etc. etc. and of a philosophical generalization of some attitudes behind these discoveries. This is a very complex process which is still far from understood. The condition of content increase may have been part of it, it may have emerged without anyone noticing what actually happened, it may have been the result of conscious effort. We do not know.[9] But we can still ask what arguments the new philosophy might have raised against commonsense and what commonsense might have replied to these arguments. Only a small part of such a debate can be presented in the present essay. The section that has just been completed gives an outline of the role of perception in Aristotle. In the next section I shall briefly explain what arguments Aristotle has against a search for greater depth of explanation and what these arguments imply. I shall also explain different notions of novelty that have not been clearly separated by Popperians. The Aristotelian arguments are continued in Section v where it is shown why Aristotle is not an inductivist. The remaining sections consider objections against a philosophy that vaguely resembled the Aristotelian philosophy. The objections turn out to be either childish, or incompetent, or irrelevant.

(ii) *On the other hand it is possible to introduce the procedures of modern science and, as a matter of fact, any scientific theory one is fond of into the Aristotelian frame. The theories will continue to make predictions, there will be discoveries and content will be increased in that sense, but this increase in content will not be accompanied by any increase in depth. It will not lead to 'novel' predictions, it will only lead to prediction of facts of the kind outlined in the basic philosophy. In this respect Aristotelian commonsense provides an interpretation of scientific theories very similar to the Copenhagen interpretation, but the basic concepts are different, and so is part of the basic philosophy. Critical rationalists, however, are not very clear as to the kind of novelty they want their content-increasing statements to have.*

The cosmology that forms the background of Aristotelian commonsense assumes (1) that man is in harmony with the world, he is well adapted to it, his perceptions and his common notions contain a true account of the universe.[10] This basic assumption is confirmed by the ease with which we carry out even complicated tasks and by the feeling of security and assurance that accompanies our actions. In Aristotle it is further confirmed by a kind of consistency proof: using perceptions and analysis of common notions we obtain a general theory of change which applied to the human organism gives us precisely the perceptions and the common notions on which the whole edifice rests. It is also assumed, (2) that the harmony between man and the world is occasionally disturbed. False notions arise, erroneous perceptions misguide us. The disturbances – and this is an essential part of the commonsense philosophy – never affect our perceptual knowledge as a whole. They do not show that our world view is fundamentally mistaken. They invade only a restricted domain and can be eliminated by a study of the problems in this domain. The elimination restores truthful perception and harmony between man and the world. This assumption, too, is confirmed by observation: sober up a drunken man, or cure a jaundiced man or a madman, and he will again see the world as it really is.

The 'methodology' that works in a commonsense world can be outlined as follows: (3) normal perception and analysis of perceptual concepts provide the basic properties, they determine what kinds of objects exist, how these kinds are related to each other and what laws are valid for each of them. The process of object-determination and – separation is a *causal* process, it occurs without any conscious activity on part of the observer (cf. again the account in *An. Post.*, 99b27ff and *de anima* 431b26ff) and the only advice to the observer is not to interfere with it. More comprehensive concepts such as the concept of change, the categories, the concepts of the actual and the potential are obtained by an analysis of the 'primary concepts' which are imprinted in man by the causal process just mentioned and are therefore nature's traces in man: by analysing the primary concepts *we* have we analyse the primary forms of *nature*.

The theories that emerge from such an investigation note common features of elements in a limited domain or of all being; they are not 'deeper' theories in the sense that they go behind appearances. 'Going behind appearances' would mean going behind the forms that are in the mind which would mean going behind the forms of nature, it would mean going 'behind' nature and, therefore, away from nature. It would mean assuming that nature is not what it is and thus introduce falsehood. Nor is there any attempt to find out whether different types of objects such as stones and stars may not be actually the same. If undisturbed perception shows them to be different, then they are different and so cannot be said to be similar.[11] 'Anomalies' can arise in two ways. They can arise by the discovery of new phenomena in a known domain or by the discovery of new domains. The last case presents no problems: a new field of research is opened and treated in the appropriate manner. In the first case there are two possibilities. We may either declare that an apparently uniform

domain was found to be two domains after all; or we may ascribe the
difference to a disturbance of normal perception and so add to our
knowledge of disturbances.[12] All these moves preserve the authority
of undisturbed perceptual knowledge which therefore continues to
have the theoretical as well as practical *certainty* that accompanies our actions in normal circumstances (this is what Aristotle
occasionally seems to mean by 'intuition').[13] Content increasing
procedures *do not* occur and *cannot* occur. For assume there is a
fact F that has just been discovered and that constitutes an 'anomaly'
in domain D. A content increasing account of the fact would be an
account of the domain, or of a larger domain that makes the anomaly
disappear while at the same time introducing a 'novel fact' G. The
'novelty' of G consists in its sharing a certain property with F which
was not known before (if there is no such property, then the
anomaly – removing procedure is nothing but an instrument of
prediction – an interpretation rejected by critical rationalists). Let P
be that property. Now P cannot be a perceptual property. In this case
it could be found by observation and there would be no increase of
content in the sense intended. Nor can it be one of the general
properties mentioned above, and for the same reason. The only thing
it can be, within the Aristotelian scheme, is a 'purely formal' property
which means, speaking the language of Simplicius as well as of
critical rationalists, it can at most be a predictive device which
again shows that no novel predictions have been made. It is therefore
quite possible to introduce content increasing stratagems into the
Aristotelian world view; however entering it, these strategies receive
an interpretation that robs them of their content increasing propen-
sities: the stratagems are interpreted as *formulae* which, by an
accident of history, predict things not yet observed, but without any
increase in depth. We see how closely (3) is connected with (1) and
(2), and how 'rational' it is, given this view of the world and of man's
position in it.

We also see how absurd the search for 'depth' behind perceptions
must appear to an Aristotelian. Any such attempt assumes that the
world is different from what it *is* (and not only from what it *appears
to be*: remember that the forms we find in our mind *are* the forms of
the world). It looks for a second, hidden world behind the real world
despite the fact that the latter is complete in itself. It looks for such a
world not because an examination of the real world has led to

problems, but in order to satisfy a philosophical dream. This is the height of irrationality as Aristotle points out more than once in his criticism of Parmenides.

This criticism, incidentally, has much in common with the objections which the followers of Niels Bohr have raised against hidden variables. The objections were twofold: first, that hidden variables would not fit into the conceptual frame of the theory and second, that any attempt to make them fit would remove them from the domain of perceptual knowledge. This is almost identical with Aristotle's criticism of Parmenides except that the concepts examined by Aristotle ('one', 'same', 'different' and so on) are much more fundamental than the concepts of classical mechanics which formed the background of the Copenhagen School.

Critical rationalists have not always been clear what they wanted when demanding content increase. Occasionally it seems that one simply wants *more* predictions, or *more precise* predictions, or predictions that *revise* already existing predictions; on other occasions one seems to demand that the predictions be both *new* and of a *new kind*. A statement such as 'all ravens are black' predicts events that have not yet been observed and in this very trivial sense gives us 'novel' predictions. Einstein's predictions about Brownian motion are not 'novel' in this trivial sense. They contain the trivial component, for they say that we shall find certain observable regularities. But the trivial component is connected with another component that is far from trivial: the observable regularities are said to be true of processes (molecular interactions with macro-objects) that not only are not found in the phenomenological world, but have no place in it. Such a transition to a 'new world' is essential for a notion of depth that goes beyond the trivial additive content-increase produced by statements of the 'all ravens are black' type and the somewhat less trivial, partly additive, partly subtractive content increase produced by statements that conflict with previous empirical regularities. It is incompatible with the Aristotelian methodology. The Aristotelian will accept the *formulae* that produce the novel predictions of the 'deep' kind, but not their *interpretation*, for he will interpret them (as Mach interpreted Boltzmann's formulae) as correlations between observable events and thus retain only their 'trivial' component which, of course, will be far from trivial for him. It is interesting to see that Popper's notion of depth which he develops in his discussion of the Kepler-

Newton case is still pretty close to what is acceptable to an Aristotelian.

(iii) *The objection that Aristotle is no good because he does not permit the right kind of content increase assumes that content increase (of the right kind) is preferable to no content increase. But that is the point at issue.*

An objection that is so simpleminded that one wonders how it could have been raised at all is: the methodology just described is unacceptable because it involves degenerating problem shifts. Anomalies are absorbed without any change in basic theory and without novel predictions.[14]

The objection consists of two parts, viz. a *description* of what is done, using special terminology ('degenerating' etc.) and an *evaluation* of the events and procedures so described. It is interesting to see that the description *insinuates* the evaluation (who is ready to praise 'degeneration'?) and thereby creates norm-fact confusion, a confusion which Popperians deplore whenever it occurs in the rhetoric of their opponents but cultivate when they need it for rhetorical displays of their own. But let us now overlook this quite legitimate attempt to forge *words* into weapons for knocking out an opponent[15] and let us ask what reasons there are for the *evaluation*: why is the Aristotelian procedure unacceptable? The only legitimate reason that might be given at *this* stage of the argument would be a difficulty for Aristotelianism that is *independent of* its being a 'degenerate research programme'. We certainly do not want to take content increase for granted, we want to argue for it and against its opponents. We want to show that it is better than a procedure that lacks depth and adds facts in an ad hoc manner. Condemning Aristotelianism by saying that it degenerates does not amount to such a demonstration. I have looked at every line of the position paper, I have tried to read between the lines, but I did not find a trace of an argument of this kind. The only thing I did find was Watkins' assertion that content increase and depth are parts of a form of science 'we would like to have', 'we' being Bacon, Descartes and Watkins.[16]

In a way the simpleminded error I have just criticised is found even in such an excellent work as Lakatos-Zahar, "Why did Copernicus supersede Ptolemy?"[17] Lakatos and Zahar try to explain the triumph of Copernicanism by showing that it was a progressive research

programme. Now quite apart from the fact that *nobody* who loved
Copernicus loved him for this reason it is assumed that people had
already adopted, or had never used anything but, a philosophy of
content increase. Just as our rationalists use content increase as a
weapon even in circumstances where the weapon is under debate, in
the very same way Lakatos and Zahar assume that content increase
was used as a weapon in the past (against Ptolemy) even in circumstances leading to a change of weapons.[18] Is that all critical rationalists have to offer for our debate? Let us see!

(iv) *A better way of evaluating a theory of rationality is to compare it
with logic. In the sciences such a comparison does not always end
in favour of logic. For example, science contains contradictions, the
attempt to remove the contradictions removes either important factual
statements or the ability of the theory to grow, or its fertility in the
treatment of concrete problems. The objection that a contradiction
entails every statement applies to special systems of logic, not to
science which handles contradictions in a less simpleminded fashion.*

As far as one can make out there are three ways in which critical
rationalists (and the philosophers they imitate) have tried to evaluate
methodologies, theories, and entire world views: logical, institutional,
normative.

The logical evaluation consists in comparing the things to be
evaluated with the laws of logic. If there is a clash, then the things
must be removed or at least changed until they agree with the laws.

The procedure is familiar from the sciences. Scientific theories are
supposed to conform to basic laws. For example, they are supposed
to be relativistically invariant, in agreement with quantum laws,
conservation of energy-momentum, entropy increase and so on.
There are however some very decisive differences between the
demands of the logicians and those of the scientists.

First, the scientific demands refer to theories that can be easily
identified: quantum theory, relativity, thermodynamics. But there is
no one 'logic' that could be used to give content to the demand that
theories, methods, philosophies must be logically adequate. There is a
whole spectrum from formally rigorous systems with non-contradiction and excluded middle via more informal systems, intuitionistic
systems without excluded middle, systems in which contradictions do
not entail every statement to Hegel's logic and beyond to logics that

are not expressed in explicitly formulated laws but through the way in which concepts are being used. In a way the demand that ideas must conform to LOGIC is very similar to the demand that they must conform to PHYSICS and equally impossible to carry out: there is no one coherent thing PHYSICS to which anything could conform just as there is no one coherent thing LOGIC to which anything could conform. Of course, what one actually means is that science should conform to one particular and rather simpleminded logical system to which philosophers have become accustomed and without which they cannot live. For them LOGIC means that system just as PHYSICS once meant mechanics. 'That' system asserts non-contradiction, excluded middle and demands stability of meaning throughout an argument. What are the reasons for preferring it over other systems? The only reason I have discovered and shall soon discuss in greater detail is connected with non-contradiction: contradictory theories entail everything and so cannot be used to say anything. This they do but only when embedded into the selected system. If we can give reasons for the usefulness of contradictions in science as I think we can then we have also reasons against the selected system. More of them below.

Secondly, the scientific demands permit approximations: it suffices if a theory is relativistic in the relativistic *limit* (assuming it goes beyond relativity) and classical in the classical *limit*. Exact agreement is not necesary. The logical demands, however, must be strictly satisfied, and not only by approximation.

Thirdly, the scientific demands are no longer absolute demands. One recognizes that a theory that clashes with basic laws may be correct, that the basic laws may be mistaken and that the mistakes may be discovered by further pursuing the theory. Not a single critical rationalist is prepared to admit that we might discover errors of logic by pursuing illogical theories. This is asserted quite generally and independently of the particular system of logic chosen. What are the reasons for this astounding dogmatism?

There are many reasons, most of them purely verbal: logic is defined in a certain way, then one states certain demands for science and concludes by showing that the demands can be satisfied only if science is built up in accordance with the laws of logic. Both the definitions and the demands lead to long chains of explanations and lemmas and so we seem to be engaged in a fruitful piece of argument, while all that happens is that we are chasing our own tails in a labyrinth of our own making. The only parts of the labyrinth that seem to have some content are the following two.

Part one: violating the laws of logic puts an end to our ability of rational discourse.

But so does systematic short-term malfunctioning of memory, computers, signs on paper (changes etc.). There are many conditions that must be met to make reasoning possible and the laws of logic (if they are among them) are only some of them. Why should they be given a special position? Besides we often find that rational discourse proceeds not only *despite* a violation of logical principles but *because of* it. Take the principle that the meaning of a term has to be kept constant throughout an argument. This means that the assumptions on which this meaning rests have to be kept constant. But the argument may be about the assumptions.

Part two: the laws of logic play a special role because they *guarantee* rationality while other circumstances only help or hinder us in the attempt to achieve rationality.

With this part we already enter the slums of a purely verbal debate and so we better become clear what we are up to. Assume somebody shows as I think I have shown (with the help of material and arguments provided by Hegel, Kuhn and Lakatos) that science often violates those laws of logic which our critical rationalists regard as a *conditio sine qua non* of rationality. For a critical rationalist the matter is clear: the parts that violate logic are not 'rational'. Assume it is also shown that certain parts of science, informal mathematics included, *progress* in the sense in which critical rationalists are happy to speak of progress. For a hardheaded critical rationalist there is no doubt what the answer must be: progress or no, the parts of science that disobey logic are not rational.[19] But it is clear that such an answer can no longer impress us. Rationality was introduced to further knowledge. Dogmatic rationality was changed to critical rationality when scientists replaced the idea of certain and stable knowledge by the idea of its growth. Rationality was changed because it was regarded as an aid to knowledge and *not* as an aim in itself. Now if one discovers that knowledge is beset by contradictions, progresses because of the contradictions and right through them, entails fundamental changes of concepts leading to incommensurability so that it is no longer possible to compare alternatives by content (though there are many other ways of comparing them) then the lesson is obvious: rationality as defined by part two no longer aids knowledge and has to be given up no matter how many lovers of antediluvian ideas of discourse keep croaking 'irrational! irrational!'

But is it really so easy to get rid of fundamental principles of logic? Has it not been shown, and then popularised that 'if two contradictory statements are admitted, any statement whatever must be admitted'?[20] And what is the force of such a 'proof'? Let us see!

The proof is based on various ideas, among them the assumption that logical rules (rules of reasoning) are valid, or must be applied, or must be obeyed no matter what the circumstances. This rule is not at all obvious and there may be reasons to reject it. Let us therefore look at the material that may provide us with such reasons.

Watch a train leave the station through a window of your own compartment in an adjacent train. The train comes to a stop. You now see it (1) moving back into the station and (2) not changing place relative to the rim of your compartment window. The phenomenologically correct description of the process is that the train changes place and does not change place. There is an experience that is correctly described only by saying that *the same thing moves and does not move at the same time and in the same respect*.[21] The rule that a contradiction while fertile must still be removed would in this case advise us to give an incorrect description of an experience. There are of course ways of avoiding the problem. One can eliminate the contradiction-causing concepts and use others. But then we cease to have the experience that led to our problem for this experience depends precisely on the interplay of concepts and percepts. So, there is no way out: remove the contradiction and with it the experience of which it provides a phenomenologically adequate description or retain the contradiction and find better ways of dealing with contradictions than are provided by the formal logic of today.

Such 'better ways' can be discovered by analysing cases that lead to even more forceful arguments against the simpleminded logic where contradictions are premises from which everything can be deduced. In these cases we do not have reasons for asserting $p\&\bar{p}$ directly, but we have a theory which, after a few steps gives us $p\&\bar{p}$. An example is Dirac's formulation of elementary quantum theory.[22] Another example is classical statistical mechanics as developed by Maxwell, Boltzmann and others. It is assumed by this research programme (e.g.

in the proof of the *H*-theorem) that the minimal elements of a gas rebound elastically from each other which means, in the case of a monoatomic gas, they are deformable. But it is also assumed (via the equipartition theorem and empirically observed specific heats) that they are not deformable.[23] A third example is the calculus of fluxions which was shown to be inconsistent very soon after Newton had used it so successfully in building up his mechanics and whose use by physicists increased rather than ceased after the inconsistencies had been revealed. The result was not disaster, but more discovery. A fourth example is the theory of the continuum which has played a role ever since the Pythagoreans' contention that it could be built on numbers entirely, but it was never put into a satisfactory shape. A fifth example is the older quantum theory. There hardly ever existed a view so incoherent and at the same time so fertile as this strange and still not too well understood research programme. In all these cases we have theories that helped in the advancement of science while being beset by internal contradictions. This clearly shows that *there are ways of handling contradictions that do not lead to the unwanted assertion of every statement but to specific and highly useful results.* Or, to express it differently, there exists a *practical logic*, used by scientists and not yet available in explicit form (except, perhaps, in some parts of Hegel's *Logic*, in Engels and dialectical materialism,[24] and in Wittgenstein's *Foundations of Mathematics*[25]) that enables us to make discoveries with the help of contradiction-infested systems. This is a serious challenge to the rather widespread belief in the supreme authority of certain types of formal logic.[26] It shows that scientific practice can overrule logic just as it can overrule 'facts' and highly confirmed laws. It can overrule logic because it contains ever changing principles of order which are far better adapted to reality and to the vicissitudes of research than the simpleminded ideas and rules that are being suggested by logicians and faithfully copied by philosophers of science.

This finishes my *general* remarks concerning the authority of logic in cosmological and methodological discussions.

(v) *The charge of inductivism, however, that has occasionally been raised against Aristotle, is misdirected. There exists no interesting theory of rationality that can be accused of inductivism. And the view which critical rationalists introduce is supersede 'inductivism' suffers from the very same faults as that fictitious theory.*

I shall conclude my discussion of logical methods of evaluation with a brief account of 'inductivism'. Inductivism is accused of trying to derive (logically!) general statements from conjunctions of singular statements. This is impossible, hence inductivism is impossible. So far the childishly simple argument against an apparently childishly mistaken philosophy. The argument is important for our case, for it is supposed to apply not only to modern philosophers, but also to Aristotle. So, let us have a closer look!

The first thing we discover is that the childishly mistaken philosophy the argument is supposed to expose never existed. At any rate, it is not found in Aristotle and it is not found in Bacon. Bacon and the early Baconians were well aware that formal logic does not give us anything new and they tried to find procedures that would produce discoveries instead of repeating what was already known. Nobody thought that these procedures would get laws from observations *with no further information given*; one assumed that observations would lead to laws in a world *whose general outlines were already known*. Like engineers who in trying to adapt their bridges to the physical surroundings occasionally argue from single observations, or collections of single observations (of ground, shore, river, material, requirements of weight) to general tendencies without claiming that the observations *entail* the tendencies, in the very same manner these philosophers, in trying to adapt their thought-bridges, viz. their theories to the psychophysical surroundings occasionally argued from observations to tendencies and laws without claiming that the observations entailed the laws. It is not easy to find out what gave them the assurance that the world was as they imagined it to be but this point does not have to be discussed in the present essay which deals with Aristotle. For in Aristotle the situation is perfectly clear. Here the attempt to solve even the simplest problem, such as the problem how individual observations are related to laws, cannot even begin without a large background of convictions, actions, perceptions, a whole world of knowledge-producing potential, only part of which enters our consciousness. Starting on our endeavour to achieve knowledge the problem is not how to get something as yet unknown from something completely separate that is already known, the problem is how to make explicit connexions that exist in nature and which the mind is capable of grasping. Proceeding in this way we seem to assume a lot before we get even started but actually we do not, for the world of commonsense that contains the connexions is given and need not be assumed. This world contains the outlines of a true cosmology. The problem is to discover the outlines, to analyse our thoughts, our actions, our perceptions in such a manner that their function is understood and lifted into the broad daylight of explicit philosophical theory: we are humans, we live and act in a perceptible world – let us examine what that amounts to. The philosophical theory Aristotle aims at is to *articulate* the known world of commonsense, not to *replace* it so that we learn how what we possess as thinking, acting, perceiving human beings can aid us in obtaining what we do not yet know. What gives Aristotle the assurance that commonsense is not an illusion, that it can be trusted, that it is not necessary to replace it by 'deeper' accounts? A survey of existing criticisms of commonsense. This survey which is a masterpiece of philosophical ingenuity starts with the strongest attack against commonsense, the attack of Parmenides. According to Parmenides, commonsense is entirely untrustworthy and has no role in philosophical debates. It must be rejected. It must be rejected because it conflicts with arguments of great force and simplicity. Aristotle takes these arguments seriously but not without examining them and the concepts they contain. He finds faults and replaces the concepts used by different concepts (*Physics*, book ii). The new concepts no longer conflict with commonsense, quite the contrary, they are perfectly adapted to it. Commonsense is now supported not only by the *practical* authority it has for us – after all, we act, live, survive in accordance with its principles – but also by the quite 'modern' authority of rational argument. It is, and it is rational, and so there are now two reasons instead of one to accept it. Aristotle soon adds a third: Extending the theory of motion he has obtained to the inter-action between man and the world he gets the result that man basically perceives the world as it is: the forms of the world are the forms that shape perception – they are *identically the same* forms, not just images. This delicate and complex fitting job[27], which confirms his original belief in the basic harmony between man and nature, enables Aristotle to give advice to those who want to explore the world in greater detail. Having found what kinds of entities there are and how they are related to each other he suggests how the knowledge may be used in the discovery of further elements of the same kind. Here is a decisive difference between the Aristotelian and the inductivist the Popperians have in mind. The Popperian inductivist makes a leap into the dark and now wants to find a way of extending the light of his starting point to the new domain in which he has arrived. He fails, for it is not possible to get something out of nothing. The Aristotelian, on the other hand, already has an account of the world, this world is given to him from birth, he lives in it, he does not need to 'assume' it and so *his* 'inference' from individual to law is not a leap in the dark but a rational step *within* a well known world. The procedure called 'induction' by the translators and by those philosophers who follow their lead (Popper among them) has therefore more than one function and none of these functions coincides with what Popperians mean by 'induction' today.[28] Common to all functions is the use of individuals for 'leading towards' a universal. The 'leading towards' is not deduction. It means, before 'reconstruction' i.e. before the Aristotelian analysis has illuminated commonsense, the process (causal process) by which the contemplation of an individual gives rise to an idea and 'after' reconstruction that individual cases can now be 'read' in universal terms "das Besondere ist das Allgemeine unter verschiedenen Bedingungen erscheinend" writes Goethe who understood these matters pretty well. Cf. also "Um zu begreifen, dass der Himmel überall blau ist, braucht man nicht um die Welt zu reisen" – both in *Maximen und Reflexionen*). The law is now directly before us and with it the reasons for its validity.[29] I conclude that Aristotle is not an 'inductivist' in the sense of Popperians and that the difficulties of 'inductivism' do not beset him

Nor can the difficulties of Popperian inductivism be overcome by the method of conjectures and refutations. The difficulties, as stated by Popperians, are that rules of induction do not work by themselves, we need further assumptions, and there is no guarantee to get anything valuable even then. But the method of conjectures and refutations is subjected to exactly the same difficulties. There is not a single method that works in all circumstances (in 'all possible worlds') and that can start without much preparation. The method of conjectures and refutations, for example gives results only in a world that is not immersed in arbitrarily distributed disturbances (for otherwise a law would be refuted the moment it is formulated and a science could never arise) just as the rule of induction gives results only in a world with clear species which are connected in all elements if they are connected in some. Now once one admits that the objections against inductivism can be raised against every methodology then philosophy or logic alone cannot decide the matter. The matter can only be decided by resolutely using the method and seeing where it leads.

A detailed examination of Aristotle's theory of knowledge therefore reveals that it does not contain any of the 'well known [speak for yourself, John!] difficulties that beset inductive theories of confirmation'.[30] The ability to defuse these 'difficulties' is Watkins' second 'major argument' against alternatives to his own point of view;[31] and there is no further argument in the position paper. Result: no logical objection has been raised against Aristotle.

To sum up the content of this section: logical objections are never strong enough to remove a blooming practice and no logical objections have been offered against Aristotelian commonsense.

(vi) *Lakatos' attempt to criticize methodologies and theories of rationality by comparing them with what has been practices by the 'scientific elite' of the 'last two centuries' fails for many reasons. The actions of the 'scientific elite' are neither rational nor uniform and the rules which Lakatos produces have no force. Besides, they are based on science, at least in intention, and are therefore parties in rather than judges of the debate.*

According to Lakatos, methodologies are evaluated by a comparison with what has been practiced by the 'scientific elite' over the 'last two centuries'.[32] I have criticised the procedure at length elsewhere[33] and I shall repeat only the main objections.

To start with the procedure assumes that science (of the 'last two centuries') is preferable to other forms of life, for example, it is preferable to Aristotelian commonsense. No argument is given for this preference. The evaluations based on it therefore involve an uncritical rejection of certain forms of life and so cannot be used as an argument against them. They *might* be used to silence dissenters claiming to do modern science (though even this use is more a declaration of faith than an argument – see below); they cannot be used as arguments against an enterprise that explicitly differs from modern science, or against changes and developments within modern science itself. All that Lakatos can possibly show is that modern science *differs from* Aristotelianism, not that it is *better*.

Secondly, the procedure assumes that science ('of the last two centuries') is sufficiently uniform and, where it is uniform, sufficiently rational to permit its use as an unambiguous authority. There is no such uniformity and no such rationality. The 'common scientific wisdom' on which Lakatos bases his standards is not very common and it certainly is not very wise.[34] If uniform standards emerge from it then they are the result of a unifying procedure which has no foundation in science itself. Thus Lakatos does not really differ from the traditional epistemologists who try to impose their ideas on science, commonsense and whatever else history offers to their formgiving mania except that the philosophers *argue* for these ideas while Lakatos does not but uses propaganda instead: he announces that he is going to support his principles by historical research but the results of this 'research' are overruled the moment they conflict with what he thinks a 'rationalist' should do. Here I definitely prefer the procedure of Watkins who says, 'in a letter to Santa Claus' that the science described by critical rationalists is the science he 'would like to have'.[35] This is not exactly the most sophisticated answer to our demand for arguments, but it is the answer that emerges whenever we ask why one should take critical rationalism seriously.

Thirdly, the reconstruction of those episodes which are to represent science as a whole (though they do not – see the second point) is often based on an arbitrary arrangement of the evidence. E. Zahar in his study of the early stages of the special theory of relativity, Peter Clark in his study of the development of thermodynamics, are faced by a *split* in the 'scientific elite'. They choose those scientists whose programme can with some effort (and often with the help of quite arbitrary adaptations) by interpreted as 'progressive' which means that 'common scientific wisdom' does not define the standards, but is constituted in accordance with them.[36]

Fourth, the standards that are arrived at in such a multiply arbitrary fashion have no force. They do not tell the scientist what to do, they merely tell him that the research programme he is interested in is related to the evidence in a certain way. And this information is not imparted in a simple manner, i.e. by saying what the relation to the evidence is, but by using biased terminology: some research programmes 'progress', others 'degenerate'. The terminology is biased because it suggests what has not been established, namely, that 'progressive' research programmes have always been favoured by all major scientists and that 'degenerating' programmes have always been rejected by them and, even stronger, that 'progressive' programs are liable to continue succeeding (problem of induction in the sense of

Hume and Popper!) and because it also suggests that the standards are strong enough to demand the elimination of what degenerates and to request retaining and elaborating of what progresses. But it is quite clear that such demands and such requests, while compatible with the standards, are not supported by them because it is 'perfectly rational' to choose any research programme one pleases.[37] This, at least, is what Lakatos says in his less rhetorical moments. But then the standards are nothing but verbal ornaments without any importance to the scientist or the philosopher of science.[38]

Considering each of these objections singly and all of them together we conclude that Aristotelianism has nothing to fear from Lakatos' manner of evaluating research programmes.

(vii) *Most of the objections which critical rationalists raise against unloved views are formal objections: the views do not agree with certain standards. But the criticism can be reversed. If the views are correct, then the standards are unrealistic. For example, if Aristotelianism is correct, then the search for 'depth' is as illusory as we now regard the search for hell. Critical rationalists give no reasons why standards can be used against views but views not against standards – and they have never considered the matter. This is another reason why their objections against Aristotelian procedures cannot be taken seriously.*

The theory of relativity gives rise to the principle of relativistic invariance which provides a standard to judge theories by: theories that are relativistically invariant are better than theories that are not. The standard is not untouchable. It may be removed when one discovers that the theory of relativity has serious shortcomings. Such shortcomings are found by the development of non-relativistic views – they are found by research that violates the standards.

The idea that nature is infinitely rich both qualitatively (there is an infinite amount of qualities in every spacetime region, however small) and quantitatively leads to the desire to make new discoveries and thus to the principle of content increase which gives another standard to judge theories by: theories that have excess content over what is already known are preferable over theories that have not. Again the standard is not untouchable. It is in trouble the moment we realise that we inhabit a finite world and we make this discovery by developing Aristotelian theories, i.e. again by research that goes against the standard.

The idea that information reaches us undisturbed leads to the standard that all knowledge must be checked by observation: theories that agree with observation are preferable to theories that do not. The standard is in need of replacement the moment we discover that sensory information is distorted and contaminated. We make the discovery by developing theories that conflict with observation and are yet excellent in other respects (in Chapters 5 to 11 of *Against Method* I showed how Galileo made the discovery).

Finally, the idea that things are well defined and that we do not live in a paradoxical world leads to the standard that our knowledge must be self consistent. Theories that contain contradictions cannot be part of science. This apparently quite fundamental standard which some philosophers accept as unhestitatingly as Catholics once accepted the dogma of the Immaculate Conception of the Virgin loses its authority the moment we find that there are facts whose only adequate description is inconsistent and that inconsistent theories are better to handle and lead to more discoveries than their decontaminated rivals (see above, Section (iv)).

In all these cases specific and well defined standards such as the standard of observational adequacy, or the standard of consistency, or the standard of content increase are defused or removed by criticising the cosmological ideas that underlie them, which criticism in turn comes from research that conflicts with the standards and has yet better results than the standard-bound alternatives – the word 'better' now referring to different standards of evaluation which are either imported from older views, or invented on the spot. And this is perhaps the most important drawback of critical rationalism: it assumes that standards are things that guide research from the outside instead of being subjected to it. But scientists not only obey standards, they also criticise them. And they criticise them by initiating research that does not agree with them and letting the results speak for themselves. They invent standards just as they invent measuring instruments and theories, they introduce new theories of rationality just as they introduce new cosmologies and new technologies. Scientific research is a process that differs widely from 'rational thought' in the sense of critical rationalists.

The tendency to replace cosmological problems by formal problems which is rather widespread among positivists and critical rationalists further stabilises the standards. The question whether there are only sense data, or whether there are also electrons, tables, pedants is a cosmological question; scientific research must decide what exists and what is wrongly assumed to exist. This was Mach's attitude. Positivists (and Popperians), turning the problem into the 'formal mode of speech' investigate interpretations of terms, consequence classes etc. instead and 'solve' the substitute problem not by physical research, but on the basis of the rationality theory they like best.[39] The only restriction imposed on the solution is that it fit this rationality theory. Solutions of special cosmological problems are thus turned from potential critics of a particular form of rationality into its most obedient servants and the practice of research is further falsified. The easy use of standards against unpopular philosophies such as Aristotelianism is therefore not a sign of intellectual sophistication, but of ignorance.

(viii) *The normative criticism which critical rationalists occasionally use against opponents is restricted to a narrow domain and is mostly circular. Much better is Mill's suggestion to judge a theory of rationality not only by its intellectual consequences, but by the consequences it has in the lives of those who adopt it. And these consequences must be seen in the widest possible way. For example, we must not forget the deterioration of theology and religion that followed the rise of modern science. Intellectuals are not likely to shed many tears over such matters, but other people are starting to take them seriously and may in the end decide to criticize their intellectuals by no longer permitting them to rule knowledge in their own way at taxpayers' expense. No doubt this financial criticism of ideas will be more effective than their intellectual criticism, and it should be used.*

The normative evaluation of methodologies consists in comparing their rules and standards with fundamental norms.

Now the difference between Aristotle and the Popperians is that they use norms of a very different degree of generality and that the Popperians hardly ever argue for the norms they try to impose on us. If one asks a Popperian why on earth one should accept his standards he will answer that this is how science proceeds (cf. the last section, about Lakatos), or that the standards are fruitful in the sense that they make us understand science (this is how Popper once presented the matter in his *Logik der Forschung*), or that they lead to a science one 'would like to have' (little John Watkins in his 'letter to Santa Claus'[40]): an entity, 'modern science' functions as an unexamined standard of all standards. Nobody asks why this entity should have such authority in fundamental matters, nobody even asks whether the unity that is insinuated by the one word 'science' actually exists and whether it is not a chimaera (on this point see again the second item in the preceding section and the details in *Against Method*, 201ff and Howson, 316ff). This trust in science or, rather, in some very special parts of the vast and incoherent enterprise which philosophers call 'science' is present already at the very beginning of the Popperian movement: Popper narrates how he was struck by a certain difference between Marx and Freud on the one side, and Einstein on the other. In the first case we have imposing theories which seem to be capable of explaining everything without there being any hint under which circumstances they might have to be given up while Einstein specified experiments which could confirm his theory, but which could also refute it. Therefore, Popper concludes, falsifiability is the hallmark of 'science' *and must be demanded of anything that claims to be knowledge.* The inference from the first part of the last sentence to the italicized part of the inference I have just criticized: it is taken for granted that science is a standard of knowledge.[41]

Even more general arguments remain firmly within the limits of intellectualist ideology. Thus Alan Musgrave has praised research programmes that 'throw up more unsolved problems than' their rivals.[42] One can understand why an intellectual loves such programs for who will pay him when every problem is solved? Also critical rationalists are fond of praising criticism and condemning dogmatism. But why should knowledge not have stable elements? The answer that stable elements are a sign of truth only if they remained stable in the face of the most relentless criticism is not satisfactory for relentless criticism will never be stopped by truth, first, because even true statements have their faults, secondly because the idea of what counts as a fault and what not may be mistaken, and thirdly because the demand to search for truth is of course also subject to criticism ('what's so great about truth?'). Wherever we look we see dogmatic assumptions and unexamined presuppositions lurking behind the great show of intellectual honesty critical rationalists put on for our (and even more for their own) benefit.

There exists a form of rationalism that is both more complete than what Popperians are offering us today and more rational in the sense that it gives reasons where Popperians have either a blank stare, or a 'letter to Santa Claus'. It is the theory which Mill developed in his immortal essay *On Liberty*. Here Mill states that knowledge is to be based on criticism, on 'negative arguments', that it is the better the harder the criticism to which it was subjected and that nothing must be exempt from criticism. This is Popper's view except for three important differences. (1) For Mill both methods and facts are subjected to criticism while Popper regards his procedure of Conjectures and Refutations as a magic wand that can change all knowledge without having to be changed itself. (2) Mill argues that criticism presupposes alternatives, the more the better and he invites us to test uncontested views by first inventing implausible and even absurd alternatives and then turning them against the status quo. Popper never made alternatives a decisive part of his procedure.[43] (3) Mill explains the point of both criticism and proliferation. He gives an answer to the questions why should we be critical? The answer is that the manifold potentialities of every single human being can be developed only in pluralistic surroundings where the order of the day is not acceptance but choice. Criticism improves not only our thoughts, but also our emotions, our imagination, it makes us not merely more rational, it also makes us better men. Now I do not want to say that Mill is right in this assertion – he may be right, he may not be right. Nor do I want to take his particular value for granted. My only point at this stage of the debate is that Mill has reasons where Popperians turn dogmatic and that the reasons are connected with the nature of man. Mill has an answer to the question "Is it not possible that my activity as an objective observer will weaken my strength as a human being?"[44] that is much more sophisticated than the barren rationalism of Popperians.[45] I shall therefore choose Mill rather than his Popperian bowdlerizers as Aristotle's opponent and I shall ask: what is the difference between the two and which philosophy shall we choose?

On the one side – Aristotle – we have progress but with the boundary condition that commonsense remain an unchanged center of knowledge. *Our knowledge must not be separated from our nature as thinking, acting, perceiving beings where 'nature' means what is common to all and not only what is accessible to a small minority of intellectuals.* Man's harmony with the world must not be disturbed by reason. Of course, the information which man acquires constantly grows, there are discoveries (such as Aristotle's discovery of a fifth element), earlier views are revised and replaced by better ones. But all these changes leave untouched the basic nature of man and those features of knowledge that are adapted to it. I have already explained how Aristotle defends this central position of commonsense. Today we can add as a defence of commonsense: commonsense-perception, -thought, and -action are results of evolution and therefore liable to be better, closer to nature than bright ideas of intellectuals. It is quite possible that abilities such as telepathy, precognition, direct interaction between mind and matter which play a large role in 'primitive' societies and explain why they have methods of healing whose success we do not understand and cannot repeat are tied to commonsense and disappear with it.[46] It is also possible that the knowledge and the moral impulses that are needed to restore the harmony of man and nature will be available only to those who again try to relate knowledge to man as a whole and not only to that small but ever growing tumour some are pleased to call his reason. But let us now see what this reason has to offer.

Mill, I have said, offers a form of rationalism that gives reasons where Popperians either have a blank stare, or a 'letter to Santa Claus'. I have also said that Mill's model of reasoning is better adapted to the sciences than Popper's – it is less formal and gives no prescriptions whatsoever for the shape research has to assume. This is left to the research process itself, i.e. to the competition of alternatives. Now this competition is not only between ideas, standards, rules, but between fully fledged *forms of life*; and it is supposed to affect not only ideas, but also feelings, intuitions, attitudes, actions, the imagination, in a word – it is supposed to affect one's whole existence. Mill seems to assume that there is not only one human nature but that there are many and that all of them are potentially contained in man. His intention is to bring them to the fore and to set them against each other so that they may be improved. The improvement of ideas is only a tiny part of this enterprise in which ideas are not viewed in isolation but as part of the way of life they engender. While critical rationalists implore us to judge only ideas and to forget their inventor, Mill realises that ideas have not just logical but also psychological, sociological and other aspects and must be judged by all of them: if the invitation to criticise makes people mean, then we shall have to restrict our criticism for it is more important to have pleasant people than to have great ideas. So the difference between Aristotle and Mill is that while Aristotle develops the consequences of the *perfection* of man Mill believes in his *perfectibility*. Aristotle assumes that man and nature are already in harmony; Mill believes that many sides of man are dormant and that the idea of harmony must be constantly explored. The modern sciences share with Mill his informality and the view that knowledge emerges from a competition of alternatives. But while Mill examines *all* sides of the alternatives chosen, while he extends them into entire forms of life and judges ideas also by their emotional etc. etc. effects, science and the contemporary *ancillae scientiae*, the Popperians concentrate on ideas and damn the rest. So they differ both from Mill and from Aristotle (and this difference explains the restricted rationality of the Popperians when compared with Mill). Let us examine some consequences of this difference.

Modern science brought about a considerable change of *ideas*. There were new laws, a new and quite abstract law of inertia among them, there were new accounts of man and matter, there was a new theory of the universe. This change of ideas was not accompanied by a corresponding change of *perception*: perception has laws of its own and does not always follow the movement of thought. A large part of perception and of the thought and actions appropriate to it had now to leave the domain of knowledge. This part was not without structure and content. It had a life of its own: we have here the arts, theology, politics and large parts of medicine. These subjects were either reshaped so that they became appendices of the new science rather than its critics, or a claim to knowledge was denied to them. Important areas of the total human existence were separated in a manner that prevented rather than furthered the emergence of a new kind of *complete man*. Of course, there were new groups of *people*, theologians, intellectuals, scientists, artists who developed fragments of their being to a high degree of perfection, invented standards that made the fragments appear complete and kept their domains carefully separated except for occasionally sniffing at each other at the boundaries. Among these new groups the scientists soon assumed a special importance. They are now regarded with awe, they can impose their quite restricted fantasies on the rest of society under the name of knowledge. They not only determine what our children are to learn, when a person is to be regarded as a useful member of society, what we may eat, how we may live, they also want to be paid for turning everyone into a slave. And the development continues until a computerized unnature inhabiting an almost destroyed nature will take care that everyone retains only the functions, the quite peripheral functions, and speaks only the language, the unpleasant and lifeless idiom that is adapted to the new idea of knowledge.[47]

So far a brief sketch of the types of man and the types of world that are collected with Aristotelian knowledge, the different types of knowledge considered by Mill, and modern scientific knowledge. What is preferable? What kind of life is preferable? This is the question to which we are finally led in our quest for arguments concerning content increase.

Now the first thing we must avoid when trying to find an answer to our question is regarding phrases that have a well defined role on *one* side as arguments against the other. The simplest example of a pseudo argument of this kind is the remark that Aristotle is no good because he is not 'scientific'. Of course, Aristotle is not scientific – but that is not our problem. Our problem is what is better: 'science' (whatever *that* is – cf. above section (iv)) or Aristotle. The same applies to the 'argument', that Aristotle does not offer any genuine 'knowledge'. Of course, if knowledge is defined in terms of unceasing content increase, then Aristotle does not offer knowledge in that sense. But, maybe, he offers something better? Nor is it good pointing out that modern science has 'superseded' Aristotle. Aristotle 'superseded' the ancestors of modern science, and yet one returned to them. Besides, it is very important to see what the word 'supersede' amounts to. First Aristotle reigned supreme, then modern science reigned supreme (to idealise matters beyond recognition). Is that supposed to be an argument? One reason might have been that modern science had many intelligent defenders while Aristotelianism had not (who would regard Popperianism as finished if Watkins and Gellner were the only Popperians alive and made a mess of things? One would want to defend Popper in a more intelligent way and *then* see what happens). But modern science has results! And better results and more results than Aristotelianism! The theory of witchcraft also had results (i.e. facts and explanations), better and more numerous results than what came immediately after it in the same area, and yet one would regard it as reasonable to work against it. Besides, the results of 'modern science' have been vastly exaggerated. There were results in astronomy and in physics, that is to be admitted – but what about the other subjects? The Aristotelian law of inertia and the theory of motion adapted to it was needed until late into the 19th century (cf. fn. 3) and certain important subjects could not have advanced without it. There is an unsolved mind-body problem in modern science, there never was a mind-body problem in Aristotle. Commonsense medicine is more humanitarian, less dangerous and in many areas much more effective than its 'scientific' rival.[48] The Aristotelian theory of the drama survived through enlightenment (when it was defended and revised by Lessing) right into our 20th century. 'But drama has nothing to do with reality!' It had for Aristotle who introduced the theory to reply to Plato's criticism that the arts not only were irrelevant to knowledge, but did damage to it. For Aristotle poetry was 'more philosophical than history' because it dealt with tendencies while history dealt with accidental results of such tendencies (here Aristotle was a predecessor of sociology). Then we have theology. For a long time it was customary to criticise theology and metaphysics because they disturbed science. I think the criticism that science disturbs theology and metaphysics is just as well taken and it must be considered now that the issue is again before us. Theology is the theory of God. It has to do with man's position in the universe, with his nature, his obligations, his salvation. Critical rationalists are not much interested in such problems, and that is their good right. But let us not forget that their lack of interest is a very chauvinistic lack, it turns into prescriptions for school curricula, it turns into standards of thought whenever the matter turns up, *it is imposed on others* via the general pressure systems which our intellectuals have been developing ever since they noticed that knowledge (their type of knowledge) might be power (for them). The much maligned Bellarmine has seen the situation very clearly: what is more important, specialist knowledge or the faith? He was prepared for change, for he did not accept the idea of a separation of faith and knowledge, but he was not prepared to change on the grounds which Galileo offered. Today Galileo's standards are praised because they are thought to be identical with those of critical rationalism and those of Bellarmine are criticised or, to describe matters better, treated with contempt. But this means that the decision I am talking about has already been made in favour of content increase. *Before* the decision the faith of the multitude is an important consideration and standards which discourage unlimited change are therefore preferable to those which make change a condition of sound knowledge. (Besides, Galileo's ideas were not even acceptable on the grounds of critical-rationalist standards – but this is a different story. Galileo was the first of a long row of scientific chauvinists who took it for granted that their ideas should overrule everything else – no matter what the consequences.)

Proceeding with the conditions of a fair comparison we must also adopt the right attitude towards the results which modern science has produced. It has produced cars and telephones and many people cannot imagine living without them. Now, to start with, the continued production of cars and telephones does not depend on a scientific philosophy. Having invented them we can produce them by memory, and without any philosophy whatsoever. It is the constant change of cars and telephones that is in need of a philosophy of change and improvement. But most changes in these fields are not improvements, they are simply incentives to keep people buying cars. Such changes can be produced under any philosophy. Real improvements of life, however, may need a fundamental change of basic philosophy, and away from content increase. So far a very brief and sketchy account of the types of argument that would enter a non-circular evaluation of the condition of content increase. It is not to be expected that the arguments will have any force with intellectuals who are not interested in the matters (faith, salvation etc.) they raise and know only little about them. But it might have some force with other people, for example with the average taxpayer (who more than once has revolted against the kind of life scientists want to impose on him and who also is often more rational in fundamental matters than are critical rationalists). And there really is no need to try to *educate* the intellectuals so that the arguments assume importance for them. Take the *money* away from them and they will soon be reasonable.

(ix) *Summary.*

Content increase is one of the basic conditions which critical rationalists want to impose on knowledge. To examine the condition it is necessary to introduce forms of knowledge that do not agree with it and yet have results. Aristotelian commonsense is such a form of knowledge. It turns out that the arguments which critical rationalists have against it are circular (sections (ii) and (vi)), non-existent (sections (iii) and (vii)) and that the 'logical' arguments are unrealistic:

they do not apply to Aristotle, and they would not be decisive if they applied (sections (iv) and (v)). Critical rationalists are not liable to listen to the reasons which are introduced by a wider discussion, but other people are and, when convinced, will refuse to continue being taxed for 'knowledge'. So unreasonable people will have to be educated by the financial measures reasonable people may soon take against them.

University of California, Berkeley

NOTES

1 According to Malinowský primitive societies always separate these elements from magic, witchcraft etc. *Magic, Science and Religion and other Essays*, Anchor Books, 1954, p. 32, J. L. Austin agrees (private communication).

2 That this is not 'inductivism' is explained in Section (v).

3 There is however a difference between the way in which a property arises in a sense organ and in a physical body. Heating a physical body involves destruction of coldness in it. Producing the sensation of heat means actualising a potentiality without destruction (cf. *de anima* 417b2ff as well as Brentano *Die Psychologies des Aristoteles*, Mainz 1867, p. 81). The reason for the difference is that a sense is not simply a physical body but a relation between extreme states (424a6f).

4 Aristotle's law of inertia, for example (things remain in their state unless disturbed from the outside), which is repeated by Descartes, *Princ. Phil.*, Section 37, has aided biologists in their research down to the beginning of this century (discovery of insect eggs, bacteria, viruses etc. etc.). Newton's law would have been completely useless in this respect.

5 This awareness may be the reason why he never revised the homocentric system and never even mentioned the observational difficulties. For these difficulties and their later use cf. Appendix 1 of *Against Method*.

6 The problem of telescopic observation is discussed in Chapters 10ff of *Against Method*.

7 G. E. L. Owen, in *Aristotle*, Moravcik (ed.), New York 1967, p. 171.

8 *Aristotle*, New York 1960, p. 57.

In his *Objective Knowledge*, Oxford 1973, p. 8, Popper writes with characteristic modesty: "Neither Hume nor any other writer on the subject before me has to my knowledge moved from here (impossibility of justifying reasoning from experienced to unexperienced instances) to the *further questions*: Can we take the 'experienced instances' for granted? And are they really prior to the theories?"

It is surprising and a sign both of the historical illiteracy of most contemporary philosophers and of their low standards of hero worship that statements such as these are taken as historical evidence and as an indication of philosophical profundity. But Newton corrected phenomena 'from above' [cf. my 'Classical Empiricism' in J. W. Davis and R. E. Butts (eds.), *The Methodological Heritage of Newton*, Blackwell, Oxford

1970], Mill required that there be a discussion of experience in order to determine both its content and its force [*On Liberty* quoted from Marshall Cohen (ed.), *The Philosophy of John Stuart Mill*, New York, 1961, p. 208], in Goethe's *Maximen and Reflexionen* we can provide the statement "Das Höchste wäre zu begreifen, dass alles Faktische schon Theorie ist" (Aus den Wanderjahren) expressing the very idea which Popper here claims for himself. Boltzmann has often quoted Goethe's dictum that experience is always only half experience [*Populäre Schriften*, Leipzig 1905, p. 222] and then there is of course Mach's observation that already the "Name 'sensation' entails a one-sided theory" [*Analyse der Empfindungen*, Jena 1900, p. 18: "Da aber in diesem Namen (der Empfindung) schon eine *einseitige* Theorie liegt ..." emphasis in the original]. Of course, all this was unknown to the followers of the Vienna Circle who wanted to start philosophy afresh, and who did start it afresh and with only minimal knowledge of earlier ideas. The Vienna Circle shares with the enlightenment an exaggerated faith in the *powers* of reason and an almost total ignorance of the *past achievements* of reason – small wonder that Popper who anxiously hovered at its periphery regarded every modification of the philosophy of the Vienna Circle as a genuine discovery. In this he is a true representative of the Viennese Neo-enlightenment. But he did write a two volume work on Plato, Aristotle and other unfortunates and so one might expect him to have familiarised himself with their philosophies. Did he notice that the empiricist Aristotle put precisely the question, the 'further question' for which he now claims prime authorship? Apparently not. Which does not prevent him from criticizing Aristotle for his 'lack of insight' (*Open Society*, 6th ed., Vol. II, p. 2).

[9] We do know that some philosophers looked for a logic of discovery and that they praised their own age for having passed beyond long established boundaries ('plus ultra' was the motto of the Royal Society and the intention was to discover new domains of knowledge just as Columbus, Dias, Magellan had discovered new parts of the earth). But this type of content-increase also occurs in Aristotle. What does not occur in Aristotle is a content increase that involves *novel* predictions and not just *additional* predictions. It is interesting to see that critical rationalists themselves are not too clear about these matters. Cf. note 13.

[10] There are many occasions on which Aristotle rejects an assumption 'because, if so, being will not be knowable', *Physics*, 189a14f. For a more general treatment cf. the essay by Owen in Moravcik, *op. cit.*

[11] In *de coelo* the 'simple bodies' are separated not only by their (perceptual) *properties*, but also by their (observable) *motions*, a procedure that is still used in elementary particle physics to separate different particles. Cf. 301b31ff as well as Solmsen *Aristotle's System of the Physical World*, Cornell University Press, 1960, Chapter 11.

[12] We may also declare the case to be a failure not of perception, but of nature itself, or a 'monster' (*Phys.* 199a38). Exceptions are classified by Aristotle as special cases and the conditions under which they occur are noted. There is no attempt to find an account that subsumes the normal case and the special case under a more general law. Just as perceptual error is supposed to leave the features of normal perception untouched, in the very same way 'errors of nature' are supposed to leave the normal case untouched. For this decisive difference between Aristotelian commonsense and modern science cf. K. Lewin's article in *Erkenntnis*, Vol. iii.

[13] Aristotle *notes* (just as Hume did later on) that our attitude towards normal experience remains unchanged by extraordinary experiences and *declares* normal

experience to be a stable foundation of our knowledge, undisturbed by the discovery of perceptual error.

[14] "A scientist *would* be pronounced 'irrational' ... by the methodology" writes John Worrall in a note (see above p. 70) showing that he regards the matter as pretty obvious "if he stuck to the old programme denying that the new programme had any merits not shared by the old one, and thus denying that his own programme needed improvement in order to catch up with the new one". The methodology of research programmes does indeed make such judgements. It supports the judgements by trying to show that they are shared by the 'scientific elite' over 'the last two centuries' (Lakatos in Howson (ed.), *Method and Appraisal in the Physical Sciences*, Cambridge 1976, p. 23). Aristotle supports his contrary judgement not by appeal to the 'scientific elite' of his time but by producing arguments (cf. his arguments against Parmenides, Democrites, and Plato). Is it not clear that a critical rationalist must prefer him?

[15] Imre Lakatos was a master in this type of verbal warfare. One could admire his skill and his artistry, one would never be put off by it. It is very different with the dreary imitators of his method who now hold the stage, who have turned art into slavish routine and who have never understood that while rhetoric without reason is empty reason without rhetoric is dumb.

[16] Cf. p. 24, this volume.

"This concludes one general argument for preferring this methodology and its appraisals to rival methodologies" writes Watkins later in his contribution (p. 32). "Although it discards some parts of the Bacon-Descartes ideal, it appears to preserve considerably more of it than do any of them". The 'argument' assumes that the Bacon-Descartes ideal is worth preserving – which is the question at issue.

Another 'general argument' of Watkins will be discussed in section (vii).

[17] R. Westman (ed.), *The Copernican Achievement*, Berkeley and Los Angeles 1976.

[18] Scientists very soon used this kind of argument against their Aristotelian opponents. They criticised Aristotle not with reasons that showed *why* content increase was more important that the preservation of commonsense, they assumed that it was and reviled everyone who thought different. Thus the Platonists of the scientific revolution regarded mathematics not just as an instrument of prediction, but as a tool for exploring depth. Combining mathematics with the mechanical hypothesis Leibnitz criticised the Aristotelians for failing to give explanations. "I should like you to think of one thing" he writes in his letter to Conring of March 19, 1687 (quoted from L. E. Loemker (ed.), *Gottfried Wilhelm Leibnitz, Philosophical Papers and Letters*, Dordrecht 1969, p. 189) 'that unless physical things can be explained by mechanical laws, God cannot, even if he chooses, reveal and explain nature to us. Or what would he say, I ask you, about vision and light? That light is the action of a potentially transparent body? [This is Aristotle's definition in *de anima* 418b9f]. *Nothing is truer even though it is almost too true.* But would this make us any wiser? Could we use this to explain why the angle of reflection of light is equal to the angle of incidence, or why a ray should be bent more toward the perpendicular in a denser transparent body, though it would seem that the opposite should happen? ... how can we hope to explain the causes of such things except by mechanical laws that is, by concrete mathematics or geometry applied to motion?'

The need of relative depth for understanding is very clearly *explained* here, but the only thing resembling an *argument* we find is a rhetorical question involving the words 'wise' and 'explain'. Now in (Aristotelian) commonsense 'explanation' has a good sense though this sense does not include depth, or content increase. It is therefore *insinuated*, though not *clearly stated that science needs depth*. But *this* demand is answered and rejected in Aristotle's arguments against Parmenides and the atomists. And his theory of mathematics has solved problems of the opposing camp which derive precisely from the assumption that the mathematical continuum has a 'depth structure' *and which are still with us* (Aristotle's theory of the continuum might therefore be of interest to quantum theoreticians and pure mathematicians). Leibnitz' rhetorical question (which was repeated by Lakatos) therefore does not advance matters a tiny little bit.

[19] This is what John Worrall says in a review of *Against Method*.

[20] For the popularization cf. *Conjectures and Refutations*, p. 317.

[21] For other and perhaps more impressive cases as well as discussion cf. *Against Method*, pp. 258ff.

[22] von Neumann, *Mathematische Grundlagen der Quantenmechanik*, Dover Publication, 1943, pp. 13, 14. Dirac uses transformations (from differential operators to integral operators) which are mathematically impossible but circumvents the contradiction with the help of functions which are 'outside the domain of customary mathematical methods' (von Neumann, p. 15). The lesson for logicians is that the disastrous consequences of contradictions can be avoided by using rules of derivation 'outside the domain of customary logical methods'. Wittgenstein, *Remarks on the Foundations of Mathematics* has discussed some such rules.

[23] For a fascinating discussion of further contradictions cf. Paul and Tatiana Ehrenfest, *The Conceptual Foundations of the Statistical Approach in Mechanics*, first published 1912, English translation Cornell University Press 1959.

[24] For details cf. Section 3 of 'Against Method', *Minnesota Studies for the Philosophy of Science*, Vol. iv, 1970.

[25] Dirac's delta function is an explicit example in the mathematical sciences.

[26] Another objection to formal logic is that it assumes stability of meaning during arguments while the most interesting arguments are those that change meanings as they proceed.

Certain developments in the sciences seem to show that the challenge, viz. to build contradiction free theories can be met. Thus von Neumann has developed a formalism that seems no longer beset by the problems Dirac tried to solve and solved by leaving the domain of 'customary' mathematical procedures, mathematicians have tried to improve the calculus to remove the unsatisfactory features already commented upon by Berkeley, and modern ergodic theory is in part an attempt to detach statistical mechanics from contradiction ridden models. *But we must distinguish carefully between the logician's approach in these matters and the scientist's approach.* A logician concentrates on the formalism of a theory without paying attention to matters that interest the scientist such as fruitfulness, testability, ease of handling, minimal amount of additional assumptions in the derivation of special cases, and especially without paying attention to those features that *point beyond* the theory to a still more satisfactory account. All he wants is a formalism that is superior to the one he sees before himself in certain formal respects. A scientist, on the other hand, always wants to have some leeway of maneuvering and "... he [will] never try to outline any finished picture, but will patiently go through all the phases of the development of a problem, starting from some apparent paradox, and gradually leading to its elucidation. In fact, he never regard[s] achieved results in any other light than as starting points for further exploration. In speculating about the prospects of some line of investigation, he dismissed the usual considerations of simplicity, elegance or even consistency with the remark that such qualities can only be properly judged after the event ..." (L. Rosenfeld about Niels Bohr in *Niels Bohr, His Life and Work as seen by his Friends and Colleagues*, S. Rosental (ed.), N.Y. 1967, p. 11. And as science is never complete, it is never 'after' the event and so simplistic elegance, 'even' consistency are never necessary conditions of scientific practice.

That they are not sufficient conditions either is revealed by a brief look at a case where logicians succeeded in imposing them.

Von Neumann's formalism of the elementary quantum theory is a monster of precision, and it may even be contradiction free, but this advantage – if it is an advantage – has been dearly bought. The relation to experience has become more opaque than it was in the correspondence principle, the calculation of concrete cases involves auxiliary hypotheses such as arbitrary cutoffs (in *Against Method*, p. 63 I have discussed such 'ad hoc approximations' in some detail. For von Neumann cf. also the literature quoted on p. 64, fn. 23 of the same volume) and the discussion of fundamental problems has been burdened by irrelevant complexities. (This is a reason way Clauser-Horne-Shimony-Holt, *Phys. Rev. Letters* **23** (1968), p. 880 preferred to use Einstein's arguments rather than those of von Neumann as a basis for their tests.) We have now lots of theorems with proofs that satisfy the highest standards of rigour but nothing that would help the physicist to improve either his predictions or the basis on which the whole enterprise rests. (This is openly admitted by R. F. Streeter and A. S. Wightman in *PCT, Spin & Statistics and all that*, New York 1964, p. 1: "The quantum field theory never reached a stage where one could say with confidence that it was free from internal contradictions ... The last 10 years have seen a number of attempts to meet the situation head on. Cynical observers have compared [physicists engaged in this activity] to the Shakers, a religious sect of New England who built solid barns and led celibate lives, a non-scientific equivalent to proving rigorous theorems and calculating no cross sections ... These efforts have not yet led to a solution of the main problem."

Another example is formalised mathematics. It was introduced with the purpose of removing certain difficulties but only succeeded in creating an evergrowing formalistic tumour whose perfection can never be guaranteed but which inhibits the development of informal mathematics. The situation has been described with excellent examples by Imre Lakatos, *Proofs and Refutations*, Cambridge 1976.

[27] For details cf. W. Wieland's excellent *Die Aristotelische Physik*, Göttingen 1970. Cf. also section (i) of the present essay.

[28] Cf. Kurt von Fritz, 'Die ΕΠΑΓΩΓΗ bei Aristoteles' republished in *Grundprobleme der Geschichte der Antiken Wissenschaften* Berlin 1971, 623ff. The essay contains a comparison of Aristotelian 'induction' with 'induction' as used by 19th and 20th century logicians.

[29] Not every reading is regarded as decisive and there is correction by the judicious use of alternatives. 'When doing research concerning principles Aristotle never refers to an

evidence that is no longer in need of arguments' Wieland, *op. cit.*, 62. In Aristotle the use of alternatives is closely connected with his gradual building up of a history of ideas. Doing history of ideas is for him part of doing philosophy. So we have a literate philosophy as well as an intelligent history.

[30] Above, p. 41.

[31] Cf. above, note 15. Aristotle, interestingly enough, developed a theory of knowledge that uses alternatives, but without increase in content.

[32] Howson, *op. cit.*, p. 23.

[33] Cf. Chapter 16 of *Against Method* as well as my article in Howson, *op. cit.*

[34] A much more detailed account is found in *Against Method*, pp. 201ff and Howson, *op. cit.*, 316ff.

[35] Cf. this volume, p. 24, as well as note 15.

[36] For details cf. again my essay in Howson, *op. cit.*, 336f.

[37] Howson, *op. cit.*, p. 16, note.

[38] Worrall (Howson, *op. cit.*, pp. 163f) "rejects [the] suggestion that if a methodology does not imply advice to scientists about which theories they should work on, then it is empty. Such a methodology will still appraise and rank theories in terms of their present scientific merits, and it will tell a scientist what general features a new theory must have if it is to be even better than any of the existing theories." But what is the point of an ethics where a thief can steal as much as he wants, is praised as an honest man by the police and the common folk alike provided he tells everyone that he is a thief?

[39] I did this once myself in my essay 'Realism and Instrumentalism' published in *The Critical Approach*, Bunge (ed.), 1963. Cf. my comments in the German edition in *Der Wissenschaftliche Realismus and die Autorität der Wissenschaften*, Vieweg, forthcoming.

[40] This volume, p. 24.

[41] Needless to say the argument has many more flaws than that. Psychoanalysis was not simply irrefutable, it was treated as irrefutable by certain people in certain problematic situations and as refutable in others. The shape it had acquired by the time Popper started quarrelling with the Vienna Circle was the result of a series of most interesting conjectures and refutations. Same about Marxism. On the other hand, treating a theory as irrefutable is not at all as criminal as Popper makes it out to be: new ideas must be given a chance to grow so that they can overcome difficulties rather than being overwhelmed by them. Finally, Einstein, Popper's star example was anything but a simpleminded falsificationist, even in the case which Popper quotes. Replying to Born who wrote, in 1952 (*Born-Einstein Letters*, New York 1971, 190, dealing with Freundlich's analysis of the bending of light near the sun and the redshift) that 'it really looks as if your formula is not quite correct. It looks even worse in the case of the redshift; this is much smaller than the theoretical value towards the center of the disk and much larger at the edges ...' he said: 'Freundlich does not move me in the slightest. Even if the deflection of light, the perihelial movement of line shift were unknown, the gravitation equations would still be convincing because they avoid the inertial system (the phantom which affects everything but is not itself affected). It is really strange that human beings are normally deaf to the strongest arguments while they are always inclined to overestimate measuring accuracies'. Of course, Einstein also sang a different tune, if the situation required it, because he was an 'opportunist'

(Einstein in *Albert Einstein: Philosopher Scientist*, P. A. Schilpp (ed.), New York 1951, 683f) and adapted talk and method to the case at hand. Popper of course, with the insensitivity characteristic for the professional rationalist turns him into the proponent of one simple (and simpleminded) philosophy. More material on this issue can be found in note 6, p. 56 and note 9, p. 57 of *Against Method*.
[42] Cf. his essay in the *Lakatos Memorial Volume*, Cohen, Feyerabend, Wartofsky (eds.), Reidel 1976.
[43] Writers ignorant of history mentioned Spinner, or me, or Popper as the inventors of proliferation, depending on their reading. None of these assumptions is correct. In modern times it was Mill who showed most clearly how knowledge can be advanced by a competition of alternatives. His conception of method is clearer than Popper's and close to scientific practice. For details cf. Section 3 of 'Against Method' *Minnesota Studies in the Philosophy of Science*, Vol. iv, 1970, "Popper's 'Objective Knowledge' " *Inquiry* 1976 as well as 'Imre Lakatos' *BJPS* 1976. After Mill came the Darwinists. They too saw knowledge emerging from a competition of alternatives. Boltzman who accepted their basic ideas criticised not only physics but even the principles of logic on this basis: proliferation had advanced beyond Popper long before he entered the scene.
[44] Kierkegaard *Papirer*, Heiberg (ed.), vii, Pt. I, No. 182.
[45] According to Popper (*Objective Knowledge*, Oxford 1972, p. 239) "what counts in the long run is a good argument". What 'good argument' was responsible for the rise of Christianity, What 'good argument' overthrew fascism?
[46] The Hopi Myth states this quite explicitly.
[47] Cf. 'Experts in a Free Society' *The Critic*, Chicago 1970.
[48] For details cf. Ivan Illich, *Medical Nemesis*, New York 1976.

ALAN MUSGRAVE

EVIDENTIAL SUPPORT, FALSIFICATION, HEURISTICS, AND ANARCHISM

My task is, I take it, to discuss some problems involved in specifying criteria of scientific progress, referring especially to the views expressed in the papers written by my friends from the LSE. I fear that profound criticisms are not to be expected from one invited here as a 'sympathizer' with the LSE position. But I will do my best.

1. EVIDENTIAL SUPPORT

Philosophers of science are, I suppose, agreed that other things being equal science has made progress if a theory which has no evidential support is replaced by a theory which has. To deny this would be to deny that empirical evidence matters in science. The philosophical problem is to provide a theory of evidential support, an account of the circumstances in which an observed fact supports or confirms or corroborates a theory. And over this problem there is much disagreement.

The reason for the disagreement is the philosophical discovery that the obvious and very simple theory of evidential support runs into trouble. The obvious theory is that *any observed fact that follows from a theory supports it.* (Those who deny that *facts* follow from theories should read 'observed fact' as 'factual proposition ascertained by observation to be true'.) And the trouble is that this obvious theory leads to two counter-intuitive results, commonly known as the *Raven Paradox* and the *Grue Paradox*. The Raven Paradox is that any observed fact which does not refute a universal hypothesis provides evidential support (thus defined) for that hypothesis. The Grue Paradox is that any body of evidence provides equal support (thus defined) for an unlimited number of different hypotheses. The obvious theory makes evidential support too easy to obtain, and makes it impossible ever to single out one theory as preferable in the light of the evidence.

A natural response to this situation is to modify the obvious theory so that evidential support becomes more difficult to obtain: to say that for a fact to support a theory it is a necessary *but not a*

181

G. *Radnitzky and G. Andersson (eds.), Progress and Rationality in Science, 181–201.*
All Rights Reserved.
Copyright © 1978 by D. Reidel Publishing Company, Dordrecht, Holland.

sufficient condition that it follows from the theory. Popper's theory of evidential support (he prefers the term 'corroboration') says exactly this. The theory drops out naturally from his falsificationist methodology, according to which observations and experiments are made to *test* theories. Evidential support can only come from a *genuine test*, an attempt to refute the theory in question. If we know in advance, from our 'background knowledge', that some observation will not refute our hypothesis, then that observation cannot corroborate it either. Only severe tests can provide evidential support, and the severity of a test is assessed in the light of 'background knowledge'.[1]

The Raven Paradox disappears because the myriads of intuitively irrelevant observations will not (on plausible construals of 'background knowledge') constitute severe tests.[2] The Grue Paradox also disappears; but before discussing that, I will comment on Watkins's discussion of that Paradox.

Watkins points out, quite rightly, that the Grue Paradox, the so-called 'new riddle of induction', is nothing but an extension of the old 'curve-fitting problem' to qualitative hypotheses. And then he solves the Paradox simply by saying that "there is no need to take into account possible hypotheses which would also have been corroborated if they had been proposed: we can concentrate upon the hypotheses actually before us".[3] Now it is true that hypotheses like "All emeralds are grue" are seldom (never?) seriously considered by scientists (the same goes for "All emeralds are green", but I shall let that pass). But such hypotheses are easy to construct. What if someone does lay before us grue-type alternatives to our best-corroborated hypotheses? Watkins would have to concede that these arbitrarily-concocted hypotheses are as well-corroborated as the existing ones. It is only because scientists happen to lack gruesome imaginations that we can ever single out one hypothesis as the best-corroborated of those available.

I wonder. Suppose "All emeralds are green" is corroborated by evidence E, and then "All emeralds are grue" is proposed. The evidence E will be part of background knowledge, and so will not corroborate "All emeralds are grue". Indeed, this hypothesis will have no corroboration *at all*. Similar considerations apply to the old curve-fitting problem, and to Watkins's fable about Mr A and Mr B.[4] In Watkins's fable, Mr B's successive curves are all drawn through data-points which already corroborate Mr A's theory. In each case all the available data is part of background knowledge to Mr B's latest theory, and hence does not corroborate it. Indeed, none of Mr B's theories is corroborated at all: in the fable each of them is refuted in its first genuine test.

What happens if we vary the fable slightly, and suppose that Mr B's last curve is proposed at the same time as Mr A's, or "All emeralds are grue" at the same time as "All emeralds are green"? We now have two pairs of untested hypotheses, and must obviously try to devise *crucial* tests between them. (I am assuming that none of them can be excluded on non-empirical grounds – which is a dubious assumption.[5]) In the first case we must make an observation in a region where Mr A and Mr B predict *different* things. In the grue case we must wait until after the year 2000 (when emeralds are supposed to turn blue) and then observe an emerald. In both cases observation will enable us to discriminate between the two hypotheses.

Let us now leave these fabulous regions, and return to the crucial notion of background knowledge. So far I have deliberately left this notion vague. What exactly *is* the 'background knowledge' to a given hypothesis, which we use to assess severity of tests of that hypothesis? Popper takes it to contain what is known to science in the given field when the hypothesis in question is proposed. And he also includes in it any auxiliary hypotheses or initial conditions needed to derive predictions from the hypothesis in question. Now I agree with John Worrall that background knowledge is doing two different jobs here, which ought to be kept apart. And I agree with him that 'auxiliary premises' should *not* be included in it because this prevents Popper from placing a premium on crucial experiments between the new hypothesis and some existing one. This means that the 'hypothesis' being considered is always a *theoretical system* containing all the auxiliary premises needed to obtain testable predictions.[6]

Worrall presents two arguments against Popper's theory of evidential support. The first proceeds from the claim, which I will not dispute, that theoretical systems are almost always inconsistent with accepted facts, and hence are 'born refuted'. Worrall then points out that Popper's formulas for degree of corroboration yield the value -1 for all such systems, leaving nothing to choose between them. Yet, says Worrall, scientists do choose between them, and prefer some to others. Popper can give no rationale for such preferences. According to Worrall, the methodology of research programmes *can* give them a rationale because it allows scientists to ignore falsifications and consider only the "verifications of excess content" which one refuted system may well have over another. Worrall concludes that Lakatos's methodology is superior to Popper's in that it is more applicable to the "messy situations" of actual science.[7]

Now I quite agree that Popper's *formulas* must be set aside in dealing with such messy situations, since they fail to discriminate between refuted systems. (I do not think Popper would disagree either: he devised the *formulas* to make a philosophical point against Carnap.) And having set the formulas aside, one obvious thing Popper could say is: "Of two refuted systems S_1 and S_2, if everything that corroborates S_1 also corroborates S_2, while S_2 is also corroborated by facts which do not corroborate S_1, then S_2 is *better* corroborated than S_1".[8] This is exactly what Lakatos said originally too: it was only when a new theory *explained all the successes of the old* that an additional success could tip the scales in its favour.[9] Only in such situations could we 'transfer the methodological spotlight away from refutations and focus it on verifications of excess content".[10] But such situations are also idealized ones, and actual science is likely to be far more 'messy'.

But Worrall, encouraged by some of Lakatos's own formulations, seems to set aside the crucial idealization in Lakatos's view. He wants us, it seems, to focus only on "excess verifications" in all messy situations. But this would be totally unacceptable. Suppose S_1 is corroborated in all tests except one, while S_2 is refuted wherever S_1 is corroborated but corroborated in the single test which refutes S_1. All the examined content of S_2 is excess content, and some of it is corroborated. But obviously S_2 is not "supported by more facts than its rival". And in reaching this verdict we must take into account the refutations as well as the corroborations of each. In this messy situation we cannot transfer the methodological spotlight from refutations to corroborations. I conclude that the claimed improvement on Popper's theory of evidential support is largely illusory. Once we set aside Popper's *formulas*, then all that Lakatos gives us is a thoroughly Popperian view of a particular idealized situation.

Worrall's second argument against Popper's theory concerns its *strictly temporal* view of background knowledge, which takes it to include everything known to science (in a given field) when a new hypothesis is proposed. Worrall objects, following Zahar, that such a view implies that the Michelson-Morley experiment cannot support Special Relativity Theory, nor the Mercury anomaly General Relativity Theory, because they were known when the theories were proposed. I accept this argument, and agree that we must construe background knowledge differently.[11]

In an earlier paper I considered two alternative construals of the background knowledge to a new theory T. First, the *theoretical view* that it consists of the best available competing theory to T. And second, the *heuristic view* that it consists of those known facts which are used in the construction of T.[12] I favoured the former, and Worrall favours the latter. It is worth noting that Worrall's arguments in favour of the heuristic view only show its superiority to the strictly temporal view. In most of the cases he cites, the theoretical view also accords with 'scientific intuitions': for example, the Mercury anomaly and the Michelson-Morley result can support Relativity Theory on the theoretical view because despite being known in advance they refuted previous theories.[13] But in other cases the two views will give different verdicts. Suppose some fact refutes the existing theory and is used to construct a new theory which implies it: the fact will support the new theory on the theoretical view, and will not support it on the heuristic view.[14] Since Newton's theory of gravity is a special case of Einstein's, facts which support the former will also support the latter on the heuristic view provided Einstein did not use them to construct his theory. But on the theoretical view, only facts which do not also support Newton can support Einstein. (The theoretical view replaces the question "Do the facts support theory T?" by the question "Do the facts support theory T better than its existing rival?".)

My original reservations about importing considerations of how a theory is constructed into the problem of evidential support were prompted by the desire to make it an *objective* relationship. Whether or not a fact supports a theory should not depend on personal or psychological traits either of the experimenter who ascertained the fact (such as whether he was sincerely trying to overthrow the theory) or of the theoretician who proposed the theory (such as whether he was aware of the fact and trying to explain it). I wondered whether Zahar's heuristic view could meet this requirement:

Suppose two scientists, A and B, independently and at about the same time, propose the hypothesis h. And suppose that A devised h to account for the known facts e_1 and

e_2 – while B, less *au fait* with current literature, devised h only to account for e_1. Does Zahar's view entail that e_2 confirms h as proposed by B, but does not confirm h as proposed by A?[15]

Worrall's answer is unequivocal:

...the heuristic considerations which led to the construction of a theory can be objectively specified ... Whether some fact was used in the construction of a theory is an objective matter – quite separate from any question about whether the theory's inventor knew of or 'was aware of' the fact ... Thus in deciding whether some fact ... supports a theory one will ask such questions as "Did x's programme give him independent reasons for fixing this parameter in this theory at this value or did its value have to be 'read off' from some observations?". And *not* such questions as "Did x know of this fact or have this fact in mind when he developed his theory?"[16]

According to Worrall we can determine whether a fact has been used in the construction of a theory by logical analysis of the theory and of the research programme which produces it. I am happy to confess that once the heuristic view is construed in this objective fashion (and I do not doubt that this was Zahar's original intention), then my original doubt about it vanishes.

I now think that the theoretical and heuristic views may be complementary rather than competing views. The heuristic view enables us to determine the evidential support of a *single* theory. But when we wish to decide whether some theory is an improvement over its predecessor, then the theoretical view comes into its own. For then we will only count those facts which do not also support the old theory. And hence, although Relativity Theory has lots of evidential support, the facts which lead us to prefer it to Newton's Theory are very few indeed.[17]

2. FALSIFICATION

So far, in considering the problem of evidential support, we have said that what is supported is an entire *theoretical system* from which a successful prediction is derived. (If we describe a single hypothesis from such a system as 'confirmed' what we mean is that theoretical systems in which it figures have been confirmed.) Empirical facts refute systems as well as confirming them. And once it has been agreed that a system has been refuted, scientists face the problem of deciding which part of the system to blame for the false prediction. (If we describe a single hypothesis from a system as 'refuted' what we mean is that a system in which it figures has been refuted and the blame pinned on this hypothesis.) Let us call this the *Duhem Problem*.

Before turning directly to this problem, I might remark that failure to distinguish theoretical systems from single hypotheses within them is endemic among critics of 'falsificationist' views. Lakatos argues, following Duhem, that only theoretical systems can contradict facts, not isolated theories.[18] He then claims that scientists ignore refutations of their theories. But scientists cannot ignore what does not exist! What 'ignore' here means is "attribute to other parts of the system in which the theory figures", a peculiar usage whose propaganda effect is considerable. Worrall follows Lakatos:

There is a well known *epistemological* rationale for scientists not getting too excited about refutations and for holding on to a theory despite clashes between it and the experimental evidence. This rationale, which was already in large part provided by Duhem, is taken full account of by the methodology of research programmes ...[19]

Now here 'theory' must mean 'theoretical system', since Worrall holds that only systems *can* clash with accepted evidence. But did Duhem (or anybody else) provide a rationale for holding on to a system which contradicts accepted evidence? Duhem provided no rationale for violating the law of contradiction. What the Duhem Problem *might* provide a rationale for is holding on to some privileged *part* of a refuted system and modifying the rest to restore consistency.

But Duhem himself did not take this view. He said that cautious thinkers will make minor adjustments to refuted systems, while bold thinkers will try to devise entirely new ones. But he recommended neither as a *general* policy: only scientific 'good sense' can say which policy is to be preferred in particular cases.[20] On this matter Popper sides with Duhem:

... we cannot at first know which among the various statements of the ... system ... we are to blame for the falsity of [the prediction] p; which of these statements we have to alter, and which we should retain ... It is often only the scientific instinct of the investigator (influenced, of course, by the results of testing and retesting) that makes him guess which statements ... he should regard as innocuous, and which he should regard as being in need of modification.[21]

Popper adds that a well-corroborated part of a system may often be retained and a more 'speculative' part modified. (He later remarks that an "unbroken sequence of refuted theories" would leave scientists at a loss in dealing .with the Duhem problem.[22]) But then he warns us that "it is often the modification of what we are inclined to regard as obviously innocuous ... which may produce a decisive advance".[23] Popper does go beyond Duhem in one important respect: he requires that however a system is modified to accommodate a refutation, the modified system should not be *ad hoc* with respect to its predecessor.[24]

So Duhem and Popper are in substantial agreement on the Duhem Problem.[25] Opposed to them are the conventionalists and Lakatos. Conventionalists argued (à la Kant) that certain basic principles (Euclidean physical geometry, conservation laws, Newton's laws of mechanics) were necessary for there to be any physical theory at all, and so should never be blamed for refutations of systems containing them. Lakatos generalizes the point: each scientific research programme is characterized by a 'hard core' of basic assumptions without which work in that programme would be impossible; adherents of that programme decide never to modify their 'hard core' (which is thus rendered "irrefutable by fiat" according to the *negative heuristic* of the programme). In this way adherents of a research programme are given additional guidance in dealing with the Duhem Problem: for them the "arrow of *modus tollens*" has a smaller target than Duhem and Popper would allow.[26]

Lakatos's argument for his 'generalized conventionalism' is based on the *heuristic power* of the 'hard core'. The 'hard core' figures in the *positive heuristic* of the programme, "a partially articulated set of suggestions or hints on how to change, develop the 'refutable variants' of the research programme".[27] For example, the positive heuristic of Newtonian celestial mechanics contains the suggestion "Attribute anomalies to masses not hitherto taken into account acting in accordance with the laws of mechanics and gravitation". Adams and Leverrier followed this suggestion when they postulated an additional planet to explain the anomalous motion of Uranus. The important point is that they *used* Newton's laws in order to calculate, from the observed discrepancy, the size and position of the new planet. To reject these laws would have deprived them of a concrete strategy for dealing with the anomaly. It is this heuristic power of the 'hard core' which, Lakatos claims, justifies its being accorded a privileged position.

This is an interesting argument. But there is one respect in which I think that Lakatos overstates his case. He says that the 'positive heuristic' not only suggests ways to deal with anomalies, but can also *predict* these anomalies. He describes it as "the strategy both for predicting (producing) and digesting" refutations. And he says that the scientist armed with a powerful heuristic can ignore "the *actual* counterexamples", for his difficulties "are mathematical rather than empirical". Refutations are taken seriously only by scientists "who are either engaged in trial and error exercises or who work in the degenerating phase of a research programme when the positive heuristic ran out of steam". This accounts, says Lakatos, for the *autonomy of theoretical science.*[28]

I do not think that a positive heuristic, however powerful, can *predict* refutations. Lakatos is led to make this surprising claim by confusing the logico-mathematical problem of deriving predictions with the empirical problem of testing them. The successive 'Newtonian models' which Lakatos describes are the result of trying to find out what Newton's theory predicts about the solar system by a method of successive approximation.[29] (If a planet moved in an unperturbed ellipse, as the 'first model' requires, Newton's theory would be *refuted* and no Newtonian could return to the 'first model' to explain this refutation.) The autonomy of theoretical science simply reflects how much activity is devoted to logico-mathematical problems of deriving specific predictions. No anti-empiricist lesson can be drawn from it: predictions cannot be tested until they have been derived.[30]

Worrall and Urbach seem to have renounced the idea that a powerful heuristic can predict refutations. Both stress that no heuristic can guarantee empirical success, which means that scientists must look at the facts to see whether the predictions of their theories are correct.[31] Yet both continue to claim that a powerful heuristic "will often point to shortcomings in existing theories ... *quite independently of any empirical difficulties*".[32] What they mean is that a scientist may lay down *a priori* requirements that a theory must satisfy, and exclude theories which do not satisfy them (Worrall's example is the Einsteinian requirement that laws be covariant).[33] But they will agree that the adequacy of such 'heuristic rules' must be judged, in the last analysis, by the empirical success or failure of theories constructed in accordance with them. So once again, no anti-empiricist lesson can be drawn from this sort of 'heuristic power'.[34]

Let me now return to Lakatos's argument for according the 'hard core' a privileged status. That argument was, essentially, that it is *easier* to devise an explanation of some anomaly which retains the 'hard core' than it is to devise a radically new system: and the more powerful the positive heuristic, the easier it becomes.[35] Is this argument acceptable?

Now I do think we should defend a methodology which is designed to give theoretical scientists an easy life. Nor, surprisingly enough, would Lakatos really disagree. To modify or renounce the 'hard core' is, in his terminology, to launch a new research programme (or to produce a 'creative shift' in the old one[36]). Lakatos insists, with Popper and Feyerabend and against Kuhn, that science must contain *competing* research programmes. He must be in favour of the proliferation of research programmes. But if so, the so-called 'negative heuristic' is an empty reed: "Do not direct *modus tollens* at the hard core of your research programme *unless* you wish to inaugurate a new programme". The upshot is that scientists can blame anything they like for refutations: it will be good if some try to accommodate them within their programme; it will also be good if others explore alternative programmes. But this is only terminologically distinct from what Duhem and Popper said about the Duhem Problem. Lakatos has not really improved upon their account of this problem. What he has done with his notion of 'heuristic power' is to give us a falsificationist account of what it is to develop a theory and defend it against criticism. He has given us, in a nutshell, a Popperian account of Kuhn's 'normal science'. (We might say that a *generalised* or *pluralistic conventionalism* such as Lakatos's is hardly worth the name 'conventionalism' at all.)

3. ANARCHISM

So far we have discussed problems arising from the confirmation and falsification of predictions from theoretical systems. Lakatos claims that such systems are produced within research programmes. And he uses their empirical success or failure to appraise the programme *as a whole*. This enables him to say, in some cases at least, that one research programme is superior to another.

As Lakatos emphasises, his rules for appraising research programmes are less stringent than previous rules for appraising single theoretical systems. A programme can (intermittently) progress theoretically even though some systems within it are *ad hoc*. A programme can (consistently) progress empirically even though every system within it is refuted. But this does not mean that Lakatos does not care about *ad hoc* adjustments or refutations. It is simply a consequence of the switch from appraising theories to appraising research programmes.[37]

Now this should already make us a little sceptical of Feyerabend's claim that Lakatos's real position is an *anarchistic* one, that "scientific method, as softened up by Lakatos, is but an ornament which makes us forget that a position of 'anything goes' has in fact been adopted".[38] But let us examine this claim in more detail. Do Lakatos's rules of appraisal have any repercussions for scientific practice? Should appraisals in the light of these rules lead to any *advice* being given to scientists? And in particular, if one programme is judged superior to another should we then advise scientists to switch to the superior one?

Lakatos once claimed that "the appraisal of any finished product is bound to have decisive pragmatic consequences for the method of its production ... scientific standards by which one judges theories, have grave pragmatic implications for scientific method, the method of their production".[39] Later he said that his central problem was to say when a scientific research programme was to be *eliminated* and work on it cease.[40] He claimed to give *"rules for the 'elimination' of whole research programmes"*, and hoped that once applied they would enable us to stem 'intellectual pollution'.[41]

But Lakatos changed his mind pretty radically. He declared that his methodology "only *appraises* fully articulated theories (or research programmes) but it presumes to give advice to the scientists neither about how to *arrive* at good theories nor even about which of two rival programmes he should work on".[42] And he said that his only piece of advice is that scientists should be *honest* about the problems they face, and that "a public record should be kept ... of known anomalies and inconsistencies".[43] As Richard Hall put it, for Lakatos an honest policy is a rational policy.[44] And Worrall seems to take this view also.[45]

This change of view means that Lakatos's earlier talk about eliminating research programmes and stemming intellectual pollution was just hot air. Provided intellectual polluters pay lip-service to

Lakatos's standards, solemnly declaring from time to time how bad their programmes are, they act rationally and will be left in peace. It also means that the persistent talk about "*rationally* reconstructing" the history of science is also hot air. Lakatos claims to explain the *rationale* behind the elimination of one programme in favour of another.[46] But according to his latest view, it would have been just as rational if the opposite had happened!

I think that this retreat into what we might call *honest anarchism* ("Anything, except dishonesty, goes") is unwarranted. Let us distinguish *epistemological anarchism* (the thesis that any theory or research programme is as good as any other) from *methodological anarchism* (the thesis that any method or procedure or research-policy is as good as any other). Obviously, Lakatos is no *epistemological* anarchist, since he provides a whole battery of standards for judging theories and research programmes. Nor, though less obviously, is he a *methodological* anarchist, since his standards are associated with a great deal of (rather obvious) advice. Here are a few examples (none of them are peculiar to the methodology of research programmes): "Try to explain refutations in a non-*ad hoc* way"; "Make severe tests, including crucial tests between a new system and its rival"; "Proliferate competing programmes, since a programme which reigns in isolation reigns supreme"; "Unexplored heuristic potential within programmes should be explored"; and so on.[47] Of course, neither individually nor collectively do such pieces of advice provide a *mechanical method of discovery* which would render ingenuity (and good luck) redundant. But we are all against *that* sort of method.

It is an illusion, then, that Lakatos defends an (honest) anarchistic position. Two things have made the illusion so compelling that even Lakatos himself was tempted sometimes to accept it. First, there is the preoccupation with an admittedly difficult but *special* question: can we advise scientists which programme they ought to pursue? Second, there is Lakatos's own anti-empiricist excesses. As we will see, these two are not unconnected.

Suppose one research programme is unambiguously superior to its rival when judged by Lakatos's standards: should we then advise scientists to pursue only the superior one? There are good arguments against such a simple-minded rule. First, the inferior programme may be newly-introduced, and not have had time either to progress or degenerate; clearly, it should be given a chance.[48] Second, some individual scientist may be very ill-advised to switch from the inferior programme: he may have mastered only its techniques, his equipment may be specific to it, and so on. Third, Lakatos's appraisals are themselves *fallible* judgements: further research may show that the programme judged inferior was not really so (for example, that theories within it were wrongly deemed *ad hoc*, or that refutations were based on faulty experiments). Fourth, and most important, even where they are correct Lakatos's appraisals are *backward-looking*, they tell us only that one programme has been superior to another *up until now*. It is possible that if both are pursued further, the progressive programme will start to degenerate and the degenerating one to progress. Since this is *possible*, any advice based only on past performance is totally arbitrary.[49]

For these reasons Lakatos (and Worrall) conclude that appraisals of research programmes can yield *no* advice about which is to be pursued: as far as future research is concerned, anything *does* go. Is this particular anarchistic conclusion warranted? The exception made for newly-introduced programmes is obviously desirable: in the interests of *proliferation* some scientists should work on such programmes to give them a chance. The exception made for individuals who, for one reason or another, could not do good work if they switched to the superior programme is also desirable. A *general* methodology can hardly issue advice to *individuals*, simply because it cannot take account of the host of idiosyncratic factors which determine what it is rational for each individual to do. As Kuhn already emphasized, a general methodology can only be expected to issue advice to the community of scientists *as a whole*.[50] Community-directed advice like "Devote more research-energy to the progressing programme if its rival has degenerated" condemns as irrational neither individuals who have good reason to persist with the degenerating programme, nor individuals who try to develop a new rival programme. What it does deem irrational is *wholesale* persistence with a degenerating programme, or a *wholesale* switch to a new but hitherto inferior one. It was not irrational for Joseph Priestley to remain a phlogistonist; it would have been irrational had the community of chemists done the same after about 1785. We can "explain the rationale" of the switch to Relativity Theory without implying that nobody should have worked within the Newtonian programme after about 1920. So much for the first two arguments.

There remain the arguments that no advice can be based on Lakatos's appraisals because they are *fallible* and *backward-looking*. I do not find these arguments decisive. Why cannot reasonable advice be based on *fallible* judgements? "Arsenic kills people" is fallible, but I think it sensible to advise my friends not to try to refute it. Since we lack pre-cognitions, *all* appraisals are backward-looking. Thus I look backwards when I judge "Arsenic kills people" to have fared better than "Arsenic is good for you". It is *possible* that this judgement will be reversed by future research. But I still think it reasonable to advise my friends not to undertake such research. Such advice is not rendered unreasonable by the mere *possibility* that it could lead us astray.

But when it comes to research programmes, Lakatos claims that there is more than a mere possibility that any degenerating programme will begin to progress if vigorously pursued. He claims that it is quite likely to happen:

If two teams, pursuing rival research programmes, compete, the one with more creative talent is likely to succeed ... The direction of science is determined primarily by human creative imagination and not by the universe of facts which surrounds us. Creative imagination is likely to find novel corroborating evidence even for the most 'absurd' programme, if the search has sufficient drive ... A brilliant school of scholars (backed by a rich society to finance a few well-planned tests) might succeed in pushing any fantastic programme ahead, or, alternatively, if so inclined, in overthrowing any arbitrarily chosen pillar of 'established knowledge'.[51]

According to Lakatos, then, the future success of a research programme depends primarily on the creative imagination of its adherents and the amount of money they possess; and this holds no matter how badly the programme has fared in the past. If this anti-empiricist view is correct, then the past performance of a research programme has little to do with its future promise. Lakatos's anti-empiricism supports his concession to anarchism.

Is Lakatos's anti-empiricist view correct? Creative imagination (and research money) are surely important. But did Ptolemaic astronomy fail primarily because its adherents were not imaginative or rich enough? Could some clever chemists backed by a rich society revive phlogistic chemistry *progressively* and overthrow a few pillars of established knowledge? There is not a single convincing example of this kind of thing having occurred.[52] And that suggests that "the universe of facts which surrounds us" exerts a greater influence on the direction of science than Lakatos allows.

Part of the trouble here derives from Lakatos's exclusive concern with "novel corroborating evidence". Lakatos is right (as were Bacon and Popper before him) that we are likely to find the odd confirmation for the most absurd idea "if the search has sufficient drive". Bacon and Popper concluded that we must seek, and pay attention to, *dis*confirmations. But Lakatos, despite his rather austere definitions of what it is for a programme to progress, seems to have slipped back into the view that only the confirming instances are to count. I have already criticised this view (see *above*, p. 184). On any acceptable definition of 'progress', it is much *harder* to make a research programme progress than Lakatos here suggests.

Suppose the 'hard core' assumptions of a research programme are *mistaken*: is it naive to suppose that this will reveal itself eventually in the track-record of that programme? Consider just one example (of a far-from 'absurd' programme): the struggle of Newtonian astronomers to deal with the anomalous motion of Mercury's perihelion. For fifty years Newtonian astronomers, lacking neither ingenuity nor money, tried to explain the anomaly without modifying Newton's laws. All such attempts failed, and several prominent astronomers concluded that Newton's law of gravity could not be quite correct, that their 'hard core' had been refuted. The suggested modification to it was artificial and *ad hoc*, and no Newtonian could rest content with the situation. The time was ripe for a new programme, with a few outstanding successes and a largely unexplored heuristic potential, to come to dominate the field.[53]

It is the same with any other programme which has degenerated. The mathematical problems involved in it have been solved. Anomalies prove to be more and more stubborn: ways of dealing with them suggested by the 'positive heuristic' have been tried and failed. As Urbach says, "A programme may, after all the instructions of the heuristic have been followed, finally reach a 'point of saturation'. At this point, scientists engaged on the programme will have nothing further to do".[54] *Ad hoc* devices for solving outstanding problems begin to proliferate. Suppose there is a rival programme which is relatively unexplored, which poses unsolved mathematical problems, and which promises to resolve the outstanding difficulties facing the existing one. In this situation, is not the community rationally advised to switch most of its energy to the new programme? I think that it is.

There is no guarantee that following such advice will lead to success. For all we know, all programmes may degenerate equally badly. But this does not mean that at any point they are all equally worth pursuing, or that all methods of pursuing them are equally worthy of adoption. No methodology can guarantee scientific progress: the most we can say is that the adoption of anarchistic views will endanger it.

University of Otago, New Zealand

NOTES

[1] See Popper [1959], note *1 to Section 28, note *4 to Section 80, and note *1 to Section 82. Popper occasionally suggests that whether or not a test is severe depends upon whether the man who performs it is sincerely trying to overthrow the theory being tested. In Section 3 of my [1974b] I objected to this psychologistic theory of severity, and pointed out that it was unnecessary since Popper had also given a perfectly objective theory. In his [1974], p. 1079, Popper replied that this was a merely verbal criticism because when he talked about the sincerity of testers he really meant the objective severity of their tests. Popper and I are agreed, then, that the severity of a test does not depend on psychological facts about the man who performs it. This means that 'background knowledge', on which severity depends in the objective theory, must also be construed non-psychologistically. I will return to this point *below*, pp. 185–186.

[2] For details see Watkins [1964], especially pp. 108–9, or my [1974a], pp. 3-6.

[3] Watkins [1978], *this volume*, p. 39.

[4] Watkins [1978], *this volume*, pp. 38–39.

[5] Mr A's linear hypothesis contains only two adjustable parameters, Mr B's complicated hypothesis many more; Mr A's hypothesis is therefore easier to falsify than Mr B's (using Popper's dimension method for comparing degrees of falsifiability: see his [1959], Section 38); and therefore Mr A's hypothesis is *simpler* than Mr B's (on Popper's account of simplicity). Similarly the grue hypothesis is harder to refute than "All emeralds are green", for to refute it we must find an emerald, note its colour, *and* note the time at which it is being observed.

[6] See Worrall [1978], *this volume*, p. 66, note 6. The anomaly in Popper's theory was first pointed out by Lakatos in his [1968], p. 375, note 2 (see also my [1974a], p. 7, note 1). It was Lakatos who proposed to avoid it by removing 'auxiliary premises' from background knowledge and incorporating them in the 'theory under test': see his [1968], p. 382 (and my [1974a], p. 17).

[7] Worrall [1978], *this volume*, p. 52.

[8] Popper has said this in so many words: t_2 supersedes t_1 if "t_2 takes account of, and explains, more facts than t_1" or if "t_2 has passed tests which t_1 has failed to pass" (Popper [1963], p. 232); again, "even after t_2 has been refuted in its turn, we can still say that it is better than t_1 [if] t_2 has withstood tests which t_1 did not pass" (Popper [1963], p. 235).

[9] See, for example, Lakatos [1970], p. 116.

[10] As Worrall puts it in his [1978], *this volume*, p. 52.

[11] For this objection see Worrall [1978], *this volume*, pp. 47–48; also Zahar [1973], Part I, pp. 102–3; and my [1974a], pp. 11–12.

[12] See my [1974a]. The theoretical view was adopted in some places by Popper (though elsewhere he hints at a strictly temporal view); it is also to be found in Lakatos (who says elsewhere that he held a strictly temporal view). The heuristic view is due to Zahar.

[13] Compare Worrall [1978], *this volume*, pp. 47–48.

[14] For an actual case, see Worrall [1978], *this volume*, pp. 47–48.

[15] See my [1974a], p. 13.

[16] Worrall [1978], *this volume*, p. 51 (I have amalgamated the text and part of note 23, p. 68).

[17] [Added in 1977:] I now think that the compromise suggested in this paragraph will not do (for reasons given by Worrall in his reply, *this volume*, pp. 330–332), and that I was too hasty in renouncing the theoretical view. For one thing, the heuristic view seems to land us squarely back into the Raven Paradox: logical inspection of "All ravens are black" will not reveal that any of the intuitively irrelevant facts which follow from it were used to construct it, and hence they will all have to be deemed to support it. For another thing, the fact that some adjustable parameter of a theory has a certain value cannot support the theory simply because the experiment which ascertained the value *could not have refuted the theory in the first place* (and this remains so *whatever* the existing theory is which the new theory challenges). Ascertaining the value of an adjustable parameter is not a *test* of a theory at all, but a completion of it.

[18] See, for example, Lakatos [1970], pp. 137–8. Popper denies this: he says, for example, that Newton's theory would be refuted if the earth increased the velocity of its motion while the other planets proceeded as before (Popper [1974], p. 1004). However, this possibility is not excluded by Newton's laws *alone*, but only by these laws *together with* additional (though unproblematic) assumptions about the solar system. Lakatos here merely repeats Duhem; and Popper himself acknowledges elsewhere that Duhem is right (see, for example, his [1959], pp. 76–7).

[19] Worrall [1978], *this volume*, pp. 32–33.

[20] See Duhem [1954], Chapter VI, Section 10.

[21] Popper [1959], p. 76, note 2.

[22] Popper [1963], pp. 233–4.

[23] Popper [1959], p. 76, note 2.

[24] Popper writes (in his [1959], pp. 82–3): "As regards *auxiliary hypotheses* we decide to lay down the rule that only those are acceptable whose introduction does not diminish the degree of falsifiability or testability of the system in question, but, on the contrary, increases it". A minor point: 'does not diminish' is not the same as 'increases', and it is the former which Popper should demand. The revised Newtonian system of Leverrier and Adams, which led to the discovery of Neptune, was not *more* falsifiable than its predecessor (which implies that no new planet would be found at the calculated position). The difference between the two systems lay in their degrees or corroboration, not their degrees of testability. And we certainly do not want to exclude the revised system as being *ad hoc*, so we must adopt the weaker definition of *ad hocness*.

[25] Worrall claims that Popper "twice denies that [the Duhem Problem] exists" (Worrall [1978], *this volume*, p. 69, note 33). And he refers us to Popper's remark that Duhem's

"famous criticism of crucial experiments" showed only that they cannot verify a
theory, not that they cannot falsify it (Popper [1959], p. 78, note 1, and [1963], p. 112).
But as I see it, this remark has nothing at all to do with the Duhem Problem. The
'famous criticism' mentioned is Duhem's criticism of Baconian eliminative induction *as
a method of proof or verification* (as inspection of p. 188 of Duhem's [1954], to which
Popper refers, will reveal). And the 'theories' which Popper maintains can be refuted
despite Duhem are actually entire *theoretical systems* (as inspection of Popper's [1959],
pp. 76–8, will reveal). Far from denying the existence of the Duhem Problem, Popper
states it as follows (in his [1959], p. 76, two pages before the first occurrence of his
much criticised remark): "By means of this mode of inference [*modus tollens*] we
falsify *the whole system* ... Thus it cannot be asserted of any one statement of the
system that it is, or is not, specifically upset by the falsification".

Let us see briefly what else Popper has written about the problem whose existence
he is supposed to have denied. He points out that the Duhem Problem can be reduced
in scope if logical investigation reveals that a refuted prediction is independent of some
parts of a theoretical system (Popper [1959], p. 76). He says Duhem overlooked the fact
that we only assert the refutation of entire systems (Popper [1963], p. 112) – I would say
that Duhem rightly saw that scientists pin the blame for refutations more narrowly.
Duhem is also said to have overlooked the fact that there are crucial experiments
between rival systems which share all assumptions but one (Popper [1963], p. 112). But
if Popper's point is that it *must* be the assumption peculiar to the falsified system which
is false, then his logic is faulty. And if his point is merely that if the common
assumptions are thought unproblematic the one peculiar to the falsified system will be
blamed, then the point is trivial and the reference to the rival system redundant.

[26] See Lakatos [1970], p. 133. Actually, Popper *agrees* with Lakatos in one special
case: in his [1967] he argues that the 'Rationality Principle' which figures in in-
dividualist explanations of human action should be rendered 'irrefutable by fiat', and
the blame for refutations always shifted to the rest of the explanation (the 'situational
analysis'). For an interesting discussion of the research programme in economics which
arose from treating the 'Rationality Principle' in this way, see Latsis [1972].

[27] Lakatos [1970], p. 135; see also p. 175.

[28] See Lakatos [1970], pp. 135–7.

[29] For the 'Newtonian models', see Lakatos [1970], pp. 135–6.

[30] For further details, see my [1976], pp. 467–473.

[31] See Worrall [1978], *this volume*, p. 69, note 39; and Urbach [1978], *this volume*, p.110.

[32] Worrall [1978], *this volume*, pp. 59–60; for a similar claim see Urbach [1978], *this
volume*, p. 108.

[33] Worrall [1978], *this volume*, p. 60.

[34] Nor can any such lesson be drawn from the heuristic power generated by *analogical*
arguments, which play a great role in the research programmes analysed by Lakatos,
Worrall, and Clark: see my [1976], pp. 471–3.

[35] In this respect the Ptolemaic programme had a very powerful heuristic: it was almost
an 'algorithm' (Worrall [1978], *this volume*, p. 69, note 42) in that "successive theories
were almost completely determined by the heuristic and virtually no latitude was
allowed to the inventiveness of scientists" (Urbach [1978], *this volume*, p. 109). Worrall
and Urbach are embarrassed about giving high marks to the Ptolemaic heuristic. They

both claim that although it was strong in one sense it was weak in the sense that it did not *predict* anomalies (Worrall [1978], *this volume*, p. 60, and Urbach [1978], *this volume*, p. 109). I have argued that all heuristics are weak in this second sense. The example shows, I think, that one can have too much 'heuristic power': the strength of Ptolemy's heuristic stemmed from the *weakness* of his 'hard core', in that (compared with Copernicus's) it excluded very few observable states of affairs.

³⁶. On 'creative shifts' see Lakatos [1970], p. 137, note 1, and p. 164. Once such shifts are allowed, the idea of a research programme loses much of its definite character: given sufficiently many of them, there is no telling whether one programme could not be transformed into a rival one! Creative shifts are always '*ad hoc₃*' for Lakatos, since to change the 'hard core' or the positive heuristic is not to act in accordance with it. Lakatos hoped to accommodate this meta-anomaly (Lakatos [1971b], pp. 176–7). I think we should simply admit that a so-called 'creative shift' is a shift to a different programme.

³⁷ For Lakatos's rules of appraisal, see his [1970], pp. 133–4 and 154–9.

³⁸ Feyerabend [1970], p. 229 (also p. 215). Feyerabend dedicates his [1975] "To Imre Lakatos, friend and fellow anarchist" (see p. 5).

³⁹ Lakatos [1968], p. 343.

⁴⁰ Lakatos [1970], pp. 154–5 and 177. Lakatos claims that "in the methodology of research programmes, the pragmatic meaning of 'rejection' [of a programme] becomes crystal clear: it means *the decision to cease working on it*" (Lakatos [1970], p. 157, footnote 1).

⁴¹ Lakatos [1971a], p. 100, and [1970], p. 176, note 1.

⁴² Lakatos [1971b], p. 174.

⁴³ Lakatos [1971b], p. 174.

⁴⁴ Hall [1971], p. 152. Hall rightly objects that one can be honest but irrational, or dishonest but rational.

⁴⁵ Worrall [1978], *this volume*, p. 70, note 45.

⁴⁶ Lakatos [1971b], p. 174. Several case-studies have been produced showing, roughly, that scientific revolutions which other methodologies must deem irrational come out rational according to Lakatos's methodology (see Howson [1976]). But this is a Pyrrhic victory if *whatever* happens (dishonesty aside) is rational for Lakatos.

⁴⁷ For some more concrete (and more problematic) pieces of advice, see Lakatos [1971a], p. 105 (and my [1976], p. 488, note 74).

⁴⁸ See Lakatos [1970], pp. 156–7.

⁴⁹ This argument was first stated by Feyerabend in his [1970], p. 215.

⁵⁰ See Kuhn [1970], p. 238. Kuhn points out that "the community's way of distributing risk" is for different groups simultaneously to pursue different lines of research (Kuhn [1962], p. 186). Clearly, no philosopher of science can decide which line should be pursued by which individuals.

⁵¹ Lakatos [1970], pp. 187–8.

⁵² Examples are not provided by vague talk about Dalton having revived Greek atomism, or Einstein having revived the corpuscular theory of light. For a *research programme* to be revived we must have the same 'hard core' and 'positive heuristic'. (I here ignore the possibility of 'creative shifts': see *above*, note 36.)

⁵³ For further details, see my [1976], pp. 461–3.

⁵⁴ Urbach [1978], *this volume*, p. 108.

BIBLIOGRAPHY

Bunge, M. (ed.): 1964, *The Critical Approach to Science and Philosophy*, The Free Press, Glencoe.

Cohen, R. S. and Buck, R. (eds.): 1971, *Boston Studies in the Philosophy of Science, Volume VIII*, D. Reidel Publ. Co., Dordrecht, Holland.

Cohen, R. S., Feyerabend, P. K., and Wartofsky, M. (eds.): 1976, *Essays in Memory of Imre Lakatos* (*Boston Studies in the Philosophy of Science, Volume XXXIX*), D. Reidel Publ. Co., Dordrecht, Holland.

Duhem, P.: 1954, *The Aim and Structure of Physical Theory*, (translated by P. P. Wiener), Princeton University Press, Princeton.

Feyerabend, P. K.: 1970, 'Consolations for the Specialist', in Lakatos and Musgrave (eds.), pp. 197–230.

Feyerabend, P. K.: 1975, *Against Method*, New Left Books, London.

Hall, R. J.: 1971, 'Can We Use the History of Science to Decide between Competing Methodologies?', in Cohen and Buck (eds.), pp. 151–9.

Howson, C. (ed.): 1976, *Method and Appraisal in the Physical Sciences*, Cambridge University Press, London.

Kuhn, T. S.: 1962, *The Structure of Scientific Revolutions*, Chicago University Press, Chicago, (second edition, enlarged, 1970).

Kuhn, T. S.: 1970, 'Reflections on my Critics', in Lakatos and Musgrave (eds.), pp. 231–278.

Lakatos, I.: 1968, 'Changes in the Problem of Inductive Logic', in Lakatos (ed.), pp. 315–417.

Lakatos, I.: 1970, 'Falsification and the Methodology of Scientific Research Programmes', in Lakatos and Musgrave (eds.), pp. 91–195.

Lakatos, I.: 1971a, 'History of Science and Its Rational Reconstructions', in Cohen and Buck (eds.), pp. 91–136.

Lakatos, I.: 1971b, 'Replies to Critics', in Cohen and Buck (eds.), pp. 174–182.

Lakatos, I. (ed.): 1968, *The Problem of Inductive Logic*, North-Holland Publishing Company, Amsterdam.

Lakatos, I. and Musgrave, A. (eds.): 1970, *Criticism and the Growth of Knowledge*, Cambridge University Press, London.

Latsis, S. J.: 1972, 'Situational Determinism in Economics', *British Journal for the Philosophy of Science* 22, 287–297.

Musgrave, A.: 1974a, 'Logical versus Historical Theories of Confirmation', *British Journal for the Philosophy of Science* 25, 1–23.

Musgrave, A.: 1974b, 'The Objectivism of Popper's Epistemology', in Schilpp (ed.), pp. 560–596.

Musgrave, A.: 1976, 'Method or Madness', in Cohen, Feyerabend, and Wartofsky (eds.), pp. 457–491.

Popper, K. R.: 1959, *The Logic of Scientific Discovery*, Hutchinson and Co., London.

Popper, K. R.: 1963, *Conjectures and Refutations*, Routledge and Kegan Paul, London.

Popper, K. R.: 1967, 'La rationalité et le statut du principe de rationalité', in *Les Fondements Philosophiques des Systèmes Economiques*, Bibliothèque Economique et Politique, Paris.

Popper, K. R.: 1974, 'Replies to My Critics', in Schilpp (ed.), pp. 961–1197.

Schilpp. P. A. (ed.): 1974, *The Philosophy of Karl Popper*, Open Court Publishing Company, La Salle.

Urbach, P.: 1978, 'The Objective Promise of a Research Programme', this volume, pp. 99–113.

Watkins, J. W. N.: 1964, 'Confirmation, the Paradoxes, and Positivism', in Bunge (ed.), pp. 92–115.

Watkins, J. W. N.: 1978, 'The Popperian Approach to Scientific Knowledge', this volume, pp. 23–43.

Worrall, J.: 1978, 'The Ways in Which the Methodology of Scientific Research Programmes Improves Upon Popper's Methodology', this volume, pp. 45–70.

Zahar, E. J.: 1973, 'Why did Einstein's Programme Supersede Lorentz's?', *British Journal for the Philosophy of Science* **24**, 95–123 and 223–262; reprinted in Howson (ed.), 1976, pp. 107–179.

HANS ALBERT

SCIENCE AND THE SEARCH FOR TRUTH

*Critical Rationalism and the Methodology of Science**

1. INTRODUCTION

I should perhaps begin by noting my fundamental agreement with the general thrust of the position paper, although I would have some bones to chew with various details. As I see it, critical rationalism exhibits three basic characteristics which are intimately connected with one another: a consistent fallibilism, a methodical rationalism and a critical realism. Each of these components plays a role in the solution of the problems we are concerned with. I should also like to emphasize that this philosophical conception has consequences that are of importance for problems of all kinds, not merely for problems pertaining to knowledge.

A *consistent fallibilism* results from the impossibility of maintaining the fusion of truth and certainty implied by classical rationalism,[1] i.e. it arises from a critique of the solutions to epistemological problems offered by the rationalist tradition. The adoption of a general principle of justification which implies a guarantee of truth involves a trilemma of infinite regress, vicious circle or recourse to dogma.[2] The only practical solution – recourse to a dogma – involves a suspension of the principle of justification itself. If one wishes to avoid this kind of *dogmatism*, and yet retain the idea of a guarantee for truth which is implied in the Aristotelian definition of knowledge, there remains the option of *scepticism*, which declares the whole enterprise of seeking knowledge to be pointless. Another possibility involves abandoning the demand for a secure foundation in favor of a critical epistemology of the kind outlined in the position paper, i.e. the idea of conceding the fallibility of human cognition without giving up the search for knowledge.

A consistent fallibilism of this kind is compatible with a *methodical rationalism*. Fallibilism denies that it is possible to justify problem-solutions in a manner that excludes doubt and error, but we can expose our hypotheses to a critical examination in order to find out if and to what extent they are preferable with respect to other hypotheses.[3] In line with the fallibilism outlined in the last paragraph, this

203

G. Radnitzky and G. Andersson (eds.), Progress and Rationality in Science, 203–220.
All Rights Reserved.
Copyright © 1978 by D. Reidel Publishing Company, Dordrecht, Holland.

version of rationalism is a general view: it deals with problem-
solutions in general, and is not restricted to cognitive or scientific
problems. Our presentation of this rationalism can take as its starting
point the fact that there is no field of human concern – the cognitive
enterprise of science being merely a special case – in which we can
demonstrate that we have perfect solutions, that our answers are free
from all possible weaknesses and thus immune from all criticism. The
idea of an absolute justification, which has prevailed not only in the
realm of knowledge, but also in other spheres – e.g. the political
justification of the social order, – presupposes this utopian possibility, a
possibility which experience in all spheres of human concern has
demonstrated to be finally unacceptable. Even in mathematics[4] and
logic,[5] fields in which the ideal of certainty was simply taken for
granted until quite recently, we have been forced to abandon the idea
of absolute justification. The attempt to provide a secure foundation
for knowledge is no longer a tenable enterprise. The only rational
alternative is to submit our proposed solutions to critical examination,
i.e. to evaluate them with a view to possible improvements, to
compare them with alternative solutions and to search for new and
better solutions. This kind of examination and evaluation presupposes
standards – criteria for evaluation – which are relevant to the type of
problem which is to be solved.

Critical realism takes the aim of knowledge to be the comprehen-
sion and representation of reality – or of aspects of reality, cognition
always being selective – in opposition to conceptions which take the
aim of science to be the construction of systems of signs – conceptual
apparati, systems of propositions, calculi – which have no represen-
tational function, but which are useful in a certain manner for
practical life.[6] That a critical realism of this kind involves the regulative idea of truth has been constantly emphasized by Sir Karl Popper.
Thus, it may be of some value to consider this idea and its
significance in this connection.

2. CRITICAL REALISM AND THE IDEA OF TRUTH

In our everyday life we are accustomed to using terms such as 'true'
and 'false', 'fact' and 'reality', 'illusion' and 'error' in a rather free and
easy manner, even when we have philosophical ambitions or are well
known as representatives of certain philosophical opinions. Of

course, these terms are not normally used in a very precise manner, but their employment is sufficiently precise to allow people with varying opinions as to which propositions or theories are to be characterized as 'true', or what is to be acknowledged as 'fact' in various fields and what is to count as 'illusion' and 'error' in detail, to understand one another. Scientists generally exhibit this kind of naturalness in the use of such terms even in connection with their own research activities. They speak of a hypothesis which has been surprisingly well corroborated, although it must be assumed that it is false; they talk about illusions of a certain kind – e.g. perceptual illusions – to be produced in certain experiments; and they point out without any hesitation that certain persons have made errors in predicting certain developments in nature or society. It is an interesting fact that they sometimes cease to use these expressions in this very natural manner when they get into a 'philosophical' mood. They may even be prepared to laugh at the naivity of taking the word 'true' seriously, to characterize propositions about reality as 'impossible' or 'unscientific', and to call colleagues who admit that they are interested in knowing the factual conditions in a realm of objects 'metaphysicians'. A milder and more inconspicuous version of the same general attitude turns up in the assertion that the truth of a proposition consists in its utility for certain purposes, that truth is to be identified with confirmation, or perhaps with the factual or hypothetical consensus of certain people – maybe under ideal conditions –, or even with relevance to the class struggle or some other cause to which they subscribe.

Now it goes without saying that neither philosophy nor science may be too tightly bound to common sense, since progress in knowledge has its root in the fact that we put the self-evident truisms of daily life into question. But it does not follow that a highly subtle philosophical doctrine must always be superior to common sense. And it by no means follows that the realism of the scientific attitude must turn out to be 'naive' when subjected to philosophical scrutiny.[7] Philosophical conceptions which involve such evaluations may be much more sophisticated and complicated than the straightforward realism of the scientific attitude, but they nearly always exhibit a certain one-sidedness and artificiality which makes them useless for the adequate interpretation of human knowledge. Of course, they might have the great merit of throwing light on some difficulties which received and more plausible views will have to deal with. And it is certainly possible that a confrontation with such theories will demonstrate that there are metaphysical and epistemological assumptions at play in common sense and scientific practice which have hardly been noticed, and which can now be made accessible to a critical examination.

Common sense thinking is generally thought to involve a tacit commitment to so-called 'naive realism', a view according to which reality by and large is as it appears to be in our sense perceptions. The qualities of sense are thought to be the properties of objects, generally with some qualifications concerning the influence of unfavorable situations on perception. It is well known that this kind of view leads to various contradictions and absurdities which make a revision of this general interpretation of human knowledge necessary. Thus, already in ancient Greek thought we find a critical realism, which concedes the subjectivity of sense perception[8] and attempts to liberate our knowledge as far as possible from its subjectively conditioned limitations, being substituted for naive realism. This critical thrust makes good use of the progress of the sciences, a progress which begins in the criticism of common sense. Critical realism is by and large the interpretation of knowledge which is native to the empirical sciences, and it is only rather recently that there have appeared tendencies in the sciences themselves to replace critical realism by other views.

Given the degree to which scientific knowledge has gained a foothold in common sense thought, we are no longer entitled to characterize common sense as permeated by a 'naive realism'. It is much more likely that considerably modified versions of realism enjoy a wide circulation today. On the other hand, it is much less probable that the more or less sophisticated variants of anti-realism – idealism, positivism, pragmatism, phenomenalism – are widely accepted, except by certain philosophers, representatives of some sciences (e.g. quantum mechanics), and by certain educated people who are much impressed by subjectivistic interpretations of some results of scientific research as formulated by famous scientists. Thus, these doctrines may for some time enjoy the support of scientists who are experts in their special fields, and this fact seems to promote their public prestige. These specialists claim to be able to reduce the realism of an educated common sense to an absurdity, and combine this claim with an interpretation of scientific knowledge as a structure of more or less useful fictions, conventions or constructions which cannot be connected in a meaningful way with an independent reality or with the received idea of truth.

At the same time, it is a most curious fact that the representatives of anti-realistic positions generally attach great importance to the experimental testing of scientific propositions, and that the opinion seems to prevail that a positive result of such tests under certain conditions justifies the acceptance of such propositions and systems, whereas a negative one justifies their rejection. One often finds a complicated methodological theory, which gives advice concerning the arrangement of tests and the evaluation of results, being proposed. But it is not quite so easy to discover just why one should accept the methodological code, for the naive idea that such procedures have something to do with the search for truth in the traditional sense, with the endeavor to discover the structure of reality or of certain parts of reality, seems to be hopelessly old fashioned.[9] Science is a game with certain rules, and it seems that the discussion of the precise specification of these rules is to be considered as a very important activity. But the original point of the game – the aim of scientific inquiry – seems to have been lost sight of.

How are we to understand this anti-realistic tendency in philosophical thought? Perhaps we should once more recall the traditional intimate connection between the search for truth and the idea of certainty. Even today there are many people who assume that genuine knowledge can only be secure, irrefutable, undoubtable knowledge. All methodological procedures are to be evaluated in terms of their contribution to the attainment and identification of such knowledge. The problem is to give knowledge a secure foundation which can serve as a guarantee for truth. Thus, we need a criterion for identifying this truth, for without the possibility of such an identification there can be no certainty that we have indeed arrived at genuine knowledge. The idea of a guarantee for truth and the concept of a criterion of truth are closely connected. If the search for a criterion of this kind turns out to be hopeless, sceptical consequences seem to be unavoidable, and modern anti-realism may be a version of scepticism in scientific garb.

But let us take a closer look at the search for truth involved in cognitive activity. The *idea* of truth involved is quite plausible. It need not be connected with the idea of a *criterion* in the above

mentioned sense, and it does not seem necessary to give a formal
definition of it. The idea is presumably very old. It can best be
elucidated with respect to a language suitable for representation, that
is, a language which has not only expressive and signal functions,
such as the communication means used by animals, but also exercises
a representative function.[10] In this context, the idea of truth can be
elucidated as the *idea of an adequate representation* of any state of
affairs referred to by linguistic means.[11] This regulative idea is a
moment in the representative function of language – whether speakers
of a language are aware of it or not – and it refers to the adequacy of
linguistic products with regard to this function. The transmission or
reception of information is hardly possible in the absence of an at
least inchoate understanding of this idea, although it is often extremely difficult to evaluate the adequacy of a given representation.
The idea that a representation can be more or less adequate is also
readily understandable, even if nobody is in a position to give such an
evaluation in a given instance. We generally operate with such ideas
in a more or less naive and successful manner, without giving them a
second thought. We don't worry about the problem of finding a *secure*
sign of truth, i.e. a criterion on the basis of which we can determine
with certainty the adequacy of a proposition or conception.

The epistemologist, of course, does not enjoy the benefits of this
naivity. He wants to know more about these things. He demands
precision. It is, however, of the utmost importance that we remember
that epistemological considerations, important as they may be, have a
secondary character. Epistemological questions can be raised only on
the basis of our everyday commerce with the idea of truth and related
ideas. The philosopher has the task of focussing on these ideas in a
critical and constructive manner, and his activity is motivated by
problems which can arise in our everyday commerce with them, for
example, the paradox of the liar. The epistemologist's raw material
comes from the life of common-sense knowledge and scientific
research, and he has the task of analyzing this material.

Roughly, there are two main goals of epistemological investigations:

(1) clarity as to just *what* this adequacy of representation *consists
in*, which is summed up by the word 'truth' when it is used in certain
contexts; and

(2) a solution to the problem of *how* we can *find out* if a given
representation is – more or less – adequate.

These two problems are by no means identical.[12] On the contrary, it is extremely important to distinguish between them. The first question concerns the *meaning* of 'truth', i.e. an appropriate concept of truth, while the second question deals with a *criterion* of truth, or, better, with appropriate methods for identifying the truth. Only an advocate of operationalism will not be able to, or not want to, distinguish between these two questions. It should also be noted that the word 'criterion' tends to arouse associations which are not fruitful for a discussion of the second question, since they tend to prejudge the answer in favor of scepticism. Thus, it is perhaps better to speak of methods of identification, since this allows us to ask how successful these methods can be, and *if* they are liable to lead to a *guarantee* of truth or not. Those who begin by demanding a *criterion* of truth are generally looking for something *more* than a method of identification. They would like to have a *secure sign* of truth, a mark warranting the truth of a proposition or a system of propositions, and it is by no means obvious that the search for such a sign will be successful. So we cannot simply take it for granted that the failure of this search involves sacrificing the idea of truth.[13] It may be that we can find procedures we can live with, even live with quite nicely, even though they offer no certainty and little security.

It is certainly conceivable that the search for an Archimedian point of knowledge and the correlative notion of an absolute truth-warranting criterion might steer us away from an adequate solution to epistemological problems, and the history of modern philosophy seems to bear out this suspicion. The above-mentioned trilemma, which arises in the classical conception of knowledge on the basis of its demand for justification, suggests dogmatism as the only practicable solution. But even this last-gasp attempt at a solution is not acceptable, since it leads to a suspension of the demand for justification itself. If we wish to hold onto the demand for justification at all costs, the only alternative will be scepticism, the relinquishment of any claim that we can acquire genuine knowledge. Thus, the same basic idea is the common root of both reactions. But whereas the dogmatist manages to sustain the illusion that he has achieved what he set out to do, the scepticist takes what seems to be the only path left which leads at least to consistency: he gives up.

There is, however, yet another possible method for overcoming the failures of classical rationalism without adopting a consistent and consequent criticist position. We might develop a conception which fuses dogmatic and sceptical components, e.g. by maintaining the search for a secure foundation and putting up with a dogmatic closure to the procedure of justification, while at the same time calling into question the cognitive character of the justified propositions or theories. In this case, there is no claim of substantial truth for these propositions.[14] A science without knowledge along these lines is the embodiment of modern scepticism, but in a dogmatic form which replaces the realism of the old scientific outlook by a version of pragmatism. This flight from critical rationalism involves saving precisely the most questionable elements from the bankrupt estate of classical rationalism. The most inconspicuous manner of accomplishing this shift involves a suitable redefinition of the concept of truth, such that it is possible to make the commitment of science to extraneous goals at least superficially plausible.[15] Viewing science as a method for dealing with practical problems can always count on a high degree of public support, and the perversions of an instrumentalism which seeks to reinterpret the cognitive foundation of the practice of life in a technological manner will usually pass unnoticed if it is sophisticated enough. Yet, the immediate connection between science and the satisfaction of human needs loses its initial plausibility when we take note of the rather surprising fact that precisely those needs which are characteristic for our species are usually forgotten: the aspiration to knowledge and to orientation in the world.

This form of pragmatism is also to be found in the theory of knowledge-constitutive interests developed in the frame of neo-marxism (Habermas) and transcendental hermeneutics (Apel). Here the problem of truth is solved in terms of the so-called consensus theory of truth, a theory which either simply does not take the idea of adequate representation into account at all, or treats it as a rather unimportant and unproblematic moment.[16] Truth is defined in terms of the consensus of a hypothetical ideal communication community, and this consensus is in turn taken to be the criterion of truth. The 'transcendental pragmatical' considerations made in this context have the task of saving the classical idea of a secure foundation for at least a certain level of knowledge, although it is not, and cannot, be shown how the epistemological trilemma is effectively dealt with.

It goes without saying that providing arguments against the classical conception of knowledge and the alternatives which appear to be consistent with that frame-dogmatism, scepticism, and a mixture of dogmatic and sceptical elements – does not commit us to providing a classical 'justification' for the criticist framework defended here and in the position paper. We can at best show how such an alternative can be elaborated, demonstrating that it has certain properties which recommend it as an acceptable solution. We can take as our starting point the fact that the idea of truth is not necessarily tied to the idea of certainty, as was characteristically assumed in classical rationalism. In other words, we need not, indeed cannot, take the demand for an absolute justification for granted. Thus, the problem of delineating adequate cognitive procedures is separable from the idea of certainty and thus from the problem of a truth-warranting criterion, without sacrificing the idea of truth. If we replace the classical principle of justification by the principle of critical examination, then we can avoid falling back into either dogmatism, as do the defenders of classical rationalism, or into scepticism, as do the disappointed rationalists who see themselves forced into irrationalism, or into those sophisticated blends of dogmatism and scepticism which are so seductive to modern philosophers. Thus, the critical stance allows us to maintain a realistic view of science, in which the idea of truth, and the idea of an approximation to truth,[17] are acknowledged as regulative ideas which can serve as an orientation point for methodological considerations. Thus, the critical realism of the received view is brought into contact with the fallibilism introduced above. Since fallibilism also refers to the methods applied in scientific research, it is not only scientific propositions, theories and explanations which are always, in principle, revisable, but also the method and criteria by means of which they are constructed and tested.

It is a currently rather popular thesis that we can dispense with notions like truth and the approximation to truth in considerations pertaining to the methodology of science. Now it goes without saying that, if we watch our step, we can avoid the use of the words 'truth' and 'falsity', and their more conspicuous substitutes, in methodological contexts. But it is a bit more difficult to discuss the significance of methodological stipulations without reference to the *idea* of truth, if only implicitly. It may well be that one can expect little sympathy if one answers the question as to why we should avoid or eliminate inconsistencies in theoretical systems by noting that the self-contradictory propositions which can be deduced in such systems are false.[18] Many philosophers will perhaps be more comfortable with the answer that any proposition can be deduced in an inconsistent system, that these systems do not discriminate between propositions at all. This answer is surely acceptable, as far as it goes, but it is not sufficient as long as there is no indication as to the properties relevant to the desired discrimination. And a discussion of these properties leads very naturally to the notion of truth.

We might also wonder why we should test theories by means of experiments and other empirical investigations at all. The vague but thoroughly adequate idea generally connected with such investigations concerns the way in which the 'resistance of reality' is utilized in the evaluation of systems of propositions. One can also identify a realistic view at the root of efforts to eliminate 'errors in observation' as far as possible.[19] It is certainly true that the refutation of a theory on the basis of 'facts' which plays such a central role in Popperian philosophy of science often involves a very roundabout process, but that does not justify the claim that efforts to test a hypothesis are in principle superfluous. The first small step out of the criticist frame rules out any rational answer to the question as to why we should not admit any attempt at immunization we please.

Any attempt to identify the difference between the 'efficiency' of a theory and a machine cannot avoid a specification of the performances expected of them respectively. To the extent that 'explanations' are demanded from theories, it is hardly possible to avoid recourse to the cognitive-informative meaning of the propositions involved – their representative character – and in these terms to the difference between a more or less adequate representation of a state of affairs. According to the usual model, explanation involves the deduction of adequate descriptions of certain states of affairs. Even philosophers partial to an instrumentalistic interpretation of scientific theories cannot avoid the realistic viewpoint in dealing with problems of prognosis – in the evaluation of the practically useable results of prediction – since the usefulness of predictions is, among other things dependent on the extent to which they give an adequate representation of the events to be expected. But if we are forced to make concessions at this point, it is difficult to see an even half-way rational motive for rejecting the idea of truth as useless in relation to the theoretical systems involved in making predictions. Such a rejection, e.g. the interpretation of nomological statements as mere 'rules', may, of course, be the result of arbitrary presuppositions such as the verifiability postulate for 'all propositions which are to be characterized as meaningful, but I hope that a discussion of early logical positivism is superfluous here. Almost everything which is to be said with regard to test procedures, the corroberation of theories, and the adequacy of criteria appears meaningless if it is not connected with the idea of the possibility of approximating truth, i.e. the possibility of arriving at a more or less adequate representation of certain aspects of reality. The fact that we cannot produce a guarantee of truth does not make the search for criteria and methods for the fallible evaluation of propositions and theories superfluous. The usual objections to the Popperian conception of truth – apart from arguments directed against certain flawed formal definitions, e.g. of 'versimilitude' – depend on ideas from the classical rationalist tradition, and the impossibility of fulfilling utopian demands generally leads these critics to an open or concealed scepticism.

3. THE CRITICIZABILITY OF METHODS AND THE NATURE OF EPISTEMOLOGY

Critics often emphasize the absence of an epistemological foundation for the methodological position of critical rationalism, putting in question the possibility of an adequate manner of formulating, evaluating and revising the criteria of knowledge in the absence of this sort of foundation. The first thing to be noted in this connection is that the problem of adequate criteria is a very general problem. It is to be found in every field of social activity – in every kind of problem-solving activity; in law, morals, politics, literature, the arts, etc. –, and not merely in the enterprise of acquiring knowledge in science. As to the evaluation of the comparative adequacy of problem-solutions, the requisite criteria will of course have to be differentiated according to the *kinds* of problems involved, e.g. the Pareto-criterion for the comparative evaluation of institutional arrangements in relation to individual preferences vs a criterion for the explanatory power of theories. Again, *within* the realm of knowledge, there may be different criteria – according to different *kinds* of problems or aims – which may be thoroughly compatible with one another.

If we are to be consequent in our fallibilism, then criteria must be fallible and revisable too. If we also wish to be consistent with the

tenets of critical realism, then it must in principle be possible to use
our fallible knowledge about reality – about real conditions – in criticizing and revising criteria, even criteria for knowledge. Our knowledge of history shows that such criteria have de facto changed, often
in connection with changes in the ideal of knowledge[20] and there have
even been situations in which people applied different – and
incompatible – criteria to the same kinds of problems. Is this compatible with the view of science as a rational enterprise? Certainly, on
the assumption that it can be shown that rational argument is relevant
for decisions concerning such criteria. Of course, we can always
simply conceive of science as a 'game' which is defined by certain
rules, and it will then be immune to criticism in this respect. Rational
discussion about the adequacy of the rules and criteria is ruled out,
and this gambit allows anti-realism (or any other favored position) to
be maintained in the face of any attempts at criticism. It is, however,
equally obvious that the price is the acceptance of a form of dogmatism. There are those who reproach Popper – the author of *The
Logic of Scientific Discovery* – with a view of this kind, but I simply
cannot see that this critique does justice to the real mode of
argumentation in this book, and it is certainly inconsistent with his
stated view in later works such as "The Aim of Science".[21] Any
consistent defense of the possibility of rational discussion in this field
will have to reject the 'autonomy' of science in the sense of the
'game'-view.

Thus, we must face the problem of *how we are to criticize* rules and
criteria. In dealing with this problem, we must presuppose an interpretation of the aim of science, as well as certain views about the
possibility of achieving this aim, e.g. about the real knowledge-
situation of man. As far as I can see, this means that we can conceive
of the *methodology* of science neither as a *normative* discipline in the
usual sense, nor as a *descriptive* treatment of the behavior of certain
experts – the scientists –, but rather as a kind of *technology*[22] related
to a presupposed goal of cognitive problem-solving activity. Thus,
methodology is dependent on background assumptions about what is,
e.g. about the structure of language, the possibilities of perception,
and the lawful structure of the Universe. Such assumptions often play
a pivotal role in methodological discussions, even if they are not
explicitly stated. It seems to me that we have to agree with Paul
Feyerabend on this point.[23] Of course, these discussions often focus

on the aims of scientific activity, on the significance and value of science in the context of cultural life,[24] but even here we must take care not to forget the technological aspects, for they are involved in the problem of realizing these aims.

Just like every other technology the methodology of knowledge must have a basis in reality: it must be appropriate to the relevant structural traits of reality. This means that we must look for an appropriate *theoretical basis* for it in an adequate theory of knowledge, an epistemology which explains or accounts for knowledge – particularly the cognitive enterprise of science – or explains how we can learn and solve problems. An epistemology of this kind is relevant for a critique and improvement of the methodology of science. But is there such an epistemology? Is a theory of this kind even possible? What would it look like? Critical realism itself is a rather metaphysical view, much too incomplete to provide an explanation of the kind we are looking for.

On the other hand, anyone who reproaches critical rationalism for the failure to develop an adequate epistemology has to confront the question of his own alternative. Hinting at a transcendental philosophy which examines the "conditions of the possibility of knowledge" is of no avail if there is no indication as to how such a conception is to be interpreted in contemporary terms and as to the exact role which such a philosophy is to play.[25] Kant himself, squarely in the Aristotelian tradition on this point, seems to have thought that the task of transcendental philosophy is to give knowledge a secure foundation. This view is no longer acceptable, but if one reinterprets the *transcendental enterprise* – in the sense of a *critical realism* such that it is understood as an *attempt at explanation*, or as an attempt to sketch an explanation of the possibility of knowledge, then the assumptions needed for this attempt will have to be conceived of as assumptions concerning real structures, and epistemology is seen to be a part of cosmology (a point emphasized by Popper, whom Feyerabend follows at least on this point). If we are to explain the possibility of knowledge or cognition as a real process, we have to assume e.g. the existence of laws of certain kinds and the existence of certain kinds of circumstances. It is thus obvious that a consistently developed critical realism must include the possibility of criticizing methodological proposals on the basis of cosmological assumptions, assumptions about the character of reality.

There remains, however, one major difficulty which seems to threaten the possible consistency of critical realism. Since our cosmological views are influenced by the development of the empirical sciences, it seems a bit paradoxical to use them in criticizing methodological proposals – rules, criteria, etc. –, the result of scientific research being themselves evaluated on the basis of proposed criteria. Is this whole enterprise based on a vicious circle?

Before falling into a sceptical sleep, we should take note of the fact that this kind of criticism is nothing new. Far from being a desperate attempt to save a flawed philosophy, it is a rather normal and unproblematical procedure. For example, neo-classical empiricism has been effectively criticized, by Feyerabend, Kuhn, Hanson and others, on the basis of research in the psychology of perception.[26] On the basis of this research, it has to be admitted that a secure, theory-neutral basis for knowledge simply cannot be produced, and that empiricist methodology must be modified accordingly. Incidentally, the usual conception of hermeneutics is open to a similar line of criticism.[27] If we are not prepared to admit the possibility of this kind of criticism, then it is hard to see how epistemology can be saved from becoming a dogmatic, aprioristic discipline. Not even methodology can be a part of logic, as some formalists would have it, and even that would not make it immune to criticism, e.g. with respect to the results of linguistic investigations. Thus, we can easily admit that epistemology is not a closed and aprioristic discipline floating *above* the sciences. It is rather itself a realm of scientific knowledge, a field in which all other sciences may legitimately interfere. There is no sharp demarcation between science and philosophy, e.g. with regard to certain ranges of objects. A glance at the history of human cognitive endeavours is enough to show that there may be feed-back between all kinds of disciplines. A conception of knowledge which attempts to exclude this kind of creative feed-back by cutting up knowledge into neat, isolated fields is simply not realistic. A consistent fallibilism, along with a critical realism, thus leads to the view that it is possible to criticize methodological proposals and epistemological ideals on the basis of the results of the positive sciences. Philosophical problem-solutions *as such* are not *immune* from criticism in this respect. At best (or worst!), they can be *immunized*, but here as in the case of scientific theories it is hard to see why such a strategy is either useful or rational.[28]

If one concedes that epistemology and methodology – and thus the criteria, rules and ideals of this field as well – are subject to change in the course of history, then it is no longer possible to think that a current methodological position can provide adequate explanations for developments in the history of science. We cannot explain anything by showing that certain scientists *did in fact use* certain criteria or rules, at least in the usual sense of adequate explanation. To this extent, it seems to me that purely 'internal' explanations are of dubious value: they are at best descriptions. But if we are to show that the application of certain rules or criteria was in fact *successful*, in the sense of producing new and/or better knowledge, then we have to presuppose an *independent* criterion for what is to count as success, and this leads directly to the technological conception of cognitive activity mentioned above. This might lead to a genuine *explanation* of progress or the evolution of knowledge, and I simply do not see how such an explanation might be formulated in the absence of a cosmological base. Certain methods are successful *because* reality has a certain structure, because certain laws hold. It is possible that very *general* structural traits of reality are involved here. By the same token, certain properties of our cognitive faculties are presumably involved, and this leads back to the question of the conditions of the possibility of knowledge, and to a realistic interpretation of Kantian philosophy in the sense of critical realism.[29]

Universität Mannheim

NOTES

* I am very grateful to Claude Evans for his help with the translation.

[1] Cf. the position paper, p. 3ff above. This fusion is at work in Aristotle: cf. von Fritz, Kurt, 'Die ARXAI in der griechischen Mathematik', *Archiv für Begriffsgeschichte* I, 21ff. (1955), for a discussion of the Aristotelian definition of knowledge. The fusion is accentuated by Descartes, and is still at work in various strands of twentieth century philosophy, e.g. the work of Hugo Dingler.

[2] For a detailed discussion, see my *Traktat über kritische Vernunft*, third edition, Mohr, Tübingen, 1975.

[3] Following Popper, it has been above all William Warren Bartley who has done great service in demonstrating in detail the possibility of separating the idea of rational criticism from the idea of absolute justification. See Bartley, W. W., *The Retreat to Commitment*, Augustus Kelley, New York, 1962, p. 134ff.

[4] Cf. Lakatos, I., 'Infinite Regress and Foundations of Mathematics', *Suppl. Vol. Aristotelian Society* XXXVI, 155ff. (1962).

[5] Cf. Barth, E. M., *Evaluaties*, Vam Gorcum, Assen, 1972, pp. 5–18.

[6] Cf. e.g. Külpe, O., *Einleitung in die Philosophie*, 10th edition, S. Hirzel, Leipzig, 1921, p. 183ff. and passim; for a comparative analysis and criticism of other views cf. Külpe, O., *Die Realisierung. Ein Beitrag zur Grundlegung der Realwissenschaften*, 1st vol., S. Hirzel, Leipzig, 1912, 2nd vol., 1920, 3rd vol. 1923. Külpe elaborated a critical realism in the Kantian tradition, taking up the transcendental question in a realistic interpretation. In a similar manner, Popper began with a criticism and revision of Kantianism. cf. his book written in the early thirties. Popper, K., *Die beiden Grundprobleme der Erkenntnistheorie*, Mohr, Tübingen, 1976. Popper's first criticism of Kant and his interpretation of the transcendental viewpoint is to be found in this work. This reference may be of particular interest to those who believe that critical rationalism has no connection to this kind of problem, e.g. Wellmer and other members of the Frankfurt School.

[7] For a defence of the realism of common sense and of science, cf. Popper, K., *Objective Knowledge. An Evolutionary Approach*, Clarendon Press, Oxford, 1972, p. 32 and passim, cf. also Agassi, J., 'Sensationalism', in J. Agassi, *Science in Flux*, D. Reidel Publ. Co., Dordrecht, Holland, 1975, p. 92ff.

[8] Cf. Külpe, O., *Einleitung in die Philosophie*, 10th ed., S. Hirzel, Leipzig, 1921, p. 191f., where critical realism is presented as the dominating philosophical view from the Presocratics to the 18th century.

[9] During a walk during the Kronberg Conference, the proceedings of which are presumably in this volume, Wolfgang Stegmüller, whose work on the problem of testing scientific propositions is well known, tried to convince me that the idea of truth has its proper place in theology rather than the philosophy of science, and this in spite of his interest in problems of this kind prior to his conversion to a sneedified Kuhnian view, cf. note 13 below.

[10] Bühler, K., *Sprachtheorie. Die Darstellungsfunktion der Sprache*, 1934, 2nd ed., Gustav Fischer, Stuttgart, 1965; and even before: Bühler, K., *Die Krise der Psychologie*, 1927, 3rd ed., Gustav Fischer, Stuttgart, 1965, where Bühler analyses – among other things – problems pertaining to meaning and understanding. As far as I know, these important investigations have not been taken into account by German hermeneutic philosophy at all. Philosophers of this stripe tend to prefer the pretentious and foggy tales of Martin Heidegger, which now seems to be finding an eager audience even in the United States, in spite of the warnings of competent scholars such as Walter Kaufmann. cf. Kaufmann, W., 'Heidegger's Castle' and 'German Thought After World War II', in W. Kaufmann, *From Shakespeare to Existentialism*, Doubleday, Garden City, 1960, Anchor Book.

[11] Cf. Bühler, K., *Die Krise der Psychologie*, op. cit., p. 49: "Der *Begriff und die Kriterien der Wahrheit* oder Richtigkeit sind wesensgesetzlich aus der Darstellungsfunktion zu entnehmen, und umgekehrt bestimmt das Ideal der zutreffenden und richtigen Darstellung weitgehend die Produktion sprachlicher Gebilde bis in die Wortwahl und Struktur der Sätze hinein".

[12] Cf. above all Kraft, V., *Erkenntnislehre*, Springer, Wien, 1960, p. 181f., where the difference between these two questions is clearly laid out. This book also contains a clear analysis of other aspects of the problem of knowledge and truth.

[13] Tarski is correct when he claims that the concept of truth doesn't differ from some other concepts in logic, mathematics, theoretical physics and other disciplines in this respect. Cf. Tarski, A., 'The Semantic Conception of Truth and the Foundations of

Semantics', *Philosophy and Phenomenalogical Research* **4**, (1944). It is interesting to note that many modern critics of the idea of truth act just like operationalists when it comes to epistemological and semantical concepts, even if they have long since understood that this conception is no longer defensible with regard to science. This remark refers to certain views expressed during the Kronberg Conference. I do not know if they will be defended in the pages of this volume.

¹⁴ The views of Hugo Dingler might be characterized in this manner. cf. Dingler, H., *Die Ergreifung des Wirklichen*, 1955, Chapter I till IV, with an introduction of K. Lorenz and J. Mittelstraß, Suhrkamp, Frankfurt, 1969. Cf. also my criticism in the book mentioned in note 2 above. Karl Popper dealt with Dingler's views as early as 1935 in his *Logik der Forschung*, comparing them to his own strongly opposed views. Cf. Popper, *The Logic of Scientific Discovery*, Hutchinson, London, 1959, p. 78ff. Something like Dingler's 'method of exhaustion' has been resuscitated in Kuhn's 'normal science'. On the contemporary scene in Germany, the school of Paul Lorenzen explicitly goes back to Dingler's views, although some of its members have recently moved away from the priority which Dingler assigned to the idea of certainty. Cf. Janich, P., Kambartel, F., and Mittelstraß, J., *Wissenschaftstheorie als Wissenschaftskritik*, Aspekte Verlag, Frankfurt, 1974. For a reply to their criticisms of critical rationalism – which are maintained in spite of their recent movement – cf. my postscript to the book mentioned in note 2, p. 190ff.

¹⁵ Cf. Bertrand Russell's criticism of Dewey's conception of 'warranted assertability' in B. Russell, *An Inquiry into Meaning and Truth*, Allen and Unwin, London, 1940, p. 318ff. Cf. also Kraft, V., *Erkenntnislehre*, op. cit. p. 175f.

¹⁶ This doctrine has been elaborated by Jürgen Habermas and Karl Otto Apel, cf. Habermas, J., 'Erkenntnis und Interesse' 1965, in J. Habermas, *Technik und Wissenschaft als 'Ideologie'*, Suhrkamp, Frankfurt, 1968, and Apel, K. O., 'Szientistik, Hermeneutik, Ideologiekritik. Entwurf einer Wissenschaftslehre in erkenntnisanthropologischer Absicht', in K. O. Apel, *Transformation der Philosophie*, Vol. II, Suhrkamp, Frankfurt, 1973, p. 96ff. and other articles in this and the first volume of this work; for a criticism cf. Albert, H., *Transzendentale Träumereien*, Hoffmann und Campe, Hamburg, 1975.

¹⁷ The viability of the idea of approximation, which has long been effective in scientific thought, does not seem to be seriously threatened by the failure of certain attempts to provide formal definitions. The situation is similar to that of the idea of truth, the acceptability of which is not impaired by the fact that certain definitions may be controversial. By the same token, representatives of the empirical sciences are certainly entitled to search for laws in their field, although it seems that there is as yet no adequate explication of the concept of law. We might at this point recall the secondary character of epistemological investigations once again.

¹⁸ But see Tarski, op. cit., who prefers this answer. At any rate, the fact that one may have to put up with contradictions *for the time being* does not make them any less undesirable, but I cannot go into this question here.

¹⁹ Külpe correctly remarks that this fact is essential for the development from naive to critical realism; cf. his above metioned book: *Einleitung in die Philosophie*, op. cit., p. 192f.

²⁰ Cf. Elkana, Y., 'Boltzmann's Scientific Research Programme and its Alternatives', in Y. Elkana (ed.), *The Interaction between Science and Philosophy*, Humanities Press,

Atlantic Highlands, 1974, p. 243ff., where he points at the role of different 'images of science'.

[21] Cf. Popper, K., *Objective Knowledge*, op. cit., p. 191ff.

[22] I am using the word 'technology' (*Technologie*) in the German sense of a discipline dealing with the means for achieving a certain end, thus demonstrating various possibilities of action.

[23] In the tradition of critical realism, the reinterpretation of Kant's transcendental philosophy mentioned above hints at this direction. Cf. Külpe, O., 'Festrede zur Kantfeier der Würzburger Universität 1904', in Kopper/Malter (eds.), *Immanuel Kant zu ehren*, Suhrkamp, Frankfurt, 1974, p. 185f.: "Es verhält sich demnach die Erkenntnistheorie zu dem wirklichen Forschen und Arbeiten in den Wissenschaften ähnlich wie die Theorie des Mikroskops zur Anwendung desselben. Unser Erkenntnisvermögen gleicht einem Instrument, desser Leistungsfähigkeit und Tragweite, dessen Grenzen und Fehler main einigermaßen muß beurteilen können, wenn man sich nicht der Gefahr einer Täuschung aussetzen will. ... Wenn wir die Theorie eines Instruments entwickeln wollen, so geschieht das an Hand seines Baus und seiner Leistungen. Seine Elemente und deren Zusammensetzung werden betrachtet und auf ihre Gesetze zurückgeführt. Genau so verfährt Kant bei der Ausbildung seiner Erkenntnistheorie ...". There is, of course, no question but that Kant's own transcendental approach is still a part of the tradition of classical rationalism in that he attempted to provide a solution of the problem of justification in the classical sense of finding a foundation. But if one dispenses with the claim to *justification*, then there is the possibility of viewing the approach as an attempt to *explain* scientific knowledge by means of a *hypothetical recourse* to the structure of our cognitive faculty.

[24] The critical remarks of Paul Feyerabend usually flow into comments about such questions. But also in the work of Max Weber we find analyses of this kind in the context of philosophy of science, cf. especially: Weber, M., *Gesammelte Aufsätze zur Wissenschaftslehre*, 2nd ed., Mohr, Tübingen, 1951, p. 566ff.

[25] This remark refers to objections during the discussions of the Kronberg conference.

[26] One of these results is the significance of socalled 'set' – and so also e.g. of possibly existing theoretical assumptions – for perception. Incidentally this result has its origin in the investigations of the Würzburg School of the psychology of thinking, one of the most influential psychological schools, whose founder was the repeatedly mentioned philosopher and psychologist Oswald Külpe. The views of this school in the last analysis go back to Kant's conception of the active character of cognition (knowledge).

[27] Cf. my article: 'Hermeneutik und Realwissenschaft', in H. Albert, *Plädoyer für kritischen Rationalismus*, 4th ed., Piper, München, 1975.

[28] This gives me the welcome opportunity to make a short remark about one aspect of Imre Lakatos' methodology. I agree with Alan Musgrave (cf. his 'Falsification and its Critics', in P Suppes *et al.* (eds.), *Logic, Methodology and Philosophy of Science IV*, North Holland Publ. Co., Amsterdam, 1973, p. 393ff.) that Lakatos' concessions to the Kuhnian critique go a bit too far. I see absolutely no reason to immunize the core of a research program against criticism, and I see even less reason for turning this immunization into a triviality by means of the 'non-statement view'. The problem of anomalies can be dealt with without recourse to such a strategy. Should someone like 'hard cores' for one reason or another, he can certainly *produce* them, but he will have great difficulty in persuading other people that it is unreasonable for them to use their power of imagination for the purpose of improving upon this arbitrary hard core.

[29] Cf. notes 6 and 23.

ERNAN McMULLIN

PHILOSOPHY OF SCIENCE AND ITS RATIONAL RECONSTRUCTIONS

When the philosopher of science stands back from his field in order to try to situate the rapid developments of the past two decades, a variety of ways suggest themselves as to how one might best characterize these. In a lengthy article "History of science and its rational reconstructions",[1] the late Imre Lakatos proposed a four-fold division of what he regarded as the major 'rival methodologies': inductivism, conventionalism, falsificationism, and his own 'methodology of scientific research programs' (MSRP). It will be the task of this essay to evaluate this way of viewing the contemporary scene, and to propose a rather different one.

Lakatos' aim was not merely to produce a classification, but to suggest a means of deciding between the alternatives. History of science, he argued, must be the testing-ground of an adequate methodology: if a theory of science does not 'fit' (i.e. is incapable, in his terms, of 'rationally reconstructing') the actual historical practice of science, then it has to be rejected, or modified at least. The tendency on the part of the philosopher of science will be to attribute to 'external history' (i.e. to socio-psychological causes) whatever does not fit his pattern; thus the line he draws between 'internal' and 'external' history in the course of his reconstruction is likely to reveal where his theory of science is unable to account for what happened in the history of science. 'Internal' history of science, in this view, is the reconstructed version presented by a philosopher of science in support of his theory of science. 'External' history, which is left to the historian proper, is the residue which does not lend itself to exhaustive reconstruction in terms of a theory of rationality, presumably because particular psychological or social (or other historical) factors have intervened. The best methodology will thus be the one which goes furthest in reducing history of science to 'internal' history.[2]

A third feature of Lakatos' presentation is his suggestion that an adequate methodology ought also function reflexively at the meta-level to decide between methodologies themselves. Validation in

221

G. Radnitzky and G. Andersson (eds.), Progress and Rationality in Science, 221–252.
All Rights Reserved.
Copyright © 1978 by D. Reidel Publishing Company, Dordrecht, Holland.

epistemology ought, as far as possible, follow the same form (exhibit the same sort of rationality) as validation in science. The main difference between the two cases would be that at the first level, where one is trying to decide between rival scientific theories, recourse will be had to systematic observation as a means of discrimination, whereas at the meta-level where rival methodologies are in contention, one will test against the 'basic normative judgements' of scientists, discovered by means of carefully drawn historical case-studies.

This gives Lakatos two criteria for evaluating competing 'methodologies'. One is to test them against the history of science; the other is to determine how far they are applicable at the meta-level as well as at the first level. Not too surprisingly, his claim is that the MSRP comes out best on both counts: it is closest to the actualities of successful scientific practice (shifts fewest incidents over to 'external' history), and is readily applicable (unlike its rivals) to the meta-level of decision between methodologies themselves. The MSRP is thus not only a methodology of *scientific* research programs but also provides "a methodology of scientific research programs of second order, or if you wish, a methodology of historiographical research programs".[3] It must thus be construed as a 'research program' in its own right, striving to overcome 'rivals', making 'predictions' as to what to expect in the history of science, developing successive theories of science in 'progressive problem-shifts'. Lakatos ends his article with a denunciation of 'a prioristic' methodologies (Bacon, Carnap and Popper are the three names mentioned) which attempt to legislate the rationality of science on the basis of logic alone ('statute law') without taking the history of scientific practice into account ('case law'). It is *hubris* for the philosopher to attempt to dictate the theory of science from the logical standpoint alone. Yet scientific practice can sometimes become "corrupted', can 'degenerate', and in these cases (it is not altogether clear how they are to be identified), the statute law of the logician can issue helpful correctives.

As a first approach to this complex and ambitious reconstruction of the philosophy of science, let us take a look at the term 'methodology' itself. Like so many other '-ology' terms (think of 'technology', for instance), it is highly ambiguous, and this ambiguity can become quite troublesome if one shifts levels, as Lakatos does.

1. METHODOLOGY

Among the many senses of that vague term, let us single out two that are particularly important to hold separate if one wishes to speak of 'rival methodologies'.[4] 'Methodology' can mean an account of the methods to be followed in a particular field, its procedures of discovery, its characteristic modes of appraisal and so forth. In order to characterize these elements of science, it will be necessary to clarify and stipulate senses for the loose methodological terms ('law', 'theory', 'explanation' ...) used by scientists to describe their work. The resulting method-schema is the sort of thing one might find in the introduction to an undergraduate science text, or in the concluding chapters of an introductory work of logic. In its intention, such a methodology is both descriptive and prescriptive. It can be presented to the student as a guide to successful practice. But it is not a *theory* of science, an '-ology' in the strict sense. The grounds for its judgements are not made explicit, its relationship to the more general problems of knowledge is not developed. It purposes are descriptive and pragmatic rather than theoretical or controversial.

Of course it need hardly be said that young scientists-to-be do not really learn their trade from such method-schemas. It is rather from wrestling with the paradigm problem-solutions that transmute theory into practice, with (it must be admitted) a dangerous neatness and rigor that may tend to conceal the messiness of 'real' science. Method-schemas are of use mainly to the non-scientist who wishes to know what goes on in science. It is an external sense of 'know' to be sure; methodology falls short of practitioner's knowledge and yet has the advantages of explicitness and reflective character that the practitioner's skill does not necessarily entail. It is not necessary to have discovered an empirical law to know what the status of such a law is, or how it differs from a theory.

A question about the "criteria of scientific progress", the topic of this book, would therefore be addressed in the first instance to methodology. What criteria does the scientist call upon in establishing that a new theory represents an advance over an older one? One can list a number of these: predictive power, coherence, explanatory potential, and so on. These would be called on to show that 'progress' has occurred in a given instance. In situations of theory-conflict or rapid theoretical change it could be important to have these criteria and some notion of their relative weighting clearly in mind. But suppose there is some uncertainty or even disagreement in their regard. Where does one turn then? The methodology of science is not self-certifying.

If the question be carried to this second level, we can speak of a 'metamethodology'. This is no longer a descriptive/prescriptive method-schema, but a theoretical account of the sort of warrant such a schema needs, of the quality of knowledge it provides and of how that knowledge fits in with other sorts of knowledge. Much of the loosely-knit area called 'philosophy of science' would come under this rubric; it comprises a wide array of problems, from the nature of explanation to the logic of confirmation. What marks it off from 'methodology' in the sense proposed above is a concern for second-order issues of justification and coherence rather than for simple specification of procedure. Two people might agree in their accounts of the methodology of natural science while disagreeing radically in their metamethodologies (i.e. in their views of how their methodologies should be warranted or what their implications are). And two people who agree broadly in their metamethodology might still come up with different accounts of how science operates (although this latter sort of disagreement would obviously be much easier to arbitrate than would the first). In methodology one provides a way to evaluate scientific theories; in metamethodology one discusses how methodologies themselves are to be tested and compared.

In practice, of course, the two enterprises cannot be sharply separated; they will usually be carried on together and will react upon one another. But there is a difference of thrust worth noting. Methodology can be a relatively empirical undertaking. Whereas metamethodology is like any other part of philosophy, and there is no ready referent (like scientific practice) to serve as falsifier of the claims made. Aristotle derived his methodology of science from a prior theory of knowledge rather than from a survey of practice; his metamethodology suggested that this was the proper method of procedure, even though there were clear discrepancies between the methodology so obtained and the methodology he himself practiced as a biologist. Descartes in the *Regulae* followed a similar metamethodology; in the *Discourse* one sees the beginning of a transition not only to a different methodology but also a different metamethodology. As science itself developed and became more confident of its methods, the older order was reversed: it came to be assumed that methodology should be based on a careful study of the practice and of the history of science. But the metamethodological implications of this reversal remain unclear, and the logical significance of the (admittedly successful) methods followed by the scientist has proved as difficult a problem for contemporary philosophers as it was for Hume.

The topic, "criteria of scientific progress", will evidently raise metamethodological issues also. One might ask what kind of warrant such criteria should have. What sort of role ought case-studies from the history of science play, for instance? A discussion of such second-level issues would not necessarily lead to methodological proposals bearing directly on scientific practice. The debates between Kuhn, Toulmin, Lakatos and other recent theorists of science have to do mainly with metamethodology; the disagreements between them in regard to methodology, i.e. as to how science is (or ought to be) carried on are not for the most part basic. Their fundamental second-order disagreements would of course lead to differences in emphasis in regard to the criteria of theory-appraisal also. Even the wide gap between logical positivism and its critics had little to do directly with methodology; though the critics rejected the positivist account of scientific method on the grounds of oversimplification, the burden of their attack was on the logicist metamethodology which made such oversimplification inevitable.

2. 'RIVAL METHODOLOGIES'?

In what sense can one speak of 'rival methodologies' of science? In the natural sciences, at least, there has been substantive methodological agreement for the past several centuries, though there are still controversies about specific regulative principles, for instance, or about the status of models in microphysics, or about the weight to be given to falsifying instances. These latter differences occur between scientists as well as between philosophers of science. But they do not amount to anything like 'rival methodologies', i.e. incompatible accounts of how natural science is (ought to be) carried on. The situation is otherwise, of course, in the social sciences, notably in psychology, where radical disagreement in regard to methodology *does* exist, a disagreement which seems incapable, for the moment at least, of being settled by recourse to long-term successful scientific practice or to a metamethodological arbitration. If there are rival methodologies anywhere in the field of science today, it is there that they should be sought.

The major debates in contemporary philosophy of science have to do with metamethodology rather than methodology, as we have defined those terms. The Kuhn-Toulmin-Lakatos-Feyerabend group have much in common, for example, when considered as critics of logical positivism, but they differ among themselves on the nature of the rationality exhibited in scientific change. There are so many cross-currents, is it possible to catalog the major rival metamethodologies? What would be the most appropriate basis of division to use?

Lakatos, as we have seen, proposed three 'rivals' to his own MSRP: inductivism, conventionalism and falsificationism. Now while there are undoubtedly inductivist, conventionalist, and falsificationist themes in the logical positivist accounts of science that are his main target, these are surely not three separate theories of science, especially if they be defined as he defines them. Inductivism is said to maintain that "only those propositions can be accepted into the body of science which either describe hard facts or are infallible inductive generalizations from them"[5] (did *anyone* in the recent history of philosophy defend this view?). Conventionalism is characterized as holding that "major discoveries (in science) are primarily inventions of new and simpler pigeonhole systems"[6] (not many takers here either). While falsificationism is of two sorts, a 'dogmatic proto-version', which assumes the definite disprovability of theory by a single falsifying fact (only for those who like their methodologies clean-cut, and haven't heard of Duhem), and a more sophisticated version, attributed to Popper, which reduces to a 'revolutionary' form of conventionalism.[7] It is obvious that these are *not* the main options open to philosophers of science today. Scarcely anyone would qualify as inductivist, conventionalist, or falsificationist, as these labels are defined by Lakatos. The scenario he sketches is in no way helpful as a rational reconstruction of contemporary philosophy of science.

What he has evidently done is to create three 'pure positions' for his own dialectical purposes, so that the MSRP can appear as a synthesis of what is best in each. Furthermore, the competition between the four 'rivals' can then be construed by him as analogous to the competition between four first-level scientific theories. And this in turn leads to the tempting suggestion that an adequate methodology ought to discriminate not only between rival scientific theories but also between rival methodologies themselves. That is, a good methodology ought to be self-referential in the sense that it would provide the criteria for deciding its own adequacy in the face of its rivals. Or in our terms, a good methodology ought to be capable of being transposed so as to serve as a metamethodology also. The MSRP, Lakatos argues, not only describes how rival scientific research programs can be judged, but also can itself be regarded as a developing (philosophical) research program in competition with the rival metamethodologies.[8] It surpasses these, not only in the consistency of the reconstruction it provides of historical instances of successful scientific practice, but also in its ability to overcome challenges by modifying old features or incorporating new ones in ways that later prove fertile in all sorts of unexpected ways.

The MSRP is thus both methodology and metamethodology, one of the strongest arguments (Lakatos believes) in its favor. Falsificationism (the rival he takes most seriously) can be refuted by means of this argument, because there are clear instances in the history of science where falsifying anomalies were (properly) disregarded. Thus falsificationism would be falsified if it were self-applied as a methodology.[9] He does not try the same move with inductivism and conventionalism, where it would not work so well. Instead, he assumes that inductivism has been straightforwardly refuted by Popper and then argues that conventionalism can be attacked in a non-circular way by a 'new theory' of how to appraise (partially conventionalist) metamethodologies of science by "criticizing the rational reconstructions (of the history of science) to which they lead."[10] The more of the 'basic value judgements' scientists have displayed in the accepted science of the past the metamethodology is able to 'forecast' correctly, the better an account it is. The test is to see how much of 'actual great science' comes out as rational if one adopts the proposed metamethodology.

One implication of this criterion is that metamethodological differences are assumed to lead to differences in methodology also. It is by means of these latter (i.e., by testing them against the history of science) that the metamethodologies themselves are to be evaluated. But as we have seen, this is not a reliable assumption. It is equivalent to holding that each metamethodology will deliver different advice to scientists as to how to proceed in cases of theoretical disagreement. An inductivist and an MSRP proponent will not, however, differ much in regard to the advice they offer; the burden of the differences between them lies not so much in the methodologies they propose (between which the differences are, for the most part, differences of emphasis) but rather in the reasons they put forward to explain why the methodology of science works as it does.

But now let us for the moment adopt a different tack. Instead of assuming (as Lakatos does for his own purposes) that there are a certain number of well-defined metamethodological 'research-pro-grams' in philosophy of science, each with its own adherents, its own history of success and failure in accounting for the history of science on the basis of its *own* view of how science ought to proceed, let us ask what are the basic alternatives open to someone who seeks to construct a metamethodological 'justification' of how scientists pro-ceed (i.e., to explain how their methods achieve the results they do). In particular, since theory-appraisal is the most important and most complex locus of scientific method, we must discover what the most basic different sorts of warrant would be for a specific mode of theory-appraisal. Put most simply, when a scientist makes use of a particular criterion in evaluating a proposed theory (i.e., in his methodology), how should this criterion in turn be justified by the philosopher (in his metamethodology)?

Only when the answers to this question are systematically laid out can one decide whether they correspond to different 'schools' in the philosophy of science. It may well be that *real* philosophers of science for the most part avoid the 'pure positions' and combine the alternative types of second order criterion-warrant in flexible and perhaps tentative ways. It is notable that only the MSRP, in Lakatos' account, comes out as pluralist in this way, admitting a variety of different sorts of criterion-warrant. Perhaps it is because its purported 'rivals' are all pure positions that they are such easy game, and that the MSRP appears as a new 'research program' supplanting them.

3. WARRANTING THE CRITERIA OF THEORY-APPRAISAL

There appear to be at least three major types of (metamethodological) warrant for the (methodological) criteria of theory-appraisal.

(1) *Logical*:   The criteria may have the status of logical rules. This has always been the 'dream of theorists of science; if scientific theories could be validated by derivation from evidence or principle by means of the ordinary rules of logic, there would be nothing problematic, nothing opaque. Science would be a matter of applied logic. Of course, the status of logical rules themselves has given rise to a certain amount of controversy. Are they *a priori* principles carrying full conviction in their own right, once clearly understood? Or do they require some sort of pragmatic sanction? Why do we accept *modus ponens* as an infallible rule? Because of its own intelligibility or because it has always worked so far? It is a familiar debate, and there is no point in rehearsing it here. Suffice it to say that the first and simplest sort of criterion is a logical rule, whatever status we later decide to attribute to logical rules.

(2) *Historical*:   A criterion might be accepted on the grounds that it has proved its value in the past. Thus, one would look to the history of science to see whether, in what circumstances, and with what results, the criterion has been used. The assumption is the pragmatic one that if it has consistently worked in the past, in the estimation of the scientific community, then it has been validated by the practice of science. There are some problems here about what it is for a theory to 'work', and who the qualified community of referees should be in case of disagreement about the status of a particular criterion. One might try to avoid these by merely asking whether the criterion has *in fact* been widely used during the history of science, but then there will be the awkward question about procedures that scientists have used which have *not* proved successful. One must be careful not to make history so strong a warrant that one is implicitly assuming that scientists have never made mistakes.

(3) *Conventional*:   The criterion may rest simply upon an act of decision on the part of scientists. It may be an agreed 'convention' of procedure; no further warrant might be demanded of it than this. This third category is often treated as a sort of 'residue'; if a criterion does not have any specific theoretical warrant in its favor, it may be assumed that its presence in scientific practice is due to a practical decision on the part of scientists, in terms of convenience, for example.

There is a type of methodological criterion the warrant of which does not seem to fit easily into any one of the three categories above.

It may be called a *systemic* criterion; such criteria have played a very important part in the history of science. Among them are consistency, coherence, simplicity, elegance, the elimination of unexplained coincidence. These are formal or quasi-formal properties of theory. As criteria they are often assumed to be conventional in their basis. Lakatos indeed tends to define conventionalism in terms of the systemic criterion of simplicity; what marks off the conventionalist, he suggests, is his use of this criterion in comparing one theory with another.[11]

But this is unacceptable. The various systemic criteria may or may not be conventional in their warrant. One cannot determine this in advance. When Einstein, for instance, made use of the criterion of simplicity, he usually associated it with a specific metaphysics, a view of the universe which would make the simple theory more likely to reflect the (ultimately simple) real. Thus, the criterion was not for him a conventional one. If one persists in regarding any use of a systemic criterion as evidence of a conventionalist orientation, one is emptying the term 'conventionalism' of any clear meaning.

It is necessary, then, to take a closer look at these criteria and see how they relate to the three types of warrant defined above. If simplicity be equated with convenience, then it would reduce to a conventional warrant. If coherence or elimination of coincidence be valued because of their significance in guiding scientists of the past, then their warrant is a historical one. If consistency be sought, it can be construed as a logical requirement, though not (be it noted) as a rule of inference relating evidence and theory, after the fashion of the other criteria listed in the first category above. So that the systemic first-order criteria can fall under any one of the second-order types of warrant we have described, depending on the case.

But there do seem to be instances where these criteria are imputed a warrant which is not reducible to any one of the three above. Einstein's appeal to simplicity would be a case in point. The significance customarily given to the elimination of troublesome coincidences (like the unexplained equality of inertial and gravitational mass in Newtonian mechanics) would be another. Most scientists would argue that such coincidences are defects in a theory, even though the theory be perfectly adequate from the point of view of accurate prediction. They would tend to fall back if challenged on this, not on any one of the three types of warrant above, but on an appeal to the function of theory as explanation, or something of the sort. This is not a simple matter of logic, nor ought it be regarded merely as a convention. There is an epistemological principle involved, one which is integral to the aim of science as a human activity.

Our goal here is not to defend or further analyze this and similar principles, but only to note their prevalence in scientific practice. They are not always associated with systemic properties of theory, although it is in that context that they are most easily documented. The 'ideals of natural order' of which Toulmin speaks, or the 'regulative principles' that a variety of writers invoke, the criterion of interest developed by neo-Marxist theorists of science would be further instances of the same kind. The warrant for these would be broadly philosophical in character as a rule (although on occasion, recourse might be had to historical case-studies in support of such regulative principles as invariance or conservation, or the like).

(4) *Metaphysical*: This suggests that we should add a fourth category of second-order warrant, which may be loosely called 'metaphysical'. It would comprise both epistemological and cosmological types of consideration. The warrant here would be whatever warrant is appropriate to the general philosophical system from which the first-order criterion is drawn; this would obviously vary from system to system. But it is sufficiently distinct from the formal, historical, and conventional considerations already discussed to make it clearly constitute a separate category. In some special contexts, the distinction might become blurred: if the philosophy in question were a pragmatic one, for instance, recourse to the history of science in support of methodological claims made might be a quite natural move. But in general, the distinction will be clear-cut; the warrant one would invoke in support of a metaphysical claim would not be reducible to formal, historical or conventional terms.

It might be argued that one could again reduce the categories to three by conflating the first and the last, the logical and metaphysical, under some such general heading as '*a priori*'. But this is undesirable, because the warrant possessed by a logical rule like *modus ponens* is of significantly different kind to that claimed for a metaphysical principle requiring an entire metaphysics in its support and open to challenge from alternative metaphysical points of view. There would be more to be said in favor of such an identification if the logic in question were inductive rather than deductive. Inductive logic commonly calls upon broadly metaphysical principles (like the principle of limited variety). Nevertheless, it seems preferable to retain the distinction, and make allowances for special cases when they arise.

In the next few sections, we shall discuss these types of warrant in more detail, asking in particular in each case how the distinction of levels between methodology and metamethodology operates in its regard, and whether a metamethodology could be based on that type of warrant alone.

4. LOGICAL MODES OF THEORY-APPRAISAL

We have talked in a general way about elements of methodology that would have the status of (and thus the warrant appropriate to) a rule of inference. It is time now to survey the logical strategies available to scientists for the purpose of theory-appraisal. The major ones appear to be four in number. First there is the *intuitive-deductive* (ID) model, where one moves deductively from a set of axioms, themselves intuitively warranted, to a set of theorems. This was the Aristotelian ideal, and it was still defended by Descartes as an ideal for some parts of science, at least. Second is the *inductive* strategy, where one goes from a set of observation-statements to a generalization of the same kind, without introducing any concepts that were not present in the original statements. This is what has often been called 'Baconian method', though Bacon did in fact stress other modes of inference also. It will be noted that there are no precise formal rules in this case, and that the generalization can never be better than plausible. The third strategy is the *retroductive* (or hypothetico-deductive) one. Here an hypothesis is held to be warranted on the strength of the correctness of the deductions drawn from it. This strategy, like the inductive one, can never suffice to eliminate all doubt in regard to the assertion made. Finally, there is the *falsificationist* (or negative hypothetico-deductive) strategy which consists in the use of *modus tollens* to eliminate hypotheses. This, like the first strategy, operates deductively to permit the (negative) assertion to be made with certainty. But unlike the other three, it does not provide a way of making positive assertions about the world, of the kind science is supposed to contain. The first three strategies are verificationist in intention; the last is falsificationist. Only the first strategy could lead to science in the original sense of necessary knowledge. But this strategy depends on the ability to formulate intuitively evident first principles, an ability (it would now seem) we do not possess.

The verificationist strategies can all be regarded as modes of *warrant-transfer*. The use of a logical rule presupposes that one already has a statement or set of statements whose warrant can be assumed. The scientist's problem is to 'transfer' this warrant, so to speak, to the general statement (theory or law) he is trying to establish. Deductive modes of transfer are of only limited use, from a theory already to some degree validated to a theoretical implication whose warrant will then be its derivability from the theory (not its independent testability on observational grounds). Indirect modes of transfer are much the more important, especially in theory-appraisal. One does not derive the theory from a more general theory. Rather, the inference works backwards; because certain implications from a hypothesis are verified, one has some reason to accept the hypothesis as a 'good' one. Retroduction normally provides the principal means of appraising a scientific theory. The theory which 'accounts for' the widest range of the observational evidence is ordinarily given preference.

But 'accounting for' is not as simple a notion, methodologically, as it might appear. As a first approximation, it can be taken as equivalent to implying. The more of the 'evidence' the theory implies, the more adequate it is taken to be. But retroduction is far more complex than this HD scheme alone would suggest. What will carry special weight is the diversity of the verified implications, their novelty, their being part of the original *explanandum*, the degree to which unification of adjoining theories is made possible (systemic coherence), the degree to which anomalies can be met by means of small and independently suggestive modifications of the theory, and so on. These features of retroductive inference are not conveyed by the simple HD logic of correct prediction: "a theory is corroborated to the extent that the predictions it makes are verified".

Logicism is the metamethodology which would maintain that the *methods* of theory-appraisal have the status of logical rules, and that methodology reduces therefore to applied logic. It is an alluring doctrine, and has had a profound influence on the history of logic as well as of philosophy of science. The syllogistic of Aristotle was the first step along a road that led in our own day to Carnap's 'inductive' logic, a logic of comparative theory-appraisal, not a logic of induction

in the narrower sense in which that has been defined here. Logicism
has nearly always taken the form of an emphasis on one of the four
methods above to the virtual exclusion of the others, not necessarily
in the sense of denying their validity but rather of viewing appraisal
predominantly through the lens of one type of inference-schema only.

   Thus logicism can take two main forms, *verificationism* and
falsificationism. The former in turn has three principal variants,
depending on whether ID, inductive or HD methods are stressed
(Aristotle, Mill, and Carnap, roughly). Sometimes a broadly
verificationist methodology will also stress the importance of falsifying instances, but it will, of course, be in the service of verification in
the long run of one of the competing theories. Strict (or what Lakatos
calls dogmatic) falsificationism is a logicist metamethodology. Its
supposed virtues are its conclusive character, and its reliance on
modus tollens alone, without need for any of the supposedly dubious
strategies of direct verification. Popper's modified form of
falsificationism (which admitted a conventional element in the designation of the basic observational statements serving as potential
falsifiers) is not logicist. A logicist metamethodology cannot allow any
conventional elements at the methodological level. Besides, Popper
was forced to introduce notions of verisimilitude and corroboration as
his concessions to the overwhelming verificationist thrust of working
science, and these notions (as he insisted) are not reducible to formal
logical terms.

5. THE POVERTY OF LOGICISM

The logicist family of metamethodologies has been a major target for
historical philosophers of science and others since the late 'fifties. It
is scarcely necessary at this point to categorize its inadequacies. Few
would now defend a full-fledged logicism, whether of the
verificationist or the falsificationist type. Efforts are still being made
to construct a 'logic of confirmation', a reconstruction in formal terms
of the intuitive notion of confirmation used by scientists. But even if
these efforts were to be successful (so far they have not been), the
resulting 'logic' would still not carry the weight of a logicist
metamethodology. The logician is explicating a concept whose major
warrant would still come from its employment on the part of scientists. The explication would not have the force of a logical law or rule.

It would appeal to various plausible principles, some of which would probably require metaphysical warrant were they to be treated as the actual grounds for theory-confirmation.

The point of such an explication is not to show that confirmation possesses an implicit structure in which every axiom and every rule is self-warranting (after the fashion of *modus ponens*). Rather it is to take it as given that confirmation works in a certain way, and then provide a hypothetical reconstruction by means of a set of epistemic principles and logical rules that would explain *why* it worked that way. It is not the reconstruction which would ultimately validate the scientist's mode of confirming theories, but rather the mode of confirming theories which would lend credence to the elements of the model which successfully explicates it. This is not a logicist warrant; in fact, it is not a warrant at all, strictly speaking. The question of how the manner of confirming theories is to be warranted would still be open, though the explication would help one to answer it. One would have to consider both historical and metaphysical types of warrant, as well as logical and possibly conventional ingredients also.

The difficulties inherent in a logicist theory of science can be summarized as follows.[12] A deductivist (ID) theory will not do, because there are no secure general principles from which the deduction might begin. A falsificationist account is inadequate because it does not represent scientific practice (theories are rarely abandoned at the first sign of anomaly; the whole process is *far* more complicated than this), and because there are no falsifying facts in the 'hard' sense a logicist falsificationist would require. Even more fundamentally, one has to ask *why* falsification should be of any interest other than as a game; if there is not *some* sense at least in which science is progressing towards truth, the whole business seems pointless.[13]

Popper's preoccupation with the Humean problem of induction led him to an exaggerated rejection of any suggestion of verificationist elements in proper scientific methodology. But unless there is *some* sense in which a theory can be 'confirmed' by progressive successes in accounting for new sorts of data, one's metamethodology serves to undercut rather than warrant the historical practice of science; Hume's critique has been allowed to triumph, and scepticism must be the result. If on the other hand one concedes some sort of 'verisimilitude' to theory on the basis of a 'corroboration' by evidence, and especially if one asserts a progressive growth in the verisimilitude of theory (all of which Popper does), then one's position is not purely falsificationist and one must take seriously the positive criteria of (partial) verification that scientists have always relied upon In short, though techniques of falsification are important elements in scientific methodology, they are only part of the story and in any event, they are hardly ever allowed to operate as a neat *modus tollens*.

The remaining two types of inference-rule are the inductive and HD ones. Inductive generalization, proceeding from particulars to generals of the same sort, is an indispensable technique of experimental science; it is the means by which empirical laws (and even usable empirical facts, which have to be stated in potentially generalizable form) are arrived at. Its most characteristic product is the smooth curve by which two well-defined physical parameters are correlated mathematically, while other causally relevant factors are held constant. Induction cannot generate theory, nor can it be used as a direct means of theory-appraisal since theories are not empirical generalizations. And there are two further reasons why induction of itself could not provide the basis for a logicist theory of science. There are elements of decision involved in the exclusion of possibly relevant physical variables (it is simply not possible to test out all possible concomitant variations that *might* affect the correlation one is attempting to establish), in the termination of the experiment after a certain (limited) number of "points" of the curve have been constructed, in the curve-fitting techniques utilized, and so on. In addition, one may be calling on some sort of regulative principle of simplicity, and, of course, there are the theories taken for granted in the definition of the measure-variables themselves. For all of these reasons, a purely logicist warrant would not suffice to account for what goes on in inductive generalization.

The most versatile of the four methods is the HD one. It does not assume secure axioms (as the ID method does); it bears on theory (unlike induction); it allows one to treat the issues of verification (confirmation) that are central to the appraisal of theory, which the method of falsification does not. But once again, there is no way to erect it into the 'single' method of theory-appraisal and at the same time represent it as purely formal in structure, possessing the sort of analytic warrant a rule of inference does. As a rule of inference, HD procedure is a simple fallacy, the fallacy of the consequent. What allows it to become a valid mode of theory-confirmation is not the mere fact that theory and evidence are deductively related. It is rather the way in which over the course of time the theory accounts for facts not forming part of the original *explanandum*, the way in which it suggests plausible theoretical extensions, the way in which it copes with the challenge of anomaly and so on. Retroduction takes account not only of the original HE premiss, but of the continuing career of H, faced with different sorts of E as it goes. This is how *ad hoc* theories can be eliminated. The HD method treated as straight rule of inference of itself has no means of distinguishing between valid and *ad hoc* theories. Time plays an essential role here. Nothing (or almost nothing) happens suddenly in science, as a logicist metamethodology would lead one to expect it should.

On what sort of warrant, then, *does* retroduction (if we can give this name to the entire complex process) rest? Several alternatives are open. One could simply choose to base it on its evident success in the past history of science, and hold off Hume and his legions as best one could. Or one could call upon a realist metaphysics, in which the fertility of the theory would be a sign of the realist implications of the model associated with the theory.[14] The argument here would run: if the model can be taken to afford a partial insight into the structure of the real, then one would expect the theory in which the model is defined and its implications explored, to show its effectiveness in certain ways over the course of time. And these are precisely the ways that retroduction is, in fact, found to operate.

A combination of these two types of warrant might seem ultimately best. It is the history of science that serves to bring to our notice the intricacies of retroduction as it is practiced; it is the realistic metaphysics of models that gives us confidence that our reading of history is correct. One cannot in the end separate history and metaphysics as types of warrant, unless one is willing to defend a view of metaphysics that would make it in principle entirely prior to practice and to history. In this paper, we have suggested a distinction between the sorts of warrant that a metaphysics and an appeal to historical practice might provide. But this distinction, whose value has already shown itself above, ought not be taken to commit us to a separation of the two in the construction of an adequate metamethodology for science.

6. HISTORY AS WARRANT

The foregoing is sufficient to suggest why historicism cannot serve as a self-sufficient metamethodology either. One could certainly construct a descriptive account of what scientists do by relying on carefully-analyzed case studies from the history of science, or for that matter by scrutinizing contemporary scientific practice. But this is not enough; it is an immediate invitation to a further (third-order) question: "Why *should* the history of method be taken as warrant for correct method?" After all, scientists can make mistakes, not only of fact or of theory, but also of procedure. One might simply respond that whatever procedures have 'worked in the long run' in the history of science define what it is that science is, and what sort of truth it can attain. But history alone is an unsatisfactory warrant somehow; one must ask: *why* did they work?

Historicist writers like Kuhn and Feyerabend are not, however, using history as warrant for methodology. Rather, they are using it either to reject the notion of a methodology entirely (Feyerabend) or to argue that revolutions in science do not come about through the application of an accepted methodology (Kuhn). Their point is a negative one, and for a negative metamethodology history must bear all-important testimony. Both make use of epistemological arguments concerning incommensurability; both note the theory-laden character of observation statements, and so on. But the weight of their case is supposed to be borne by historical case-studies.

Feyerabend is saying: despite the rationality that philosophers and scientists alike have tried to attribute to the methodology of theory-appraisal in science, a closer look at the history of science will show that appraisal is a chaotic and wilful affair where persuasion and propaganda often carry the day. Here would be a conventionalist methodology par excellence, though Feyerabend would prefer to think of it as a non-methodology since it does not lead to progress, assurance, and truth, as methodologies have been expected to do. And the metamethodology is a predominantly historical one. Kuhn, on the other hand, though he stresses the decisional elements by using metaphors of conversion and *gestalt*-switch, also stresses the role played by anomaly in ultimately forcing a theory-change. And, of course, he is perfectly prepared to admit that normal science has a well-defined methodology. But since he rejects the notion of theoretical progress in science and the realist metaphysics that underlies it, the thrust of his metamethodology appears more negative than positive; this is certainly the way he has been understood by his readers, many of whom were puzzled by the more positive-sounding tones of his 1970 postscript. Our concern here is not, however, with the claims he is making in regard to the methodological 'structure of scientific revolutions'. Rather, we are interested in the metamethodology that underlies his analysis. And this is altogether clear-cut: since science is a complex activity, its structures are best understood by the techniques of the historian and the sociologist. To discuss theory-appraisal, therefore, one must start from what history tells us about theory-changes.

Critics of this historicist trend in recent philosophy of science have, of course, been more worried by the associated conventionalist challenge to orthodox notions of methodology that they have by the historicist standpoint of itself. They see a threat to rationality in the easy acceptance of conventional elements by Kuhn, Polanyi, Toulmin; the response of these writers is to say that rationality ought to be defined by what is found in the history and practice of science rather than set out formally in advance and imposed upon history.

Lakatos is attempting a middle course here. His proposals for the methodology of science (the MSRP, taken at first level) are well on the side of traditional notions of rationality, yet he repudiates the logicist assumptions on which the classical account of rationality was always based. Though he rejects much else in Popper's philosophy, he believes that Popper's Humean critique of verificationist principles of method still stands. Falsification is also in his view a broken reed. It has to be augmented by all sorts of conventionalist strategems, so that there is a real question as to whether the resultant methodology is much better than a disguised scepticism. There is only one alternative open if rationality is to be saved, and that is to allow verificationist 'inductive' principles to operate at the *second* level even though their invalidity at the first level has been shown (he assumes) by Hume and Popper. His plea to Popper to allow a 'whiff of inductivism'[15] amounts to asking him to concede a warrant-role to history of science at the level of metamethodology. This allows him to reintroduce a basically verificationist first-level methodology (of the sort Popper had rejected) but now on a historical rather than a logicist warrant. He hopes in this way to concede as much to Hume as Popper demands, while not allowing the rationality of science to be undermined, as he believes Popper is in danger of doing.

This is an uneasy compromise. It is one thing to reject a logicist principle of verification which purports to be independent of the *praxis* of science. But if one rejects *any* sort of verificationist principles of methodology, any sort of 'inductive principle' whereby experience could count *positively* for a scientific claim, then there is no way to resurrect such a principle at the second level to make experience (historical experience) count for a methodological claim. If one is prepared to countenance the latter, one has to reconsider the entire set of assumptions within which the anti-inductivism of the first level took on whatever plausibility it has to begin with. Lakatos is correct in perceiving that if history cannot count as warrant in metamethodology, the only alternatives are absolute *a priori* ones of a logical or metaphysical sort or else one or other form of conventionalism. His sense of what science has been prevents him from accepting either alternative. But his Popperian residue makes him unduly apologetic about this, and leads him to look over his shoulder at Hume when the issue rather is how the warrants of history, logic and metaphysics are to be combined to do justice to the reality of what we know science to be.

The philosopher must take account of the historical dimension of science in two separate ways. He must, as we have seen, allow history to serve as partial warrant at least, for methodological claims. But in addition, he must recognize that the entity to which his methodological criteria will be applied is itself a historical one; it is theory taken over its entire career and not just at one frozen moment. The impact on the 'logicality' of science of these two considerations is not the same. The first suggests that the warrant of a theory of science cannot be a wholly formal one, as the classical logicist views supposed. The second implies that whatever the methodological criteria chosen and whatever the warrant thought appropriate for them, the unit for appraisal is itself a historical one. To evaluate a scientific theory, one will need to exercise at least something of the skills of the historian; one will have to be aware of its successes and failures over a period of time. The scientist evaluates retrospectively, and thus he is calling upon the testimony of history, though of such recent (and to him familiar) history that it is easy to overlook the philosophical impact of what he is doing.

Is it necessary to have recourse to the *history* of science? Would it not suffice to look at its contemporary *praxis*? Is there something to be gained by tracing the *historical* record, as such? Several answers now suggest themselves to this often-asked question.[16] In establishing the role of history as (partial) warrant in metamethodology against logicist claims, it will be important to discover whether the criteria of scientific appraisal have themselves developed over the years, and if they have, in response to what sorts of critique, what sorts of failure, what sorts of new initiative. Secondly, the appraisal of a theory must *necessarily* take the historical dimension into account. It will not be enough to scrutinize its role at the present moment in the *praxis* of science. The scientist will, explicitly or implicitly, take into account its past record and the past record of its competitors (this may require quite a bit of historical information) and make an estimate of its resources against the future. Thus history enters into the warp and woof of appraisal. There is no way to isolate a single body of propositions on which purely logical techniques can operate to produce an appraisal which would satisfy the demands made on *real* appraisal in the practice of science.

7. CONVENTION

Less has been said about convention than about the other types of metamethodological warrant. Strictly speaking, convention is *not* a warrant; rather, it suggests an absence of warrant, a practical decision to proceed where cogent reasons are not available. In what sense could conventionalism possibly constitute a metamethodology, then, since such a view would seem to imply rather that no sufficient warrant can be found? And this would surely not be adequate as a basis for a theory of science, other than a purely sceptical one, perhaps.

The ambiguities latent in the notion of convention have already been noted. A quotation from Popper will serve to illustrate these:

> The source of the conventionalist philosophy would seem to be wonder at the austerely beautiful *simplicity of the world*, as revealed in the laws of physics. Conventionalists seem to feel that this simplicity would be incomprehensible, and indeed miraculous, if we were bound to believe with the realists that the laws of nature reveal to us an inner, a structural simplicity of our world beneath its outer appearance of lavish variety ... (Consequently) the conventionalist treats this simplicity as our own creation ... (Nature

is not simple), only the *'laws of nature'* are simple; and these, the conventionalist holds, are our own creations, our inventions, our arbitrary decisions and conventions.[17]

The two facets of classical conventionalism, decision and simplicity, are clearly delineated in this influential passage. The conventionalists did not regard simplicity as a warrant; rather, it was a puzzling feature of science to be explained by postulating that the laws of science, despite appearances, are really no more than conventions, resting thus on human decision rather than reflecting a feature of the world or a structure of the human understanding. Because they *are* conventions, it is possible for scientists to choose the simplest forms and even retain them in the face of apparent refutation (this was the feature that made conventionalism a principal opponent for Popper). Why *do* scientists choose the simplest forms?

This was a question that the conventionalists themselves were not especially concerned with; aesthetic or pragmatic considerations seemed to them a plausible answer. They were concerned to deny the orthodox view of law as being supported by inductive warrant and thus as resting directly upon experience. But their stress on simplicity was not intended as a proposal of an alternative and more appropriate warrant for theory-appraisal; rather, their thesis was that scientific laws did not *need* a warrant. A decision is enough; the fact that aesthetic or practical criteria usually govern the decision would not count as providing the warrant, only as making intelligible the simplicity that scientific laws were already known to possess. One could, if one wished, aim at more complicated forms of law; the conventionalist could not object to this, strictly speaking, but would merely remark that this is an unusual course because of the unaesthetic or inconvenient results to which it would lead.

The major emphasis in conventionalism was not, therefore, in the domain of methodology. Lakatos interprets it as a theory of warrant, and makes it prescribe either keeping "the center of the pigeonhole system (of science) intact as long as possible" or else simplifying the system as far as consistency will allow.[18] These prescriptions are not identical, and indeed would often conflict with one another. But apart from that, it is misleading to construe conventionalism as a methodology, as a way of doing science, as a set of directives to scientists in regard to procedure.[19] The problematic of the conventionalists was an entirely different one; the gradual transformation of conventionalism at Popper's hands into a rival theory of warrant leads Lakatos (as it did Popper himself) to formulate it in a way that runs a grave risk of incoherence.

In its origins, conventionalism was something like a metamethodology of scientific law. It assessed the significance of the procedures already followed by scientists in the formulation of law, asked what sort of warrant they enjoyed, and argued that instead of the logico-experiential warrant it appeared to be, it was, rather, an ultimately arbitrary affair. This was not a sceptical or a relativist claim in its intention; it was anti-realist, but the analytic statements to which it reduces laws were not wholly arbitrary. It was not clear, however, to what degree they permitted assured predictions to be made. This was equivalent to asking whether conventionalism was seen as a solution to the Humean problem of induction; Poincaré and various of his successors, like Le Roy and Dingler, disagreed on this, the last-named believing that it did provide a solution.[20] To be able to use a law for predictive purposes we must know that it applies to the particular context in which we are interested. But conventionalism, though it tells us the status of law, does not help us when it comes to *applying* the law to concrete instances. And indeed it warns us that this may be a tricky affair, since no matter what the outcome, stratagems may be employed to retain the original law unchanged, if one wishes (for whatever reason) to do so.

One critical comment is in order here. The plausibility of the conventionalist view depended in large measure upon the rather special scientific contexts to which it was applied, and especially upon the restriction of the discussion to law-like statements. Poincaré utilized it for the discussion of mechanics (specifically of the status of physical geometry) and of classificatory types of definition/generalization ("phosphorus has the properties P, Q, R ..."). But what of the structural types of theory proper to chemistry, geology, biology, where some form of realism of the associated models is relatively easy to defend? It would have been far more difficult to defend a straight conventionalist interpretation of the Bohr theory of the H-atom, say, where the considerations governing appraisal are evidently far more complex than those affecting either empirical generalizations or mechanical explanations.

A conventionalist metamethodology of science as a whole has never been tried, and it is not even clear what it would look like. The arguments in favor of construing law as a convention of language simply do not work for theoretical models. The tendency on the part of those who follow Popper in making conventionalism a major option in metamethodology is to shift the focus from *convention* (of language) to *decision*. Popper himself attributed "conventional" status to the basic observational statements around which his method of falsification hinged; testing them has to stop at some point, even though there is always a finite chance that they might have to be modified in some way:

Basic statements are not justifiable by our immediate experiences, but are, from the logical point of view, accepted by an act, a free decision ... a decision reached in accordance with a procedure governed by rules.[20]

This he regards as a 'conventional' element in his own system, not because the basic statements are conventions in the sense in which Poincaré believed scientific laws to be conventions, but because of the element of 'open' decision involved in their designation. It might be supposed, then, that he would attribute similar status to theories, since after all "the choice of any theory is an act, a practical matter".[21] But he does not, on the grounds that the choice of a theory is 'decisively influenced' by the success or otherwise of the theory in escaping falsification; no decision has been made which would render the theory temporarily immune from challenge (Lakatos would make some distinctions here). A 'conventionalist' move for him is thus not just one involving decision on non-cogent grounds ('acceptance without belief' in Lakatos' term)[22] but one that arbitrarily suspends the falsifiability criterion which constitutes the demarcation criterion separating science and non-science. He needs to allow this to happen somewhere in his system in order that the logic of falsifiability should begin to function; what prevents the basic statements from becoming non-science in consequence is that the *possibility* of a later test is held open, even though for the moment the statements are being treated as foundational.

It is not unexpected that what 'conventionalist' means for Popper should be conditioned by his falsificationist perspective. There are two positions from which he wants to distance himself: verificationism (because of its supposed inability to handle the problem of induction) and conventionalism (because of its implicit rejection of any clear-cut criterion of demarcation). Although he allows a 'conventional' element in his system, he limits it to the basic statements only, so that from *his* point of view, his system is not a conventionalist one.

For Lakatos, however, it is: Popper's theory of science reduces to a 'revolutionary' form of conventionalism in his view. But now the definition has shifted once again; the stress is on *decision* only, that is, decision which is not dictated by logical or experiential considerations. Thus, the acceptance of a theory involves a 'conventional' element, since it rests on decisions of all sorts. Lakatos discusses five ways in which decision is involved in a naive falsificationist account of science; three of these still recur in the 'sophisticated' account, where one is forced to make decisions, "at least occasionally and temporarily", about the truth-value of a theory.[23] The line which is drawn between 'background knowledge', not up for test, and the theory under appraisal, is necessarily a 'conventional' one. In Lakatos' own MSRP, the distinction between 'hard core' and 'protective belt' requires a decision, admittedly not one without guidelines, but nonetheless one where there is a considerable latitude, and disagreement between competent scientists is thus possible. This is what leads Lakatos to say that the MSRP in a sense 'rationalizes classical conventionalism'.[24] The rationality to which he alludes derives from the positive fertility criteria which diminish the conventionality involved in theory-appraisal.

If this shift in the notion of 'convention' be accepted, then one can characterize many of the themes in Feyerabend and in Kuhn as 'conventionalist'. What they stress is the importance of the part played by non-coerced decisions in scientific change. Feyerabend's rejection of orthodox methodology could be called 'conventionalist' in its inspiration. He goes far beyond Popper's mild (and very unrevolutionary) 'conventionalism' to extend the decisional elements into every aspect of methodology. Kuhn's stress on science as a community enterprise, Polanyi's thesis of personal knowledge, are so many ways of bringing out that decisional elements play a far larger role in scientific appraisal than either logicists or scientists themselves would ever have been inclined to admit.

Conventionalism in this broader sense belongs primarily to the level of methodology. The decisional elements pertain to the problematic of the scientist. He is faced with choices of all sorts, for only some of which he can produce compelling reasons. He must, for example, decide when in his analysis of causal relationships he has sufficiently excluded the possibility of other relevant causal factors; when an anomaly shows itself, he has to decide where to begin and end the questioning, and so on. The methodology of science must, therefore, include reference to conventional/decisional components, as it already does to deductive rules, regulative principles, and the like.

Conventionalism is the view that stresses the presence in scientific method of these components. It *stresses* them, but it does not exclude the presence of other non-conventional constituents. If one were to insist on a 'pure' conventionalism which would exclude all procedure-with-justification from science, it would make the label a useless one. But what sort of warrant does conventionalism itself rest on? What kind of justification would someone offer for saying that there are decisional elements in scientific method? It is most likely to be a combination of historical testimony and logical analysis, of the sort one finds in Kuhn's work, for instance.

This brings out an important point: that the *defence* of conventionalism cannot itself be a conventionalist affair. Poincaré gave all sorts of reasons to back up *his* conventionalist analysis of law; he would never have conceded that his analysis itself rested on conventionalist principles. He was proposing a conventionalism of science, not of philosophy. What confused this issue is that the proposal of a conventionalist science is itself called a 'conventionalist' philosophy. But the sense of 'conventionalist' in the two cases is not the same. Scientific methodólogy is 'conventionalist' because it contains elements of convention; a philosophy (or metamethodology) would be 'conventionalist' not because it contains elements of convention but because it defends the presence of such elements at the *lower* level.

This is why the question: "can conventionalism be a metamethodology?" is such a troublesome one. If by 'conventionalism' one means that philosophy which argues for the significant presence within scientific method of decisional elements, then it is clearly a metamethodology. It differs from the first-level descriptive 'conventionalism' by being an argument, a presentation of warrant, a justification of one theory of science over against another. If on the other hand 'conventionalism' is taken to mean (as it does at first level) a methodology containing within itself decisional elements to a significant degree, then it does not belong to the level of metamethodology. This is the level at which reasons are given and warrant is sought. To invoke 'decision' here would have an altogether different significance, and would not ordinarily be an adequate meta-consideration.

Convention enters into what scientists do; that is agreed. But what about philosophers? Are decisional elements present in the same sense in the procedures of philosophical analysis? Most philosophers would deny this, just as logicist theorists of science once denied it of science. Suppose, however, that a metamethodologist evaluates his account of method against the historical praxis of scientists, ready to modify his claims where significant discrepancies occur. Such a philosopher will be prepared to admit not only that scientific theories evolve, but also that methodology itself evolves, as particular methods, particular principles, are tried and proven effective, or else prove unhelpful and are modified or dropped. Would not this imply that his methodology, as it is formulated at any given time, would have to be taken to have a provisional character, to some extent at least? Is there not a decisional element involved, then, in adherence to a particular methodology, knowing that it is itself under test and may ultimately have to be modified? And does this not lead one to conventionalism in metamethodology (philosophy)?

To answer this, it might have been helpful had we in the first place drawn a three-fold distinction between method-schema, methodology and metamethodology, instead of the simpler two-fold distinction that has sufficed so far.[25] The decisional element is involved in *adherence* to a theory, in testing it out, and retaining it perhaps in the face of some failures. The philosopher who puts forward a methodology tentatively, with the proper qualifications, is making a decision at the level of methodology, but at the metalevel he is ready to provide reasons for saying that this interim decision (though provisional) was an entirely rational one.

It would not be worthwhile to carry this complicated point further. But enough perhaps has been said to make the point that the shift from the level of methodology to that of metamethodology which has been relatively straightforward to visualize for the other types of warrant proves equivocal for convention, in part no doubt because it is not properly a warrant in the first place.

8. CONCLUSION

Our survey has proceeded at two different levels. First, there were the different factors involved in theory-appraisal in science: logical, historical/practical, metaphysical, conventional. Singling out one of these could give rise to a 'pure' position which would attempt to reduce methodology to that factor alone. Second, there were the different types of warrant appropriate to the support of methodological claims, i.e. the factors involved in the appraisal, not of scientific theories, but of theories of method. These also could be classed as logical, historical/practical, metaphysical; the fourth category, 'conventional' becomes ambiguous at this level and is consequently unhelpful.

Philosophers of science generally work on both levels at once. They list and discuss the criteria that play a part in the appraisal of scientific theory. In addition, they concern themselves with the kind of warrant that would be appropriate to the defense of the proposed criteria. 'Pure' positions, at either level, are rarely defended; they exist mainly as useful pedagogical or dialectical devices. The only really plausible views are 'mixed' or pluralist ones, in which a variety of factors or of warrant is recognized, and the issue becomes the more complex one of how they should be harmonized.

Looked at in this light, the MSRP appears quite sensible as a pluralist methodology. It attempts to do justice to the variety of factors, logical, metaphysical, historical, conventional, which are intertwined in the sophisticated practice of science. It is not especially new; one can find similarly pluralist accounts all the way from Whewell to Campbell. What makes it *seem* new is the contrast Lakatos draws between it and the more-or-less 'pure' logicist and conventionalist methodologies cast as its rivals. The most important single point that he makes is that the unit of appraisal is the theory (or as he prefers, research program) taken over an adequate period of time and not just as a timeless interconnected set of propositions. This is important, not because it has not been said before (once again, back to the solid Whewell-Campbell line), but because the logicist counter-assumption still surfaces now and then.

A fairly broad agreement appears to have developed over the past decade among philosophers of science, however, that the logicist monolith has been shattered. Writers as diverse as Hempel, Stegmuller, Salmon, Grünbaum, Toulmin, Lakatos, are in agreement on that. The differences between these writers, in regard to *methodology* at least, would lie in the relative strength they would attribute to the different strands that go to make up, in their view, the strong rope of scientific rationality. These differences would not constitute sharply-different schools, well-defined 'rival methodologies'.

At the level of metamethodology, the matter is perhaps otherwise. When it comes to assigning the warrants appropriate to methodological claims, the divisions between the (broadly) logicist school and the (broadly) historicist school are still deep. One might also cite a (broadly) metaphysical school, comprising those who attribute primary significance to regulative principles of a phenomenological or a constitutive sort or who would reduce the warrants of logic or of history to those of metaphysics. But to trace these divisions would be a new, and equally arduous, task.

One moral still remains to be drawn, however, from our tortuous tale. It is that the project of testing the adequacy of a methodology by noting how it (or something vaguely like it) functions as a metamethodology is a misguided one. Not only does it lead to a blurring of the categories utilized, but it takes for granted an unargued assumption that for consistency a good methodology ought somehow 'transfer' to the second level. One might grant that in some very broad sense, if inductive elements are admitted at one level, one might expect that inductive elements would appear at other levels too. But Lakatos tries to argue a far more specific thesis than this: that the MSRP functions equally well as methodology and metamethodology. This is what leads him, as we recall, to his scenario of rival theories of science, with hard cores, protective belts, predictive tests, anomalies, and all the rest, just as though they were rival scientific theories ready for the falsificationist axe.

What gives this suggestion whatever plausibility it enjoys is that he is introducing historical factors at both levels (though not in the same way). This is in opposition to the logicist view which would exclude such factors at both levels. He is insisting that scientific theories have to be tested 'over the long run', and that theories of method are not to be imposed arbitrarily but have to be tested against the actual history and practice of science. But these points have only a superficial similarity; one might maintain one without the other even though they both might very loosely be summarized as 'taking the history of science seriously'. He is aware that a purely logicist *a priori* is likely to do violence to the realities of science; it always has. But he is also aware that history needs *some* sort of 'reconstruction' in terms of logical and metaphysical principles if it is to yield a methodology at all. His problem (and indeed that of most recent philosophy of science) is how to do justice to what he calls the resulting 'pluralistic system of authority'. Whatever of the 'historiographical meta-criterion' that he proposed, one thing is certain: it will not work if it is construed after the pattern of a criterion for *scientific* theory-appraisal. The sense in which theorists of science must take account of the history of science is, in the last analysis, quite different from the sense in which scientists have to take account of observations. Only a preoccupation with the Popperian problematic of induction would make them seem sufficiently similar to allow any sort of likelihood to the project of treating them as basically the same strategy.

In this essay, we have attempted only to lay the groundwork which would permit the needed taxonomies to be constructed. The purpose was not to propose a methodology, nor a specific metamethodology, of science. Rather it was the more modest one of providing some of the analyses needed by those who set out to construct a rational reconstruction of the somewhat bewildering and rapidly changing pattern of agreements and disagreements in contemporary philosophy of science.

University of Notre Dame

NOTES

[1] *Boston Studies in the Philosophy of Science*, Vol. VIII, R. C. Buck and R. S. Cohen 1971, pp. 91–134 (abbreviated below as HSRR).

[2] This way of making the internal/external cut is open to serious objection, especially from the side of the historian, but we shall not engage on this topic here.

[3] *Op. cit.*, p. 116.

[4] Lakatos himself in his 'Popper on Demarcation and Induction', in P. Schilpp (ed.), *The Philosophy of Karl Popper*, 1974, pp. 241–270, abbr. PDI; paper written in 1969) makes use of a distinction between 'methodology' and 'metamethodology', but does not elaborate on it.

[5] HSRR, p. 92.

[6] HSRR, p. 96.

[7] HSRR, p. 97.

[8] "Progress in the theory of rationality is thus marked by historical discoveries; by the

reconstruction of a growing bulk of value-impregnated history as rational. This idea may be seen as a self-application of my theory of scientific research programmes to a (nonscientific) research program concerning scientific appraisals", PDI, p. 251.

[9] See HSRR, p. 109 seq.; PDI, p. 246 seq.

[10] HSRR, p. 109.

[11] HSRR, p. 96.

[12] These are documented in the works of Kuhn, Polanyi, Toulmin, Feyerabend, and other critics of the logicist orthodoxy of yesteryear; McMullin, 'Logicality and Rationality', *Boston Studies in the Philosophy of Science*, Vol. XI, 1974, pp. 415–430, and 'Empiricism at sea?', *Boston Studies in the Philosophy of Science*, Vol. XIV, 1974, pp. 21–32.

[13] Lakatos has developed this critique of falsificationism very clearly in his PDI.

[14] McMullin, 'What Do Physical Models Tell us?', in B. van Rootselaar and J. F. Staal (eds.), *Logic, Methodology and Philosophy of Science*, North-Holland Publ. Co., Amsterdam, 1968, pp. 389–396. See also R. Harré, *Principles of Scientific Thinking*, University Press, Chicago, 1970; and McMullin, 'The Unit for Appraisal in Science', *Boston Studies in the Philosophy of Science*, Lakatos Memorial, Volume XXXIX, 1976, pp. 395–432.

[15] PDI, p. 256.

[16] See R. Giere, 'History and Philosophy of Science: Intimate Relationship or Marriage of Convenience', *British Journal for the Philosophy of Science*, 24, 282–297 (1973); McMullin, 'History and Philosophy of Science: a Marriage of Convenience?', *Boston Studies in the Philosophy of Science*, Vol. XXXII, 1975, pp. 515–531.

[17] *Logic of Scientific Discovery*, Hutchinson, London, 1959, p. 79.

[18] HSRR, p. 94.

[19] Lakatos inherited from Popper a way of speaking that makes it uncertain who the 'conventionalist' is: "The conventionalist decides to keep the center of such a pigeonhole system intact as long as possible" (HSRR, p. 94). Is the 'conventionalist' a scientist who is acting in the way described? (do scientists always act this way? Did the conventionalists think they should?) Or is it a philosopher reflecting on what the scientist does and *interpreting* it in this way (without presuming to suggest any alteration of procedure)? But then this mode of speech is misleading. Or is it a philosopher reflecting on his *own* approach in philosophy of science, i.e. construing conventionalism as a second-order methodology? Popper's way of describing conventionalism seems to make it hover on all these levels at once, and of course Lakatos explicitly makes it both a methodology and a metamethodology.

[20] LSD, p. 109.

[21] LSD, p. 109.

[22] PDI, p. 261.

[23] 'Falsification and the Methodology of Scientific Research Programmes', in I. Lakatos and A. Musgrave (eds.), *Criticism and the Growth of Knowledge*, University Press, Cambridge, 1970, pp. 91–195, abbr. FM; see p. 131.

[24] FM, p. 134.

[25] Method-schema: a description of procedures; methodology: a reflective theoretical account of a particular method-schema; metamethodology: a philosophic discussion of the warrant(s) appropriate to methodological argument, and an appraisal of the rival theories of science based on them. 'Method-schema' is used here in roughly the sense in

which 'methodology' has been used above; what has been called metamethodology above would correspond to both 'methodology' and 'metamethodology' here. A distinction between these two would be helpful to this present discussion. The notion of metamethodology used throughout this essay covers several rather different activities, but for the purposes of the essay, it is usually not necessary to distinguish them. It will be obvious to readers of G. Radnitzky's book, *Contemporary Schools of Metascience* (Akademieforflaget, Goteborg, 1970) that it corresponds to what is there called 'metascience', and that many of the issues raised in this essay find a place in the very similar perspective of Radnitzky's work. He too is attempting to provide the categories needed to make a taxonomy of the major options in contemporary philosophy of science. The main difference between his approach and ours is that he devotes a great deal more attention to the hermeneutic approach than we have done.

[26] HSSR, p. 121.

Written for the conference on 'Progress in Science and Methodology', sponsored by the von Thyssen Stiftung at Kronberg, July 1975, and revised and enlarged (October 1975) in the light of the helpful discussions at that conference.

NORETTA KOERTGE

TOWARDS A NEW THEORY OF SCIENTIFIC INQUIRY[1]

1. DESIDERATA FOR A THEORY OF SCIENTIFIC INQUIRY

An instructive model for the philosopher interested in scientific method is a novel entitled *Zen and the Art of Motorcycle Maintenance*. The narrator (who is a dropout from Mortimer Adler's program at the University of Chicago) stresses the importance of good external and internal working conditions: Loud music is distracting; you will need good light and room to lay out your tools; if you are feeling impatient, knock off for the day – or at the very least, don't attempt a sensitive job; try to remain flexible in your diagnosis of what is wrong; investigate various alternatives in a systematic way.

There are also general strategy suggestions: Check on the *most accessible* sources of trouble first. (In case of ignition failure, test the spark plugs before dismantling the carburetor.) Check on the *most probable* source of trouble early on. (If your carburetor has a history of causing trouble, you will move it up on the check list.) Etc.

But surely this is all just common sense, you are thinking. I agree – though one often finds even such simple methodological maxims breached in practice. It is just the sensible, reasonable nature of the manual which impresses me. It does not recommend conventional decisions to protect a certain part of the system from criticism. Motorcycles have no hard cores. Neither does the novelist try to justify his theory by appeals to an elite group of motorcycle mechanics. The theory stands or falls on its own merits – it does not rest on garage sociology.

Let us begin to construct a theory of scientific inquiry which is at least as sophisticated and right-minded as that found in *Zen and the Art of Motorcycle Maintenance*. Such a theory will draw on decision theory, epistemology, and the history of science. It will recognize that both logic and luck play essential roles in the growth of science. Its injunctions will be modest, but not trivial.

The methodological theory which I envisage will have three parts.

(i) First, it will contain a *description* of the typical sorts of problem situation which arise in scientific inquiry. Some of the more interesting

253

*G. Radnitzky and G. Andersson (eds.), Progress and Rationality in Science, 253–278.
All Rights Reserved.*

problems will come from the history of science – puzzles about white swans and balls in urns do not exhaust the problems which arise in real science.

(ii) For each typical problem situation, the methodological theory will present an analysis of the options open to the scientific community and an *evaluation* of each of the strategies which could be adopted.[2]

(iii) Most importantly, the methodological theory will give *epistemological arguments* for these evaluations. If we are to prescribe a certain strategy to scientists, we should be able to argue for that prescription – to explain why that response appears to be the optimal one. *Conventional rules or mere explications of our intuitions are not enough.*

The result will be a reasoned prescriptive manual for working scientists: "If you are in problem situation P, the best solution is to adopt strategy S (and here are the reasons why)."

Given this ideal for a theory of inquiry, several points regarding the two theories discussed in the LSE position paper emerge immediately:

Popper's theory of falsification is at the very least *incomplete* because it provides no detailed epistemological theory of basic statements. *Which* statements should scientists choose as relatively unproblematic? *Why* should they do so? I am not requiring that Popper provide a justificationist epistemology – a conjectural one will do. Allowing the 'basis' of science to be purely conventional opens the door to less satisfactory conventions, such as Lakatos' decision to declare the hard core unproblematic. (I should immediately point out that no one else has provided a very good theory of the empirical basis of science – and of course Popper cannot be expected to solve all problems at once.)

Although Popper's falsificationist account and Lakatos' confirmationist account[3] would appear to be flatly inconsistent, we must look carefully to see exactly wherein the disagreement lies. It could be that they are merely referring to two different problem situations, i.e., that Popper is saying (roughly), "In situation P_1, be a falsificationist" while Lakatos is saying, "In situation P_2, count only confirmations".

If this were the case their only disagreement might be about whether P_1 or P_2 is the more typical scientific problem situation. Although this interpretation has a certain plausibility (one recalls that in the *Poverty of Historicism* Popper says that although the inter-pretative viewpoints according to which histories are written cannot be falsified, they can nevertheless be "distinguished by their *fertility* – a point of some importance" (p. 151)), I think that the conflict is much deeper.[4]

It is also important to note that apart from very general maxims, such as "Foster criticism" or "Don't fake your laboratory results", methodological advice will always depend on the details of the scientific problem situation. It may sometimes be proper to protect the core of our scientific theory from refutation, but a good theory of methodology should specify exactly when this is good strategy.

Let us now begin to construct an adequate theory of scientific inquiry by dealing with two important methodological problems which are discussed in the position paper, viz., the Duhemian problem and the problem of adhocness.

2. THE DUHEMIAN PROBLEM SOLVED

The Duhemian problem arises when a prediction from a scientific system does not match experience. Making the usual distinction between theory (T) and auxiliary hypothesis (A), it can be put as follows: Suppose $T \cdot A$ imply e, but experience suggests $\sim e$. What should one do?

There are a variety of possible responses and many examples of each can be found in the history of science:

(i) One may challenge the derivation by showing that e does not in fact follow from $T \cdot A$.

(ii) One may show that the experiment which purports to show $\sim e$ is unreliable.

(iii) One may reject A.

(iv) One may reject T.

This list is neither exhaustive nor exclusive. But let us simplify the discussion by considering a special case in which there are good reasons for not adopting responses (i) and (ii) or any of the unlisted moves (such as ignoring the contradiction or reserving judgement). The problem can now be stated very simply: In the case of a prediction failure, when is the theory itself (as opposed to auxiliary hypotheses) refuted?

I will proceed by describing two Duhemian problems which

Mendeleev faced and his response to each of them. Next we will ask what advice Popper, Kuhn, and Lakotos would have given Mendeleev and try to determine how good that advice is. At the end, I will suggest a more adequate philosophical treatment of the Duhemian problem.

2.1. *Mendeleev's Problem Situations*

Mendeleev and his contemporaries believed that the Periodic Table was more than just a convenient classification system. Rather it was based on what Mendeleev called the Periodic Law and which he summarized as follows:

"... If all the elements are arranged in order of their atomic weights, a periodic repetition of properties is obtained." (*Principles of Chemistry*, Vol. II, p. 17.)

It is well known that Mendeleev's Periodic Law brought a good deal of order into existing chemical data and also had many dramatic predictive successes. Not only were the three 'missing elements' which were needed to fill the gaps in the Table quickly discovered, it also turned out that their atomic weights and the specific gravity, specific heat, molecular volume, boiling point, etc., of their compounds had almost exactly the values which Mendeleev had predicted. He also used the Periodic Table to suggest corrections in the current values for the atomic weights of titanium, osmium and platinum and the valencies ascribed to beryllium, uranium, and indium. And these theory-based assignments were independently confirmed.

But there were also two persistent areas of anomaly. First, there was the problem of reversed pairs. Because of its valence and other chemical and physical properties, it was clear that iodine should be located after tellurium and before xenon. Yet this order did not correspond to that given by the best available atomic weight data. There were three such reversals in the table.

Secondly, there was the problem of where to put the rare-earths. In Mendeleev's Table there was room for only *two* elements between barium (II) and tantalum (V). (Hafnium (IV) was not discovered until 1923.) However, by 1903 (the date of the 7th edition of his textbook) *eleven* elements, all with valence III and atomic weights intermediate between those of barium and tantalum, had been discovered.

Mendeleev's reactions to these two different sets of prediction failure were explicit and unambiguous. He considered the reversed pairs to be anomalies – when their atomic weights were correctly determined, they would fall smoothly into place. Thus in his textbook he not only puts the elements in the order required by their chemical properties but also lists *theoretical* values for their atomic weights to make them conform to his Law. (The best available experimental values, which do not agree, are relegated to a footnote.)[5]

His stance vis-à-vis the rare-earths is just the opposite. In the Introduction to his book he writes that "... this portion of the periodic system is, in a way broken ..." (*Principles*, Vol. I, p. xvii). Writing later Soddy agreed with Mendeleev's judgement, saying that the rare-earths constituted "... a point blank contradiction to the chemical principle of the Periodic Law ..." (*Isotopes*, p. 2).

I want to stress that in both the reversed-pair case and the rare-earths case, the predictions made using the Periodic Law did not match experiment. Yet in one case, that of the reversed pairs, Mendeleev brazens it out – the Periodic Law is all right; something somewhere else is wrong. But in the rare-earths case Mendeleev simply admits that his Law is refuted.

The historian of science must ask, "Why *did* Mendeleev respond differently in these two seemingly similar cases?" The philosopher of science must ask, "*Ought* Mendeleev to have responded differently to these two cases of prediction failure?" For example, we should determine whether the reversed pair experiments also refute the Periodic Law.

We immediately notice that deductive logic alone does not force Mendeleev to consider his theory refuted in either case. Rather we have a typical Duhemian situation in which auxiliary assumptions are involved in the derivation of experimental predictions from the theory. Let us look at just *some* of the auxiliary hypotheses which Mendeleev had to use.

In the tellurium-iodine experiment, the reproducible experimental data concerning iodine are based on vapor density measurements of a reddish-brown gas prepared in a specified way. So one may accept as true the experimental report that such a gas has a vapor density corresponding to an atomic weight of 126.97 ± 0.01 but deny the auxiliary assumption that the reddish-brown gas is *pure* iodine. And as a matter of fact, Mendeleev conjectures that since the vapor was dried over anhydrous calcium chloride, maybe some of the iodine was replaced by lighter chlorine (*Principles*, Vol. I, p. xviii).

Thus one can save the Periodic Law from refutation by dropping
A_1: The red-brown vapor is pure iodine,
and replacing it by
A_1': The red-brown vapor is iodine plus chlorine.
Mendeleev follows this strategy in dealing with all the reversed pair discrepancies. He challenges very minor auxiliary assumptions about the reliability of the experimental methods used and puts forward specific suggestions for improving them. These conjectures are not only *testable* (one can find out whether iodine replaces chlorine in $CaCl_2$), but also in many cases quite *plausible* (vapor density methods *are* less precise than gravimetric techniques).[6]

The logic of the situation is this. Let $\sim e$ be the lab report and T the Periodic Law. Now $T \cdot A$ are inconsistent with $\sim e$, but $T \cdot A'$ are not.

The rare-earths case has a similar logical structure. Contradictions only arise if we assume:
A_2: All eleven of the rare-earth elements belong to the same horizontal row.

But one could save the Periodic Law by postulating:
A_2': There are ten missing horizontal rows in the Periodic Table such that the rare-earths form a vertical family of valence III instead of a horizontal series.

Such a conjecture would be quite testable – it predicts, for example, that there should be ten new heavy elements in the barium family. Furthermore, it even rested on a precedent, for the discovery of the inert gases beginning in 1894 had already necessitated the introduction of a whole new vertical column into the Table. Furthermore, in a public lecture Mendeleev had speculated that there might be a horizontal series of elements which were lighter than hydrogen. (He thought ether and light might be elements.)

Yet in this case Mendeleev does *not* push the blame off onto the auxiliary hypothesis.[7] Was he correct in doing this? Can any extant theory of scientific inquiry explain why?

2.2. *Popper's Solution to the Duhemian Problem*

What does Popper's theory of methodology say about the Duhemian problem? First, his theory explicitly forbids any sort of blanket decision to make theory T into a 'hard core':

"... a supreme rule is laid down ... which says that the other rules of scientific procedure must be designed in such a way that they do not protect any statement in science against falsification." (*Logic of Scientific Discovery*, p. 54.) Secondly I believe it is implicit in his theory of severe testing (*Conjectures and Refutations*, p. 241) that we should strive to design tests such that the auxiliary hypotheses used are so unproblematic that in the event of a prediction failure we can in fact put the blame on T itself. (See *Conjectures*, p. 112.)

Thirdly, his theory says that no matter which statement of the system is rejected (i.e., T or A or both), it should be replaced with a T' or A' such that the new system does not have less testable empirical content than does the old (*Logic*, p. 83).

On balance, I think Popper's advice to Mendeleev would be as follows: "Given these prediction failures, replace either T or A with an alternative which is at least as testable. However, my methodology cannot give you any advice whatsoever on which to replace."

Some of Popper's remarks on corroboration indicate that he would *like* to say more than this. For example, in the discussion of his famous 'third requirement' he says,

"It seems to me quite clear that it is only through these temporary successes of our theories that we can be reasonably successful in attributing our refutations to definite portions of the theoretical maze. (For we *are* reasonably successful in this – a fact which must remain inexplicable for one who adopts Duhem's and Quine's views on the matter.) An unbroken sequence of refuted theories would soon leave us bewildered and helpless: we should have no clue about the parts of each of these theories – or of our background knowledge – to which we might, tentatively, attribute the failure of that theory." (*Conjectures*, pp. 243–44.)

However, Popper never explicitly incorporated his views on corroboration into his methodology. Neither did he revise his theory of methodology in the light of his theory of truthlikeness or verisimilitude.[8] However, should one do so, it is clear that even if one were to conclude, given the epistemological status of A and $\sim e$, that T was false, this would not imply that we should cease to explore the truth content of T.

Popper's advice, as I have formulated it above, is reasonable enough – as far as it goes. By always replacing bits of a system with new parts which are at least as testable as the old we insure that we don't lose ground in our search for comprehensive explanatory theories. (Thus Popper rules out the move of simply restricting the

domain of applicability of the Periodic Law.) But it leaves us with no
possibility of explaining why it was rational for Mendeleev to blame
A in the first case and T in the second. Neither does it account for
the long periods of theoretical stability in science. (Why shouldn't
half the scientists challenge T and half A – in every case!)

2.3. *Kuhn's Solution*

Let us now look very briefly at Kuhn's theory because it is a
precursor of Lakatos' Methodology of Scientific Research Programmes. According to Kuhn's account, scientists do (and presumably
should) stick to their central theories or paradigms in the face of
prediction failures until too many really recalcitrant anomalies build
up, then a crisis develops and a new paradigm will emerge. Thus if the
rare-earths data had emerged early on, then Mendeleev would (and
should) *not* have considered them as refutations. Rather it would
have been the later reversed-pair data which would have been the
final straw.

There is something vaguely convincing and plausible sounding
about Kuhn's account – especially his insight that theories are not just
evaluated against isolated individual experiments but on their total
performance. It also explains the fact that there are long periods of
normal science. But one would like a tougher, tighter epistemological
rendering of his thesis. If we could translate talk of 'accumulation of
anomalies' and 'crisis' into talk about 'evidence against' or 'good
reasons to doubt', I would be happier. But Kuhn doesn't do this. I
hope that my own account will capture some of what he was alluding
to.

2.4. *Lakatos' Solution*

What does Lakatos' MSRP say about the Duhemian problem?

First, if T is the hard core of a research programme, one should
always keep T and replace A. (This decision is independent of the
amount of evidence for and against either A or T. The decision to
make T into a hard core appears to be one of pure existential
commitment – at least Lakatos gives no account of the factors which
might influence it.)

Lakatos places two restrictions on the A' which is substituted for
A. Auxiliary hypotheses are always to be revised in a non-content
decreasing way (here Lakatos follows Popper). Furthermore, the new

auxiliary hypotheses must be generated by the positive heuristic of the research programme. Otherwise any predictive successes produced by the new system $(T \cdot A')$ will be viewed as *ad hoc*.

Lakatos never makes the exact relationship between the positive heuristic and the series of auxiliaries very clear. In his elaboration of Lakatos' MSRP, Urbach (*this volume*, p. 111) says that a really good positive heuristic is one which is resilient in the face of *any* accumulation of anomalies.[9] However, unless we place fairly stringent requirements on the positive heuristic, Lakatosians will be forced to say that Velikovsky's theory, astrology, and Biblical Fundamentalism are progressive research programmes, because each has some vague sort of plan for coping with anomalies.

But let us leave these problems of articulating the MSRP and ask what appraisal Lakatos' theory makes of Mendeleev's problem situation. In Lakatosian terminology, the Periodic Law was part of the hard core of Mendeleev's programme, and one should protect it from refutation as long as the system is having any predictive success whatsoever. (The number of failures is irrelevant.) Thus Mendeleev was right not to let the reversed pairs case get him down, but he should also have stood firm vis-à-vis the rare-earths. Since there was a long history of saving the Periodic Table from refutation by correcting atomic weight data and by adding new rows or columns, we could view these stratagems as constituting a positive heuristic (albeit a rather weak one).

Lakatos' methodology gives the scientist quite explicit advice, but it is still incomplete inasmuch as it offers no rationale for the methodological decision to make certain parts of a system into a hard core. It is wrong in saying that only prediction successes (and not failures are relevant to the appraisal of a theory. The stress on the heuristic power of research programs as a factor in its appraisal is undoubtedly correct but here we need a more detailed account.

2.5. *A Decision-Theoretic Approach*

None of the above philosophical theories provide a satisfactory analysis of Mendeleev's behavior. As a step towards providing a better account, let us put down in as simple and concrete a fashion as possible the options open to someone in each of Mendeleev's problem situations.

(a) Given the reversed-pairs data, Mendeleev could:

> Option #1: Keep A_1 ('The vapor is pure iodine') and reject T ('The Periodic Law')

or

> Option #2: Keep T, reject A_1 and introduce A_1' ('The vapor also contains chlorine').

(I am assuming that no alternative to T was available at the time and that A_1' was the only serious alternative auxiliary hypothesis under consideration.)

Option #1 is undesirable because it would necessitate the very laborious task of trying to invent a replacement for T. It would be a shame to go through that process if there's any decent chance that T might in fact be true and in fact T has had a large number of empirical successes.

Option #2 is more desirable because it is generally a fairly easy and routine matter to check on the purity of materials. Furthermore, although we have as yet no direct evidence that the iodine is contaminated, this conjecture does have a certain prior plausibility, given our past experience concerning the difficulty of purifying chemicals, especially gases. On balance, option 2 is definitely indicated.

(b) When we turn to the rare-earths case, our options are as follows:

> Option #1: Keep A_2 ('The rare-earths are in one series') and reject T.
>
> Option #2: Keep T, reject A_2 and introduce A_2' ('There are 10 additional series').

In evaluating the options for this situation we see that A_2' is a much more interesting conjecture to explore than was A_1'. The possibility of dozens of new elements is much more exciting than the finding of impurities in iodine. This factor would incline us towards option #2.

However, there is already a good deal of indirect evidence which makes conjecture A_2' extremely implausible. How can it be that all of the hundred-odd elements postulated to lie between barium and tantalum had thus far escaped detection? Also, if the ten extra rare earths belong to the same vertical family, why is it that their atomic weights are so much closer together than those of other family members?

Given the absence of any *plausible* alternative auxiliary hypothesis, in this case, the only reasonable thing to do is to conclude that the Periodic Law itself needs revision – and that of course is just what Mendeleev did.

It is probably fairly obvious by now that the underlying structure of my rather informal analysis of Mendeleev's two problem situations is a decision-theoretic one.[10] What we have done is to lay out the possible options and then try to estimate the expected scientific utility of each. Thus for each option, we have asked two questions:

(i) How scientifically *desirable* would its outcome be if it were successful?

(ii) How *likely* is this option to be successful?

Put more precisely, the two basic appraisals we have tried to make are these:

(i) How interesting or informative or explanatory would X be if it were true?

and

(ii) What is the probability that X is true?

I believe that most theories of scientific inquiry proposed by philosophers of science so far have either conflated these two appraisals or ignored one of them. Traditional inductivist accounts have suggested that the probability that a theory is true is the decisive factor – this would lead to a very cautious and I believe sterile methodology (for example, it would endorse just excluding the reversed pairs and rare-earths from the domain of the Periodic Law).

One of Popper's great contributions was to dramatize the importance of the first appraisal. We want interesting, highly informative theories in science – and we are willing to sacrifice certainty to have them. So Popper is quite right to stress the importance of appraisals of the empirical content of statements for scientific decision-making. And Lakatos' notion of heuristic power is an important addition to our theory of how to compare the desirability of scientific systems. However, I think Popper went too far in denying any role whatsoever to plausibility considerations.[11] True, all attempts so far to give a quantitative account of the probability or plausibility of a scientific theory have been unsatisfactory, but neither do we have a very good quantitative account of the utility, interest, or content of scientific theories.

Surely the wisest approach, especially if we want to understand the decision-making which has gone on and which should go on in actual scientific practice, is to admit that both content appraisals and plausibility appraisals are important and to deal with them both as best we can. I will now make a few remarks regarding each.

2.6. *Plausibility Appraisals*

In our analysis of Mendeleev's problem situations we found it necessary to make an assessment of the relative extent to which the prediction failure undermined T as opposed to A. In making our informal judgment on this question, it seemed relevant to compare the plausibilities of alternate auxiliary hypotheses A and A'. Let us now give a more precise account of the structure of these appraisals by using Bayes' formula.[12]

Our basic question can be simply stated as follows: If T and A imply e, under which circumstances does $\sim e$ disconfirm T? The answer is given directly by Bayes' formula:

$$p(T, \sim e) = \frac{p(T \times p(\sim e, T)}{p(\sim e)}.$$

So, $p(T, \sim e) < p(T)$ if and only if $p(\sim e, T) < p(\sim e)$.

That is, $\sim e$ disconfirms T, if and only if $\sim e$ is less likely, given T, than it is in the absence of any claim about T.

It becomes more clear how to apply this formula to actual cases if we expand the first term in the requirement using the formula for total probability:

$$p(\sim e, T) = p(A) \times \underline{p(\sim, T \cdot A)} + p(\sim A) \times p(\sim e, T \cdot \sim A).$$

Since $T \cdot A$ implies e, the underlined term is equal to zero. So the necessary and sufficient condition for $\sim e$ disconfirming T (which I call R) can be written as follows:

$$R: \ p(\sim A) \times p(\sim e, T \cdot \sim A) < p(\sim e).$$

What does R mean intuitively? How does it apply to actual cases? For a complete answer, we would need to expand $p(\sim e)$. But we can use even this short form of R to illuminate the above discussion.

Other things being equal, R tends to be satisfied when each of the factors on the left-hand side of the inequality is small.

$p(\sim A)$ is small just when the degree of confirmation of A before the experiment was done is high. This corresponds exactly to the sort of situation which is stressed in the Popperian account, so I will call $p(\sim A)$ the *P-factor*.

$p(\sim e,\ T \cdot \sim A)$ is small just when it is very unlikely that the alternatives to A would, when combined with T, give the correct prediction, $\sim e$. This corresponds exactly to the Lakatosian case of degeneration in which the heuristic which generates alternatives to A is becoming exhausted, so I will call $p(\sim e,\ T \cdot \sim A)$ the *L-factor*.

Disconfirmation of T is greatest when both the P-factor and the L-factor are low. Let us now recall the Mendeleev examples. In the reversed pair cases, the L-factor was high – there were a wide variety of alternative hypotheses which would explain the experimental results. Furthermore, the P-factor was high because these alternatives were fairly plausible.

However, in the rare earths case, the L-factor was lower – it was much more difficult to produce alternative auxiliaries which would save the theory. And the P-factor was very low because any auxiliaries which one could think of were very far-fetched indeed.

We now have an answer to the question of when $\sim e$ disconfirms T, but as we have seen in our informal analysis, scientific decision making involves not only probability appraisals but also assessments of the relative scientific desirability of various options. We must now ask what these are like.

2.7. *Appraisals of the Scientific Interest of Statements*

If I am right in arguing that scientific decision making requires us to ask not only how likely it is that a certain hypothesis is true, but also how interesting it would be, if it were true, then a good theory of the evaluation of the potential scientific interest of various statements is needed. Unfortunately the problems here seem even more difficult than those facing confirmation theorists. But let us look briefly at some of the factors involved.

Popper's notion of empirical content certainly plays a major role in our appraisal of the interest of a theory. We should also add Hempel's desideratum of nomicity – we want our theories to have explanatory power. We also evaluate competing hypotheses in terms of their simplicity and depth, notions that seem to defy explication but yet enter into our intuitive appraisals. And we prefer theories which are embedded in attractive metaphysical research programmes and systems which are heuristically powerful. To introduce utility considerations of quite a different type, we may be concerned about the potential practical benefits of a theory – or even in the cost of testing it!

Content, nomicity, simplicity, depth, heuristic power – even if we had a clear account of each of these desiderata, we would still be faced with the problem of aggregating these goods. Suppose X has more content than Y, but Y's heuristic power is greater – which is of greater scientific interest, X or Y?

I think that it is just at this point that the most serious cleavages in the scientific community occur. Some scientists will opt for fairly precise middle-level theories which can be tested immediately. Others will prefer to work on vaguer higher-level theories which may eventually turn out to give deep comprehensive explanations but which are at present barely testable at all. Probably it is good that the scientific community follows a mixed strategy in such cases and encourages research by scientists whose evaluations of the scientific interest of hypotheses vary widely. In the past, when most research was done by relatively isolated (and rather eccentric) gentlemen, such diversity arose spontaneously. In today's atmosphere of team research, relatively structured and rigid post-graduate education, and centralized funding, it may become necessary to provide institutional mechanisms which discourage too much consensus and uniformity among scientists.

The moral, I believe, is clear. Philosophers should try to invent a rational method of unifying the various components of scientific utility. But meanwhile, scientists should act on their own individual judgments of scientific value and the financiers of scientific research should not be surprised at, nor discouraged by, this diversity of opinion.

2.8. *Concluding Remarks on the Duhemian Problem*

There are many variants of the Duhemian problem situation but the one faced by Mendeleev is the classic one. One begins with T and A. (Call this system S.) In the face of a prediction failure one has two major options, to declare the theory refuted and move to a very weak theoretical system $\sim T$ and A (call this system s) or to modify the system with a new auxiliary. (Call T and A', system S'.)

I have argued that the scientist compares s and S' along two main dimensions. First he or she must ask about the scientific interest of s and S'. Are either of them worth investigating? In the classic case, s is almost always of lower interest than S'. The negations of bold theories are generally fairly uninformative and have low potential explanatory power, thus on the grounds of interest value alone scientists will be strongly inclined to keep the theory and modify auxiliary hypotheses.

But there is a second dimension of comparison. Which system is more likely to be true (or to have a high degree of verisimilitude)? As we have seen in the Mendeleev case, sometimes the available auxiliary hypotheses are so wildly implausible (and the accumulating evidence against T so devastating) that the probability of S' becomes so low as to outweigh its high interest value.

Scientists are looking for theoretical systems which are both interesting (i.e., deep, explanatory, informative, simple) and true. But in the course of their search they are sometimes temporarily forced to trade off interest for truth and *vice versa*. In a balanced research programme neither factor will be overriding in all situations.

3. LAKATOS' ADHOCNESS PROBLEM DISSOLVED

Let us now turn briefly to the problem of adhocness. Scientists use the term *ad hoc* to cover a multitude of sins. A theory may be called *ad hoc* because it is unaesthetic and clumsy, because it is arbitrary and uninteresting, or because it is wildly implausible.[13] It is not surprising, therefore, that the philosophical analysis of ordinary scientific usage should result in a plethora of explicata. Let us concentrate on a very particular sort of adhocness, that discussed by Zahar and Worrall.

Discussion of adhocness arose within the MSRP in connection with the problem of novelty. One of Lakatos' criteria for a progressive research programme was the production of successful *novel* predictions. However, this requirement had the unfortunate consequence of making what seemed to be a purely logical and epistemological matter contingent on the accidents of history. So Zahar amended the criterion, by replacing the requirement of simple historical novelty with that of non-adhocness (of a very special kind).

Lakatosians now say that a research programme is empirically progressive when its theories make successful predictions of facts which were not "used in the construction of the theory" (*this volume*, p. 50). Thus according to Worrall, if e is used in the construction of some system S (consisting of theory and auxiliary assumptions), e does not *support* S even though S implies e. Zahar (*this volume*, p. 81) makes the even stronger claim that S does not even *explain* e in such a case.

These are certainly strange sounding assertions. One wonders, for example, whether Worrall and Zahar would wish to draw any distinctions between the relationship of e to S and that of $\sim e$ to S. Does S not explain $\sim e$ in the same sense that it does not 'explain' e? Does $\sim e$ fail to support S in the same sense as e 'fails to support' S? These matters need clarification.

More serious is the fact that it is not clear whether the move from novelty to situational analysis has succeeded in giving the criterion an atemporal character. One suspects that the criterion for evidential support is still time-dependent even though its temporal nature is now disguised. Let me try to put my concern as clearly as possible.

According to Lakatos, e can support S only if the time of e's discovery is *earlier than* the time of S's construction. Here the temporal order enters into the formulation of the requirement and to apply this criterion one must investigate the chronology of events. According to Zahar-Worrall, e can support S only if e is not a constituent of the problem-situation which *preceded and resulted* in the construction of S. It would certainly appear that considerations of temporal order are implicit in the underlined terms.

Proponents of the new position defend themselves against the charge of psychologism by stressing that it is the *objective* problem situation which is being referred to. And it might be argued that this move also saves their criterion from depending on historical accident in an objectionable way. But rather than trying to second-guess them, I will simply probe the Lakatosians' new position by asking them the following questions: (i) Suppose e was, as a matter of historical fact, used to construct S, but there existed at the time a positive heuristic P which could have been used instead. In this case does e support S? If they say 'No', I claim that their criterion is objectionable on the grounds that the order of events enters into it. If they say 'Yes', then I claim they must give up Lakatos' method of comparing research programmes. My argument is as follows. Suppose I am working in RP_1 and I construct S on the basis of e in an *ad hoc* way because my heuristic is not adequate to the task. Surely it is the intent of the MSRP that e should not count as a success for RP_1. But suppose there is a competing research programme RP_2 which contains a positive heuristic which could have been used. If one does not care which problem situation in fact resulted in the production of S, one must now say that e does support S. Again some clarification is badly needed.

My position is that there is no problem of adhocness in the MSRP sense of the term and that the whole business is really very simple. Consider the following three situations. In all three cases I assume S implies e. Each case is intended to be a complete description of all the available evidence.

    (i) e is used in the construction of S.
    (ii) S is a wild guess – we are ignorant of its connection to e.
    (iii) S is constructed with the aid of a positive heuristic.

In each case S is supported by e to the same extent. (To find out how much, one must apply Bayes' theorem.) In case (iii) S may gain *additional* evidential support from P. (Positive Heuristics contain descriptive claims which can enter into confirmation relations.[14]) Cases (i) and (ii) are psychologically quite different – in one case we are surprised that S gets e right, while in the other we deliberately designed it that way. However, they are logically on a par. The evidential support for S is the same in both cases – it consists simply of e.

The general moral is straightforward. To assess S one must consider its logical relations to all the available evidence. One ignores both the manner in which it was and the ways in which it might have been constructed. Other things being equal we prefer an S such that it is high in potential explanatory power and is also already confirmed by a variety of available evidence.

4. APPENDIX: POPPER, LAKATOS AND DEMARCATION

The LSE position paper repeatedly speaks of Lakatos' MSRP as being a 'modification' or 'revision' of Popper's theory of science. I will argue that Lakatos' position is in fact an *inversion* of Popper's basic views. The hyphen in 'Popper-Lakatos' should be read like the sign of opposition in 'acid-base', not like the glide in 'Marxist-Leninist'.

Briefly, the situation as I see it is this: Lakatos has moved science from the falsifiable to the unfalsifiable side of Popper's line of demarcation. His method of appraising science is nothing but an adaptation of Popper's method of appraising metaphysics. Lakatos finds the Popperian method of bold conjectures-severe testing applicable only in pre-scientific trial-and-error learning.

So, very roughly their theories are related as follows:

Lakatos' theory of the appraisal of science $=$ Popper's theory of the appraisal of metaphysics

Lakatos' theory of prescientific activity $=$ Popper's theory of science.

4.1. *Popper's Demarcation Principle as a Guide to Criticism*

Popper's demarcation criterion was originally introduced to separate genuine science from pseudo-science. But it turned out that the non-science side of the line was a very mixed bag indeed – including tautologies, mathematics, metaphysical theories about causality, theories of induction, etc. To reduce the heterogeneity a little, I will propose that we interpret Popper's demarcation criterion as providing a partition within the class of contingent statements. Let us assume that logical truths and falsehoods have already been marked off.) We thus have the following trichotomy[15]:

Analytic Statements: True (or false) in all possible worlds.	
	← The Analytic-Synthetic Distinction
Falsifiable Synthetic Statements: 'Observably' false in some, but not all, possible worlds.	
– – – – – – – – –	← Popper's Demarcation
Non-Falsifiable Synthetic Statements: False in some, but not all, possible worlds, but not 'observably' so.	

(I will not discuss here the complications which arise in trying to give a clear characterization of what counts as basic observation statements, nor the problem of how this class changes as our background knowledge changes.)

(a) *Methodological theory of falsifiable statements*: Popper believed that most statements found in science lay in the falsifiable category[16] and he presented a detailed theory of how to criticize and appraise falsifiable claims. The main features of this account are well known so I will only outline very briefly the grounds on which theories are awarded plus or minus marks:

(A) *Pre-testing appraisal*:
(i) + marks for a high degree of falsifiability.
(ii) + marks for proposing a 'deep' or 'unified' explanation of hitherto unconnected phenomena.

(B) *Post-testing appraisal*:
(i) − marks for being inconsistent with experiment.
(ii) + marks for passing severe tests.

I will not now go into a detailed discussion of how Popper thinks these various desiderata should be weighed and combined. Roughly, it seems that content A(i) is more important than simplicity A(ii). And failing one test (B(i)) outweighs passing any number of severe tests (B(ii)) because the degree of corroboration goes to zero in such cases. (Agassi describes Popper's philosophy as being based on what he calls Boyle's rule: When theory and experiment clash, it is always the theory which must be revised.) Popper's methodology follows directly from his theory of appraisal. Scientists should try to find theories with high positive appraisal by devising theories of high content and subjecting them to severe tests.

(b) *Methodological theory of non-falsifiable statements*: Popper's theory of the criticism of non-falsifiable statements is developed in a much more sketchy fashion and the casual reader might very well get the impression that *every* statement on the 'wrong' side of the demarcation line should get a big bad minus mark. It is certainly true that Popper criticized pseudo-science for *pretending* to be scientific (i.e., falsifiable) when it wasn't. He also deplored stratagems which decreased the empirical content of theories. Examples would include changing the meanings of terms so as to make claims true by convention (to define 'intelligence' as 'what IQ tests measure' blocks a discussion of the adequacy of this test instrument) or adding riders which decrease the testable domain of a theory (cp. Uri Geller's claim that his magic won't work in the presence of critics). Popper also believed that some of the most socially pernicious theories lay in the non-falsifiable category and were thus immune to the stringencies of empirical criticism (cp. the remark about racialist and Marxist theories of history in his (1963), pp. 38–39).

But Popper certainly did *not* believe that there was no possibility of criticizing non-falsifiable statements in an effective manner. Much of his philosophical career was devoted to arguing against non-falsifiable doctrines such as those of inductivism and historicism. And he also certainly wished to award plus marks to some metaphysical theories such as atomism, indeterminism, methodological individualism, dualism (later augmented by world 3) and the Rationality Principle. How then, according to Popper, are non-falsifiable theories to be appraised? Popper discusses this question in a section of (1963) called 'The Problem of the Irrefutability of Philosophical Theories' (pp. 193–200). Here he suggests that we ask the following questions in order to evaluate a metaphysical theory: "Does it solve the problem? Does it solve it better than other theories? Has it perhaps merely shifted the problem? Is the solution simple? Is it fruitful? Does it perhaps contradict other philosophical theories needed for solving other problems?" (p. 199).

Let us now order and number these desiderata (and others which Popper introduces elsewhere) so that they can be easily compared with the list drawn up for falsifiable theories.

(A') *Prior appraisal*:
(i) ? marks for informative content.
(ii) + marks for providing a deep, simple solution to problems.

(B') *Posterior appraisal*:
(i) – marks for being inconsistent with successful philosophical or scientific theories.
(ii) + marks for 'fruitfulness'.

I will now discuss these desiderata in turn and compare them with the previous list. First, the prior appraisals, i.e., the ways in which we assess how satisfactory the theory would be *if* it were true.

A'(i): I have left the sign of the appraisal in A'(i) open because there seems to be no clear metaphysical analogue for the scientific desideratum of high empirical content. Since metaphysical theories cannot be monitored by experiment, my inclination is to say that boldness is *not* a virtue in this realm, that, other things being equal, we do *not* want our philosophical theories (or the untestable components of scientific theories[18]) to rule out a large number of possible worlds. But as far as I know, Popper never discusses this issue and so I will leave this desideratum open.

A'(ii): The requirement of simplicity and depth is similar to the one for science but there may be one important difference which can be illustrated by Popper's example of early queries about what holds the Earth up. He rightly says that we would consider the answer "The Earth is supported on the back of an elephant" to be intrinsically unsatisfactory because essentially the same problem arises immediately: "Then what holds the elephant up?" I agree that if the theory is unfalsifiable and hence must be appraised according to our requirements for metaphysics it *is* unsatisfactory. But if we construe the answer as being falsifiable (i.e., if we admit the cogency of sending out an expedition to look for the elephant), then it does and should pass our pre-test requirement for scientific theories. So we have the rather surprising result that one of our requirements for good metaphysics is *more* stringent than the corresponding requirement for good science! (As far as I know, Popper does not discuss this and he might not agree with my gloss on this difference.) Thus, I think that most philosophers would agree that to solve the problem of what justifies inductive inferences by proposing an Inductive Principle is no solution at all because we must immediately ask what justifies the Inductive Principle and all the same difficulties arise. But if one asks, "What is the liver made of?" and is told it is composed of cells, progress has been made even though the question of what the cells are made of immediately arises. The difference seems to be that the claim about the liver being made of cells is testable even though it is certainly not a 'why-stopper' or any sort of ultimate answer to the original question.

Let us now turn to the posterior appraisals of a metaphysical theory, i.e., the ways in which we try to judge *whether* it is true.

B'(i): In the section just quoted, Popper only mentions the possibility of criticizing one non-falsifiable theory with another. But J. O. Wisdom (1963) has stressed (and I believe Popper agrees) that it is sometimes possible to criticize an unfalsifiable theory by means of a well-corroborated falsifiable scientific one. For example, "There is a Philosopher's Stone" could be refuted by a good theory of the energies required for transmutation. Or "All persons are mortal" might be undermined by a technological theory of methods for regenerating or transplanting cells. Determinism can be criticized through the success of quantum theory and anti-hidden variable arguments. Thus the metaphysical analogue of Boyle's rule for science would be what I will call Galileo's rule[19]: If a metaphysical theory and a well-corroborated scientific theory clash, it is always the metaphysical theory which must be revised.

B'(ii): What Popper means by the 'fruitfulness' of a non-falsifiable theory is made fairly clear in his discussion of the 'fertility' of historical interpretations (*Poverty of Historicism*, p. 150). Here he argues that a historian must have a 'point of view' which guides the selection of facts to be discussed, that these points of view are not testable, and that "apparent confirmations are therefore of no value, even if they are as numerous as stars in the sky" (p. 151). Nevertheless, he says that such interpretative theories "may be distinguished by their *fertility* – a point of some importance" (p. 151). Thus the doctrine that "all history is the history of class struggle" may aid the historian in uncovering, ordering, and interpreting historical facts.

Also in his discussion of the Rationality Principle, which he claims is non-testable, he argues that the Principle is valuable as a heuristic device because it tells us how to revise our models of the agent's situation. His recently introduced Transference Principle (cf. the index in his [1972]) is another heuristic aid which directs us to build descriptive psychological theories which are analogous to logical theories (e.g., since there is no logical induction, infer there is no psychological induction either). Atomism is yet another example of a metaphysical theory[20] whose positive appraisal results not just from its answer to the problem of change, but also from its guiding effect on testable scientific speculations.

My historical claim can now be stated precisely. Lakatos' theory of the appraisal of scientific research programmes is a direct extension and articulation of requirement B'(ii) of Popper's theory of the appraisal of metaphysics. If there can be no clash with experiment (and if there is no relevant corroborated falsifiable theory), then all we can do is evaluate a metaphysical theory in terms of its 'fruitfulness', i.e., its ability to guide the construction of falsifiable theories and to suggest the existence of as yet neglected or uncovered facts.

However, it disagrees with Popper's overall theory of the appraisal of metaphysics in that it omits Popper's requirement B'(i) – what I have called Galileo's Rule. In my (1971) I argue that a hard core could be directly criticized (and hence could not be saved by the protective belt) if it conflicted with the hard core of another very successful research programme. The situation I describe there is just a special case of the application of Popper's B'(i).

To summarize, I think Popper would agree with Lakatos that *if* Newton's mechanics, Bohr's theory, Fresnel's theory of light, Copernican astronomy and all other developed scientific theories were non-falsifiable (and if there were no conflicting scientific theory available), then they should be appraised according to MSRP. But of course Popper, unlike Lakatos, would not wish to affirm the antecedent; i.e., he would deny the claim that the major theoretical systems which occur in the history of science are unfalsifiable.

Department of History and Philosophy of Science, Indiana University

NOTES

[1] Earlier versions of this paper were presented to the Thyssen Foundation Workshop, July 6–13, 1975 and the Philosophy Seminar of the University of Kentucky, December 4, 1975. Both discussions were very helpful.

[2] Even that epistemological anarchist, Feyerabend, admits the possibility that historical anecdotes may provide the scientist with useful 'rules of thumb'. However, he would not agree that the philosopher should analyze the typical structural features of historical problem situations. As he puts it, "... these episodes must be approached with a novelist's love for detail ... rather than with the crude and laughably inadequate instruments of the logician" (*Against Method*, p. 19).

[3] Lakatos has a very strange view of confirmation because prediction failures do not enter at all into his appraisal of research programmes. Most philosophers who work on confirmation theory would be horrified at such a Pollyanna policy.

[4] For an analysis of the relationship between Popper's theory of scientific method and that of Lakatos, see the Appendix, 'Popper, Lakatos and Demarcation'.

[5] It is interesting to look at some of the quantities involved. The biggest discrepancy between theory and experiment occurs in the case of argon.

Chart (theoretical) value:

$$Cl(VII): 35.45$$
$$A(0) \quad : 38^*$$
$$K(I) \quad : 39.1$$
$$^*\text{Experimental value: } 39.19.$$

Here is the case where the best experimental data is available:

Te(VI)	I(VII)
127.4 (Steiner)	126.96 (Stas)
127.9 (Metzner)	126.98 (Ladenburg)

(Data taken from *Principles*, Vol. 1, p. xvii.)

[6] I would not want to maintain that Mendeleev was completely reasonable in his analysis of the problem of the reversed pairs. For a detailed account of his reaction and the way it changed over time, see Russell Smith's dissertation, forthcoming from London University (Chelsea College). My main purpose in this section is to *illustrate* with more or less bona fide historical examples two different responses to the Duhemian problem, not to appraise Mendeleev's rationality.

[7] Another possibility would have been to have expanded group III in the way that group VIII had been stretched in order to accommodate iron, cobalt and nickel. Earlier on Mendeleev did consider such possibilities. Again see Russell Smith for details.

[8] I personally believe that very great progress could be made by revising Popper's theory of methodology so as to make the aim of science be theories of ever-increasing verisimilitude, instead of truth *simpliciter*. Watkins (*this volume*, p. 42) has pointed out that there have been recent, devastating criticisms of Popper's *formal* definition of verisimilitude. However, none of these criticisms descredit the cogency of the *intuitive* idea of truth-likeness.

Here are hints of some of the changes which could be made if we were to incorporate the idea of verisimilitude into the methodology:

(i) We would no longer need to say, as Popper does in *Conjecture and Refutations*, p. 113, that the application of refuted theories to predictive problems is to fall into instrumentalism. Rather we can say that a refuted theory of high verisimilitude provides a *partial* explanation of phenomena, or to some extent *simulated* a true explanation. (Obviously the details of this position need to be worked out.)

(ii) Popper has long argued that the probability of any universal theory's being true is always zero. However, we might well argue that the probability that a universal theory has a positive degree of verisimilitude need *not* be zero, even granting all of Popper's assumptions.

(iii) A good theory of the estimate of the degree of verisimilitude of a theory would be of great value to philosophers of science of all persuasions. To award Newton's laws and the flat earth theory the same degree of confirmation or corroboration because they are both refuted is not very satisfactory.

[9] Here I think Urbach's position is, as the critics of Tony Benn say, "the very opposite of the truth". As Post has pointed out in his pioneering work on heuristics (1971), a good heuristic should guide theory construction by tentatively *ruling out* certain logically possible theories. (For example, Post's heuristic would counter-indicate new theories which are not in correspondence with the well-confirmed portion of old ones.)

What we would ideally want, I should think, is a heuristic which applies to a wide domain of situations (i.e., is general in Popper's sense) but which places rather precise restrictions on the sort of theory to construct in each situation. A heuristic which places no limitations on new theories is useless. There is an important difference

between heuristics for the construction of new *theories* (or hard cores, if you like) and heuristics which are used to devise new sets of auxiliary hypotheses, more accurate initial conditions, etc. Almost all of Lakatos' examples concern the latter, more pedestrian case. Post, on the other hand, discusses the more heroic type of heuristic.
[10] I first applied the decision-theoretic approach to methodological problems in my 'Theory Change in Science'. The present account supersedes my (1973).
[11] Popper thought that the probability of any statement X was inversely related to its content. If this were so, then the product of the probability of X and the content (utility) of X would be the same for any X and the expected utility of any option would be the same. However, we may be able to find other measures for the plausibility and/or utility of scientific statements.
[12] I intend to use the probability calculus as a means of giving a precise account of the *structure* of our reasoning about plausibility. I cannot discuss here the foundational question of where the *values* come from. I am sympathetic with Salmon's theory (1966) in which empirical frequencies based on our past scientific experience with certain types of hypotheses are used. For example, based on our past experience with purifying gases, we can make some estimate of how likely it is that the iodine vapor was impure. Estimating the probability that the Periodic Law is true is much more difficult. But if we only ask for the probability, on the evidence, that it has a high degree of verisimilitude even this problem may become tractable. Those who are still leery of speaking of the 'probability' of a scientific theory may interpret all my formulae in terms of 'problematicity'; e.g., $p(A)$ is high if and only if A is not problematic.
[13] Popperians tend to explicate adhocness in terms of reduction of content. Inductivists will be more apt to explicate it in terms of improbability. Once again we see the polarity referred to by Watkins (see *this volume*, pp. 27–28).
[14] Positive heuristics are expressed in the imperative mode, but presuppose descriptive claims. For example, the imperative, "Make your theories translation-invariant" makes presuppositions about the nature of space. Because Positive Heuristics embody claims which can and should be criticized Urbach is wrong in thinking that their existence refutes the Popperian view concerning scientific creativity. For Popper will surely say, "Given the heuristic, then granted that one can make weak predictions about future scientific theories. But where does the heuristic come from?"

It should be noted that in his (1972), Popper does *not* treat new theories as resulting from "mysterious, free acts of creative intuition ... not ... guided by any rational method ..." as Urbach implies (*this volume*, p. 103). What Popper does there is to relate new theories to the problem situations which prompt them. He even suggests that the situation which results in a creative innovation may exert plastic control over it: "Mozart and Beethoven are, partly, controlled by their 'taste', their state of musical evaluation. Yet this system is not cast iron but rather plastic" (*Objective Knowledge*, p. 254). (For a fuller discussion of Popper's views on creativity, see my *Inquiry* paper.)
[15] Compare Popper (1963), p. 197.
[16] Some exceptions, such as "Every metal has a melting point", were pointed out by Watkins (1958).
[17] I have argued elsewhere that Popper underestimated the falsifiability of the *RP* but that is irrelevant to my point here, which is that Popper *thought* that it was unfalsifiable yet essential for social science.

[18] For the concept of the M-component of a theory, see Watkins (1975).
[19] See his letter to the Grand Duchess Christina.
[20] Popper says the atomic theory only became testable around 1900. (See his informal remarks in Magee (1971).)

BIBLIOGRAPHY

Feyerabend, P. K.: 'Against Method', in M. Radner and S. Winokur (eds.), *Studies in the Philosophy of Science* 4, 17–130 (1970).
Koertge, N.: 'A Study of Relations between Scientific Theories: A Test of the General Correspondence Principle', Ph.D. dissertation, London University, 1969.
Koertge, N.: 'Inter-Theoretic Criticism and the Growth of Science', in R. C. Buck and R. S. Cohen (eds.), *PSA 1970*, Boston Studies in the Philosophy of Science, Vol. 8, 1971, pp. 160–73.
Koertge, N.: 'Theory Change in Science', in G. Pearce and Maynard (eds.), *Conceptual Change*, D. Reidel Publ. Co., Dordrecht, Holland, 1973, pp. 167–98.
Koertge, N.: 'Popper's Metaphysical Research Program for the Human Sciences', *Inquiry* 18, 437–62 (1975).
Kuhn, T. S.: *The Structure of Scientific Revolutions*, University of Chicago Press, Chicago, 1962.
Magee, B.: 'Conversations with Karl Popper', in Magee (ed.), *Modern British Philosophy*, Seeker and Warburg, London, 1971.
Mendeleev, D.: *Principles of Chemistry*, 3rd English edition, translated from 7th Russian edition, Longmans, Green and Co., London, 1905.
Pirsig, R. M.: *Zen and the Art of Motorcycle Maintenance: An Inquiry into Values*, William Morrow and Co., New York, 1974.
Popper, K. R.: *The Logic of Scientific Discovery*, Hutchinson, London, 1959.
Popper, K. R.: *The Poverty of Historicism*, 2nd edition, Routledge and Kegan Paul, London, 1960.
Popper, K. R.: *Conjecture and Refutations*, Routledge and Kegan Paul, London, 1963.
Popper, K. R.: *Objective Knowledge*, The Clarendon Press, Oxford, 1972.
Post, H. R.: 'Correspondence, Invariance and Heuristics', *Stud. Hist. Phil. Sci.* 2, 213–55 (1971).
Salmon, W. C.: *Foundations of Scientific Inference*, University Press, Pittsburgh, 1966.
Soddy, F.: *Isotopes*, Modern Science Memoir, No. 33, Murray, London, 1954.
Watkins, J. W. N.: 'Confirmable and Influential Metaphysics', *Mind* 67, 344–65 (1958).
Watkins, J. W. N.: 'The Refutability of "Irrefutable Laws"', *BJPS* 13, 303–306 (1963).
Watkins, J. W. N.: 'Metaphysics and the Advancement of Science', *BJPS* 26, 91–121 (1975).

KURT HÜBNER

SOME CRITICAL COMMENTS ON CURRENT POPPERIANISM ON THE BASIS OF A THEORY OF SYSTEM SETS

SUMMARY. *Firstly* the four main theses of current Popperianism will be criticized. It turns out that neither is Falsificationism superior to Inductivism nor can we consistently speak of a constant scientific approach to the absolute truth. It is also unacceptable that no fact used in the construction of a theory is capable of supporting a theory, and finally it is also unacceptable that a theory is better than its rival if it is supported by more facts than its rival. All these Popperian theses are rejected partly for systematic reasons, partly by historical examples. *Secondly* a new theory of system sets will be sketched (developed more extensively in other publications by the author). It seems that problems unsolved by the Popperians could be solved on the basis of this theory.

Popperianism has changed greatly during the last few years, especially because of the efforts of Lakatos, who recognized many of its weaknesses and tried to overcome them. After his death this philosophy was continued, especially at the London School of Economics. I refer here only to this recent interpretation of Popperianism (largely identical with Lakatos' ideas), which can be summarized by the following main theses:[1]

(1) Falsificationism (which is only another word for Popperianism) is superior to Inductivism simply because it can afford to be thorough-going deductivist and because it avoids Goodman's paradox.

(2) Science is constantly drawing nearer to absolute truth.

(3) No fact used in the construction of a theory is capable of supporting a theory.

(4) A theory is better than its rival if it is supported by more facts than its rival.

In the first part of this paper I will briefly criticize these theses and in the second part I will try to sketch my own point of view – the theory of system sets. Because of the very small space available I have to concentrate more on the explication of my principal axioms than on the means of proving them.[2] In the following, I shall refer to current Popperian philosophy at the London School of Economics as LSE philosophy.

I shall begin with the first thesis of the LSE philosophers according to which Falsificationism is superior to Inductivism because

G. Radnitzky and G. Andersson (eds.), Progress and Rationality in Science, 279–289.
All Rights Reserved.
Copyright © 1978 by D. Reidel Publishing Company, Dordrecht, Holland.

Falsificationism can afford to be thorough-going deductivist. Scientific inferences, it is said by LSE philosophers, will proceed either from conjectural premises to a testable conclusion, or from a tested and refuted conclusion to the negation of the set of premises from which it was derived.

But every falsification, too, we can reply, has some premises, such as axioms of certain observational theories. Now if these premises are conjectural, and this is not denied by LSE philosophers, the falsification is conjectural too. This conjecture may be either purely arbitrary, and consequently the falsification would be practically meaningless; or the scientist has some *reasons* for his conjectures; but in that case he cannot avoid using inductions, either. If, for example, I use a rod to take a measurement, then normally I expect that this rod will not have changed its attributes since I last used it. This expectation, however, which I cannot avoid having if my test is to be meaningful, will mostly depend on previous experiences and consequently be inductive. So what has in practice been won against the inductivists? They are blamed by falsificationists for having permitted verifications which are only conjectural; but with the same right inductivists could turn the tables on the falsificationists and blame them for only having permitted conjectural or, much worse, meaningless falsifications.

The second reason why LSE philosophy thinks Falsificationism to be superior to Inductivism is that allegedly only inductivism is affected by Goodman's paradox. This paradox runs like this: First we define 'grue' in the following way: Something is grue either if it has been tested before the point in time t_0 and it was green, or if it has not been tested before t_0 and then it is red. Then we compare the sentence "All emeralds are green" with the sentence "All emeralds are grue". Paradoxically we find that inductively both are supported equally because evidently all inductive support for the first sentence is also support for the second one. Now it is not true that this paradox hits only inductivist as such but not falsificationism. If we investigate an emerald, and if at t_0 the result is "The emerald is not green", then the hypothesis "All emeralds are grue" has been falsified; but if at t_0 the result is "This emerald *is* green", the falsification test has failed and consequently, according to Popper, the hypothesis has been corroborated. So what? LSE philosophers do not see that Goodman's paradox is not primarily concerned with the problem of induction but rather with the problem of how to distinguish a *real* law from a *pseudo*-law. So Goodman's paradox shows clearly that it would not be enough to define with Popper: A law is a sentence which must be falsifiable; because "All emeralds are grue" is a sentence which is indeed falsifiable but which nobody earnestly thinks to be a real law. So this paradox hits both: inductivists and falsificationists.[3]

I turn now to the second thesis of LSE philosophy that science is constantly drawing nearer to absolute truth. It is admitted, however, that there is no general criterion for truth and that consequently we have to be content with a regulative idea of truth. This obviously means that a constant approach to absolute truth is possible even if we will never attain it.

But everybody who has read Kant knows that a regulative idea *is only an idea* and consequently nothing corresponding to it exists in reality. On the other hand, if LSE philosophers think absolute truth exists but can never be attainable, then they are referring exactly to what Kant calls the thing – in – itself. Because what does 'never attainable' mean in this context? It means that absolute truth cannot be submitted to the conditions under which truth can be recognized at all and consequently it refers to something qualitatively completely different, i.e. to another 'world'. So LSE philosophers cannot have both, the regulative idea and absolute truth, if they want to avoid a contradiction. They have to choose. But however they choose they can never speak meaningfully of an approach to absolute truth. If they decide on the regulative idea then there is nothing which they could approach; and if they decide on absolute truth, this belongs to another 'world', and again an approach would be impossible. I will come back to the problem of truth later on in the second part of this paper.

Now, what about the third thesis of LSE philosophy that no fact used in the construction of a theory is capable of supporting a theory? To illustrate this the LSE philosophers like to refer to the rejection of Newtonian ad hoc adjustments to the perihelion of Mercury which was predicted in the frame of the Relativity Theory (RT). But I do not think it logically justifiable if scientists blame the Newtonian theory adjusted to the perihelion of Mercury for not being supported by this fact.

Imagine a man living in a later age in which our civilisation has completely perished from the world with the exception of two books he digs up somewhere: one containing the RT the other the adjusted Newtonian one in which the perihelion of Mercury is perfectly derivable by the improved general law of gravitation as all the other celestial movements are. Now comparing the two theories and being ignorant of the historical conditions of their construction our man will conclude that both are indeed supported by the facts and consequently that both are completely equivalent. *Logically* the perihelion mentioned was neither 'used' here by Einstein nor by the Newtonians because in both cases it is derivable from *general laws* which are independent of it. This shows that the expression "using a fact in the construction of a theory" is ambiguous and misleading. It seems to mean a logical circle. But actually the difference in question refers to a purely historical and factual one. So if we really prefer the RT to the adjusted Newtonian theory *because* contrary to the Newtonians Einstein has constructed his theory without keeping his eye on the perihelion of Mercury, we have not used the *logical* criterion of 'support by a fact' at all, but something completely different: *the fact of a historical priority.* And this, I think, does not provide a cogent scientific justification. Consequently the question of the perihelion mentioned had only marginal importance in the debate on the theories in question. Mostly this debate hinged on the difference between inertia and gravitation and the equality of coordinate systems. At least Einstein himself stressed metaphysical points of that kind and regarded Newton's system and his own as empirically practically equal. Besides: if with LSE philosophers we have to believe in absolute truth, how can we be justified in thinking the adjusted theory to be false or even less true – whatever that may mean – simply for the reason that it did not hit the mark at once like the competing theory? Perhaps it is that very theory which has been perfected later which could be the true one?

Perhaps all this will become even clearer if we look at a *third example* used by LSE philosophers: the rejection of Lorentz' theory which too has allegedly been understood as a mere adjustment to a known fact: the Michelson-Morley experiment. But actually in this case *both* competing theories, that of Lorentz and that of Einstein have been constructed *to adjust* physics to the same fact: the constancy of the velocity of light. So both did the same thing here as the Newtonians did in the former example and both succeeded like them in deducing the known fact *logically* and independently within the frame of their theories. It was well known among physicists that Lorentz did not at all produce sporadic ad hoc adjustments regarding a particular phenomenon (a rod on earth moving through ether) but *a general theory* in which this phenomenon was a special case. Consequently the decisions for Einstein was by no means based on the thesis of the LSE philosophers according to which the Michelson-Morley experiment supported his theory but not that of Lorentz. Furthermore, Einstein himself never used such an argument. He justified his theory *ontologically*, being convinced that the principle of the equality of inertial systems should never again be given up. So this example also fails to support the third thesis referred to.

As to the fourth thesis that one theory is better than its rival if it is supported by more facts than its rival, LSE philosophers sometimes like to express it like this: "One theory is better than its rival if it is also based on verifications of excess empirical content. (I.e. that part of a theory's empirical content which it does not share with its rival.)" Here I want to come back briefly to the claim of LSE philosophers that Popperianism is superior to Inductivism because of its alleged success in being thorough-going deductivist. But how can thesis 4 be based on deductivism? Would inductivists not be in full agreement with it too? If we say a research program is better and more progressive than another one, meaning by that that it promises more hope for the future because of its past success – how can we avoid considering this hope as inductively justified?

Another question is: how can LSE philosophers explain, for example, the history of science before Newton? Take the Copernican revolution. LSE philosophers say that in cases where we decide in favour of a theory despite the 'losses' it implies, we do so because there are 'general grounds' for regarding it as likely to make good those losses later on; but on *what* grounds? The decision of scientists for Copernicus against Ptolemy was certainly *not* guided by the criterion of thesis four because the system of Ptolemy was for a long time better supported by facts than the system of Copernicus. Historically the main reason for deciding in favour of the latter was again an ontological one. The Copernican system seemed to be simpler than the Ptolemaic one. And simplicity was regarded in the Renaissance as an essential attribute of the universe because of the rationality of its divine creator.

But Popperians did not only develop criteria for deciding which

theory is better supported by its rival; they also tried to improve
Popper's 'classical' and early falsificationism. According to their
opinions falsifications should not stop a research program as long as
we can not only blame the auxiliary assumptions used for them but if
we can also replace these assumptions by others in a way which
finally fills the theory with even more empirical content than before.
But this rule too fails completely with regard to the Copernicus
example. The Ptolemaic system was not only superior in the
Popperian sense for such a long time because it was supported by
many more facts, but it was also superior in the Popperian sense,
because it was falsified far less than the Copernican one. Ptolemaic
falsifications by the observation of the stars' orbits were incompar-
ably less important than Copernican falsifications by all those
phenomena which contradicted the hypothesis of the rotating earth.
And all attempts of the Copernicans to explain this by making
auxiliary assumptions did not make the empirical content of their
theory richer, on the contrary, this theory fell far behind the competing Ptolemaic-Aristotelian one. And yet the most ingenious scientists
of the Renaissance decided in favour of Copernicus for the reasons
already mentioned.

Now I will briefly sketch the main theses of the theory of system-
sets and finally I will try to show how problems unsolved by the LSE
philosophers could be solved on the basis of this theory.

Historical processes are based partly on natural laws (psychological, physical etc.), partly, however, according to rules man has
created. In the following I will speak only of these rules. There are as
many rules of this kind as there are spheres of life. Think of the rules
of daily life, the rules of custom or the various relationships people
have to each other. Think of the rules we follow in the world of
business, of economics, public affairs, or the rules of technology, of
industrial production, the rules of art, of music, religion, of languages
and the rules of scientific work to which the *description* of natural
laws also belongs in e.g. physical works (they too are man made).
Now, I call rule systems of the kind mentioned '*historical rule
systems*'. Obviously every human community at a given time lives
within a set of such partly contemporary, partly traditional systems
related to each other in many different ways. These relations can be
those of practical motivation or of theoretical criticism of one by the
other etc. Very often one system is only an application of another one

to a different field or a deduction of one from another one with the help of special border conditions (of which I will give examples later on). Of course there are also such systems which are incompatible with each other and which are even incommensurable with each other. I call a system set of the sort just described 'historical system set' because it dominates during a special period, a special historical situation. And now it is my principal thesis that sentences on scientific facts as well as scientific principles are determined by a historical situation, that means by a certain historical system set. The reason is that sentences on scientific facts are dependent on the theories, which are part of a system set; and principles on the other hand are principles of theories and therefore they too are part of a system set. We never can break out of a historical system set to find absolute facts or absolute principles – they all are given and definable only within the frame of a historical system set.

From this point of view the debate on the question "Falsificationism or Inductivism" is completely fruitless. On the one hand no science can operate totally without inductions or totally without falsifications; on the other hand neither inductions nor falsifications as such are essentially important; what is essentially important is the context in a historical system set within which they are used. Within this context alone and nowhere else the value of inductive verifications or of falsifications is given and in this context alone it is really clear what has to be regarded as a fact and what not. This is also the reason, why LSE philosophers fail so often in their attempt to explain the history of science. Neither the decision between Ptolemy and Copernicus, nor the decision between Newton and Einstein, nor the decision between Einstein and Lorentz has been made on the basis of this philosophy, because these decisions were based mostly on different ontologies, understandable only in the context of a special historical situation and definable only within a special system set. Consequently in some examples some scientists could say that the excess empirical content exists, some scientists, however, could not acknowledge this content as an empirical one; in some cases the question of the empirical content does not play any role at all, etc. Falsity, corroboration, fact, empirical content, truth are not as simple as Popperians want to make us believe, and we have to be well aware of the context within a system set in which alone they can be used and understood. Looking at that special context we can resign ourselves to proving in a rather sophisticated way that Falsificationism is better than Inductivism. We can make understandable and we can often even recognize as rationally founded those examples of the history of science which Popperians can never explain.

A special case of one of them is the rise and fall of certain metatheories. So LSE philosophers think that their point of view is better in harmony with what they call the Bacon-Descartes-Ideal (whatever that may mean). But how can one justify this ideal *as such* as progressive? Take for example Huyghen's criticism of Descartes. He rejects his rationalism and his contempt of empirical facts (that is, what he calls Descartes' 'κριτήιον veri'). Well, this may be right, but why? In arguing against Descartes Huyghens has to argue in favour of other scientific aims and other *normative* concepts of science than the Cartesian ones and consequently he has to explain, for example, *why* it is better to predict more empirically testable facts than to deliver more 'clare et distincte' insights. The discussion of Descartes by Huyghens reminds us that the rules of the LSE philosophers for scientific progress are given too on the basis of certain normative concepts which are those of something we could call a sophisticated empiricism. Remember too Popper's criterion for distinguishing empirical sciences from metaphysics. Obviously he was convinced that this is a normative criterion and that scientists have to avoid metaphysical sentences if they are useless for testable contexts (falsifiable contexts). Now we may accept this or not; in any case the arguments for criteria of that kind are regarded as arguments for progress concerning the *philosophy of science* and *not* concerning a *special research program* of a particular empirical science. Consequently Huyghens' claim of having made progress through metatheoretical considerations of the kind mentioned, or the claim of the LSE philosophers of having made similar progress in the same field cannot be judged on their criteria regarding empirical content, support by facts etc. That means that these criteria are not applicable to all decisions in science with regard to the question "progressive or not progressive?"; on the contrary, they miss some of the most important and most revolutionary decisions ever made. These decisions on the other hand can only be justified if we consider the system set within which they took place.[4]

But we can also show that we need no idea of an absolute truth or the idea of a paradoxical approach to it in speaking of truth at all. The reason is that we have an implicit or explicit idea of truth in different system sets on the basis of which we can find true sentences. To find sentences of that kind we need a metatheoretical 'coordinate system' which also contains a special idea of truth. And again this coordinate system is not necessarily constructed just arbitrarily but can be introduced historically in a very reasonable way. But if we do not deny that we see true things even if our observation is dependent on the special construction of our eyes, why should we deny sentences to be true even if the truth of these sentences is dependent on special metatheoretical and changing presuppositions regarding the horizon within which true sentences can be found?

Thus, for example, by studying the main turning points in the interpretation of space during the history of science we can observe with particular clarity how presuppositions render certain decisions on truth or falsehood possible at all, and how the horizons within which these decisions take place are changed, and indeed changed by reasonable arguments and certainly not by pure arbitrariness within a special system set. So in antiquity, under the influence of Aristotelian philosophy, people believed in a non-Euclidean space of a special sort; Descartes having his starting point in a special rationalism presupposed the Euclidean space; and later Einstein thinking the equality of all coordinate systems to be a true metaphysical doctrine (namely as an expression of God's harmony) introduced the Riemann space into physics. In all these cases we find some kind of a priori decisions regarding space which are reasonably deducible from different systems within a system set; within a given historical situation (Aristotelian philosophy, Renaissance rationalism, traditional concepts of God etc.) – and *then* we can get true sentences in this special frame (measurements on the basis of the presupposed geometry, physical results on the same basis as, for example, that we have gravitational forces in a Euclidean space and no forces at all in a Riemann space). So I think it makes no sense to say that one geometry is nearer to the absolute truth than another one; but on the other hand in the frame of each we can find true sentences. Different theories and different geometries give different general ideas of truth within which we can operate empirically. But all those different ideas of truth are not purely arbitrary but more or less logically mediated by a special historical situation, by a special system set. So to cancel the idea of *absolute* truth and the paradoxical idea of approaching that truth does not mean to give up an idea of truth at all. And the historical point of view stressed in the theory of system sets is not at all a relativist one, because it has nothing to do with relativism if people react differently in *different* situations. We can speak of relativism only, if people react differently in the *same* situation, and in an arbitrary way. But this is not the case if we look at things in the light of the theory of system sets.

Summing up I think that with the exception of LSE philosophy's thesis 2 all other theses deliver metatheoretical rules which *may be* very reasonable in special cases. But we should beware of generalizing these rules and of extrapolating them for every scientific situation as a general norm. They are applicable in some cases, in some cases they are not. We should beware of seeing in them some form of *the* rationality in science and a means of reconstructing *the* rationality in historical progress. On the other hand it has nothing to do with some irrational elitism, as Popperians think, if we stress the historical context of system sets in which alone rationality can be present and developed. For the logic of a situation, the logic of detecting obscurities and contradictions in a given system set, the logic of removing both, and of creating connections which are as comprehensive as possible, or in other words the logic of harmonizing a system set within which we live and in which scientific systems are included, is not only not irrational but the only one way here to operate rationally, and that means objectively and intersubjectively. Every kind of rationalism and consequently so-called critical rationalism, too, has this flaw that it is blind to the variety and the complexity of history and that it succumbs to the temptation of too simple generalisations. Historicism on the other hand has – at least in the kind presented nothing to do with relativism or irrationalism but tries to consider decisive differences. And the theory of historical system sets delivers the categories with the help of which we can find them out and detect their situative logic.

Philosophisches Seminar, Universität Kiel

NOTES

[1] At least they were the main theses put forward by the philosophers of the London School of Economics during the Symposium held at Schloß Kronberg in July 1975.

² For a more extensive presentation of this theory and its application in different fields see: Hübner, K., 'Kritik der wissenschaftlichen Vernunft', Freiburg 1978. – 'On the Question of Relativism and Progress in Science', *Man and World* 7, 394–413 (1974). – 'Philosophische Fragen der Zukunftsforschung', *Studium Generale* **24**, 851–862 (1971). – 'Philosophische Probleme der Technik', in Lenk and Moser (eds.), *Techne-Technik-Technologie*, Dokumentation, Pullach, 1973. – 'On Theories in Historical Sciences', *Man and World* **8**, 363–382 (1975).

³ Some Popperians think that "all emeralds are grue" is *not* supported by past observations of green emeralds because these observations are part of 'background knowledge' at the time this hypothesis is proposed. Hence according to this opinion this hypothesis is *not* as well supported as is "all emeralds are green", which presumably was proposed in advance of at least some of these observations. This objection, however, seems to me completely irrelevant. We can easily imagine a man proposing the sentence "all emeralds are grue" in advance of the observations mentioned and *simultaneously* too proposing the sentence "all emeralds are green", and then "All emeralds are grue" would have been equally supported in the Popperian sense, too. The Goodman paradox is so convincing *because* it is not dependent on special situations but can be constructed every time to every general sentence of the kind mentioned.

⁴ See also my article: 'Descartes Rules of Impact and their Criticism. An Example of the Structure of Processes in the History of Science', *Boston Studies in the Philosophy of Science* **XXXIX**, 299–310 (1976).

*THE PROBLEM OF VERISIMILITUDE

The aim of the theory of verisimilitude is to answer the *semantical* question of what is intended when we say that one theory is closer to the truth than another. It is not intended to answer the *epistemological* question of how we can know that a theory is closer to the truth than another.[1] Miller has formulated the problem of verisimilitude as the question, "Are some theories *closer to the truth* than others are; and if so, what is the objective determinant of their greater truthlikeness?"[2] Miller has chosen this formulation in order to stress that the problem is not merely a verbal one of defining words or concepts. But when he says that he seeks an objective determinant of truthlikeness, he of course does not mean that we can have certain knowledge about it. From a fallibilistic point of view hypotheses about relative degrees of truthlikeness are highly conjectural, but given an objective determinant of truthlikeness they can be criticized. Thus a central problem for a theory of verisimilitude is to define the concept of verisimilitude in such a way that a critical discussion of conjectures about relative degrees of verisimilitude is possible.

The thesis of this paper is that it has not yet been possible to give a satisfactory definition of verisimilitude. First I will show the main reasons why Popper's comparative concept of verisimilitude is not tenable. Then I will show three problems confronting Popper's quantitative concept of verisimilitude. The last of these problems is of a fundamental nature and has forced me to regard also the quantitative concept of verisimilitude as unsatisfactory.

1. THE COMPARATIVE CONCEPT OF VERISIMILITUDE

In order to define the concept of verisimilitude Popper introduces the idea of a consequence class. The *consequence class* (or *content*) of a statement a is the set of its logical consequences, and is written $Cn(a)$ or A.[3] In Figure 1 the circle A represents such a consequence class. In the square are all statements under consideration. The true

G. Radnitzky and G. Andersson (eds.), Progress and Rationality in Science, 291–310.
All Rights Reserved.

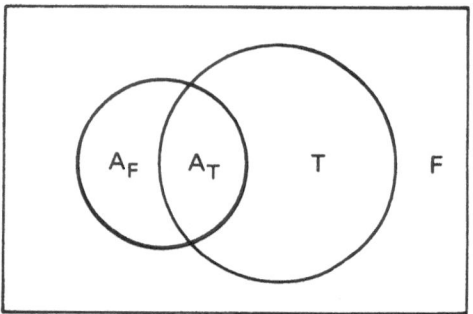

Fig. 1.

statements are collected in the circle T, the false statements are in the area F outside the circle T.

The set of all true consequences of a is called its *truth content* (A_T), and it can be defined as $A \cap T$. In the same way the set of all false consequences is called the *falsity content* of a (A_F) and can be defined as $A \cap F$. Intuitively speaking, the statement a with the consequence class (content) A is the closer to the truth, the greater its truth content and the smaller its falsity content is. In Figure 1 the circle A should cover as much as possible of the true area T, and as little as possible of the false area F. "The task of science is, metaphorically speaking, to cover by hits as much as possible of the target (T) of the true statements, by the method of proposing theories or conjectures which seem to us promising, and as little as possible of the false area (F)."[4] According to this idea the objective determinant of truthlikeness is sought in terms of the true and false consequences of a theory. In the case of more general theories we do not know all their deductive consequences and still less the truth values of them. But since we only want to answer the question of what can be *intended* by truthlikeness, not how it can be known, this need not worry us.

Can we use the ideas of truth content and falsity content in order to define a comparative concept of verisimilitude? The main problem is how to compare different truth contents and falsity contents. Suppose that we want to compare the two theories a and b with the contents A and B. Then the problem is how to compare A_T with B_T and A_F with B_F. If we assume that these sets are set-theoretically comparable, that is, that one is a subset of the other or that they are identical, then we can define a comparative concept of verisimilitude (v):[5]

DEFINITION 1: $v(A) < v(B) \leftrightarrow (A_T \subset B_T \& A_F \supseteq B_F \vee A_T \subseteq B_T \& A_F \supset B_F)$.

In Figure 2 $A_T \subset B_T$ and $A_F \supset B_F$. Hence $v(A) < v(B)$ according to Definition 1. In order to be comparable by v, two theories must have consequence classes related in a way similar to that in Figure 2.

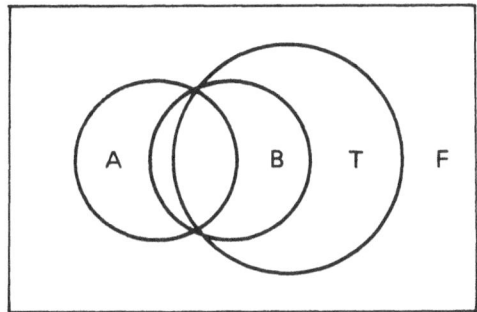

Fig. 2.

The main drawback of Definition 1 is, as Miller and Tichý have shown, that very few theories are comparable by v. It can be shown that very few consequence classes are related as required by Definition 1.

It follows immediately from Definition 1 that a false theory cannot be closer to the truth than a true theory. Suppose that A is true and B false. Then $\emptyset = A_F \subset B_F \neq \emptyset$, and it follows from Definition 1 that B cannot be closer to the truth than A.

A closer examination of the relation between truth contents and falsity contents shows that they vary rather directly with each other, if at least one of the theories is false. The following relations are valid:[6]

$$B_F \neq \emptyset \rightarrow (A_T \subset B_T \rightarrow A_F \subset B_F), \qquad (R1)$$

$$B_F \neq \emptyset \rightarrow (B_F \subset A_F \rightarrow B_T \subset A_T). \qquad (R2)$$

Proof: Let B be false and $A_T \subset B_T$. Assume that $A_F \not\subseteq B_F$. Then there is an $a \in (A_F - B_F)$ and a $b \in B_F$. $b \rightarrow a$ is an element of A_T and hence of B_T. Since b and $b \to a$ are elements of B, $a \in Cn(B)$. But $Cn(B) = B$, since B is a consequence class. Thus $a \in B$, against the choice of a. Thus $A_F \subseteq B_F$.

Since $A_T \subset B_T$, there is a $c \in (B_T - A_T)$. $b \& c$ belongs to B_F, but not to A_F, since otherwise $c \in A$, against the choice of c. Thus $A_F \neq B_F$, and $A_F \subset B_F$, which completes the proof of (R1).

Let B be false and $B_F \subset A_F$. Assume that $B_T \not\subseteq A_T$. Then there is a $b \in B_F$ and a $c \in (B_T - A_T)$. $b \& c$ is an element of B_F, and hence of A_F. Thus $c \in A$, against the choice of c. Therefore $B_T \subseteq A_T$.

Since $B_F \subset A_F$, there is a $d \in (A_F - B_F)$. $(b \to d) \in A_T$. Assume that $(b \to d) \in B_T$. Since $b \in B$, it follows that $d \in Cn(B) = B$, against the choice of d. Thus $(b \to d) \notin B_T$, and $A_T \neq B_T$, which completes the proof of (R2).

From the relations above it follows that two false theories cannot be compared with respect to v, and hence that two false theories cannot be related as in Figure 2. The reason is that if one false theory has more truth content than another false theory, then it also has more falsity content. This reduces the value of set-theoretical comparison of truth contents and falsity contents for a definition of verisimilitude.

Since Gödel we know that in every sufficiently rich language the set T of true statements is unaxiomatizable. If we furthermore assume that the theories to be compared are (finitely or recursively) axiomatizable, the following stronger versions of the relations above can be proved:

$$(A_F \neq \emptyset \lor B_F \neq \emptyset) \to (A_T \subset B_T \to A_F \subset B_F), \qquad \text{(R1')}$$

$$A_T = B_T \to A_F = B_F. \qquad \text{(R2')}$$

In the proof Popper's theorem on truth content will be used. Popper has shown that it is valid under the presuppositions that T is unaxiomatizable, and A and B are finitely axiomatizable. Miller has shown that it is valid also when A and B are recursively axiomatizable.[7] According to Popper's theorem on truth content:

$$(A \subset B \leftrightarrow A_T \subset B_T) \& (A = B \leftrightarrow A_T = B_T).$$

With its help (R1') and (R2') can be proved:

Proof: Assume that $A_T \subset B_T$. Then $A \subset B$ according to Popper's theorem on truth content, and $A_F \subseteq B_F$ according to set theory. Assume that $A_F = B_F$. Since A or B is false, $A_F = B_F \neq \emptyset$, and there is a

$b \in B_F$. Since $A \subset B$, there is a $c \in (B - A)$. $b \& c \in B_F = A_F$. It follows that $c \in A$, against our choice of c. Thus $A_F \neq B_F$, and $A_F \subset B_F$. This completes the proof of (R1').

Suppose that $A_T = B_T$. Then $A = B$, according to Popper's theorem on truth content. It follows that $A_F = B_F$. This completes the proof of (R2').

From (R1') and (R2') it follows that two false axiomatizable theories, or a true and a false axiomatizable theory cannot be compared with respect to v. For if at least one of the theories is false, then they have the same truth and falsity contents, or one has more truth content and more falsity content than the other. In both cases they are not comparable with respect to the qualitative concept of verisimilitude as given in Definition 1.[8]

The criticism of the comparative concept of verisimilitude can be summarized:

(1) a false theory cannot be closer to the truth than a true theory;

(2) two distinct false theories cannot be compared;

(3) in every sufficiently rich language axiomatizable theories can be compared only if they are true.

The first conclusion is counterintuitive. We can imagine interesting false theories with a lot of true information which we would prefer to any almost trivially true theory.

The second conclusion is serious, because one of the aims of the theory of verisimilitude is to allow us to compare false theories. This is for example important in order to understand what is meant by progress in the history of science, where we often think that one falsified theory can be closer to the truth than another falsified theory.

The third conclusion is devastating. According to it only true axiomatizable theories are comparable. But that a true theory has more truth content than another true theory means that it is logically stronger, and we would not need the theory of verisimilitude in order to prefer the logically stronger theory.[9]

2. THE QUANTITATIVE CONCEPT OF VERISIMILITUDE

One of the reasons for the inadequacy of the comparative concept of verisimilitude is the requirement of set-theoretical comparison of truth contents and falsity contents. But Popper has defined measures for truth contents and falsity contents, and with them arbitrary truth contents and falsity contents can be compared. Can the shortcomings of the comparative concept of verisimilitude be overcome with their help?

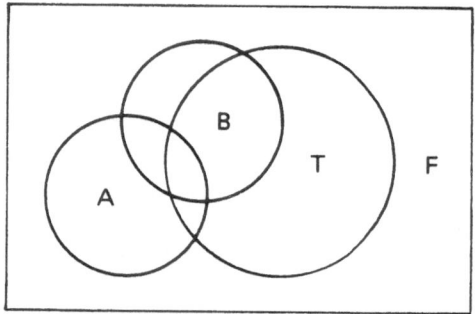

Fig. 3.

In Figure 3 A_T is not set-theoretically comparable with B_T, and A_F is not set-theoretically comparable with B_F. But since B covers more of the true area (T) and the less of the false area (F) than A does, we would intuitively think that B is closer to the truth than A is. With the help of measures for truth contents and falsity contents it seems possible to define a *quantitative concept* of verisimilitude which overcomes the limitations of the qualitative concept.

For the sake of simplicity I will limit the discussion to cases in which not only the theories discussed but also the set of true statements T are finitely axiomatizable. This means that for the theories A, B, ... and for the set T of true statements there are statements a, b, ... and t such that $a = Cn(a)$, $B = Cn(b)$, ... and $T = Cn(t)$.[10]

If $A = Cn(a)$, then the *measured content* (ct) of A can be defined as[11]

$$ct(A) = 1 - p(a).$$

If $T = Cn(t)$ and $A = Cn(a)$, it can be shown that[12]

$$A_T = A \cap T = Cn(a) \cap Cn(t) = Cn(a \vee t).$$

Thus we can define the *measured truth content* (ct_T) as[13]

$$ct_T(A) = 1 - p(a \vee t).$$

The falsity content A_F is not a consequence class, that is $A_F \neq Cn(A_F)$. (If $a \in A_F$ and $b \in T$, then $(a \vee b) \in Cn(A_F)$. But $(a \vee b) \in T$. Thus $(a \vee b) \notin A_F$. Hence $Cn(A_F) \neq A_F$, and it has been shown that A_F is not a consequence class.)[14] Hence a special definition is necessary for the *measured falsity content* (ct_F). Popper has argued that the following definition is adequate[15]

$$ct_F(A) = 1 - p(a, a \vee t).$$

Thus we get if $A = Cn(a)$ and $T = Cn(t)$:
(1) if A is true
$$ct_T(A) = 1 - p(a) \text{ and } ct_F(A) = 0$$
(2) if A is false
$$ct_T(A) = 1 - p(a \vee t)$$

and

$$\begin{aligned} ct_F(A) &= 1 - p(a, a \vee t) \\ &= 1 - p(a \& (a \vee t))/p(a \vee t) \\ &= 1 - p(a)/p(a \vee t). \end{aligned}$$

Now we can define a *quantitative concept of verisimilitude* (vs)[16]

DEFINITION 2: $vs(A) = ct_T(A) - ct_F(A)$.

By this definition the limitations of set-theoretical comparison of truth contents and falsity contents can be overcome. But unfortunately this definition gives rise to other problems.

2.1. *The Normalization of vs*

Popper has argued that if we are interested in numerical values of vs, it is preferable to multiply by a normalizing factor $1/(2 - ct_T(A) - ct(A))$. Thus normalized verisimilitude (vs') is defined:[17]

DEFINITION 3: $vs'(A) = vs(A)/(2 - ct_T(A) - ct_F(A))$.

Let *contrad* be a contradictory statement, and *Contrad* = $Cn(contrad)$. (*Contrad* consists of all statements in the language in question.) The main reason for the introduction of vs' is to give $vs'(Contrad)$ the minimum value -1, and $vs'(T)$ the maximum value $+1$. Without normalization we get $vs(Contrad) = 1 - p(t) - 1 = -p(t)$, and $vs(T) = 1 - p(t)$. With normalization $vs'(Contrad) = -p(t)/p(t) = -1$, and $vs'(T) = (1 - p(t))/(1 - p(t)) = +1$. We get[18]

$$-1 = vs'(Contrad) \leqslant vs'(a) \leqslant vs'(T) = +1.$$

The chosen normalizing factor does not, however, preserve the order of vs. Let us consider the language L_3 generated by three primitives a, b and c. The statements $a\&b\&c$, $a\&b\&\neg c$, ..., $\neg a\&\neg b\&\neg c$ are called the *constituents* of L_3. Every statement in L_3 can be written as a disjunction of (one or more) constituents. Since the constituents are mutually incompatible, jointly exhaustive and of equal logical strength, and since there are eight constituents in L_3, we get that the probability of one constituent is $\frac{1}{8}$. Suppose that $T = Cn(a\&b\&c) = Cn(t)$.

We want to compare the consequence classes of the two statements $\neg a\&b\&c$ and $\neg a\&\neg b$, and we note that $\neg a\&\neg b$ is logically equivalent with $\neg a\&\neg b\&\neg c \lor \neg a\&\neg b\&c$. $p(\neg a\&b\&c) = \frac{1}{8}$, $p(\neg a\&b\&c \lor t) = \frac{2}{8}$, $p(\neg a\&\neg b) = p(\neg a\&\neg b\&\neg c \lor \neg a\&\neg b\&c) = \frac{2}{8}$, and $p(\neg a\&\neg b \lor t) = \frac{3}{8}$.

Thus

$$vs(Cn(\neg a\&b\&c)) = \tfrac{1}{2} - \tfrac{2}{8} = \tfrac{6}{24}$$
$$vs(Cn(\neg a\&\neg b)) = \tfrac{2}{3} - \tfrac{3}{8} = \tfrac{7}{24}$$
$$vs'(Cn(\neg a\&b\&c)) = (\tfrac{1}{2} - \tfrac{2}{8})/(\tfrac{1}{2} + \tfrac{2}{8}) = \tfrac{1}{3}$$
$$vs'(Cn(\neg a\&\neg b)) = (\tfrac{2}{3} - \tfrac{3}{8})/(\tfrac{2}{3} + \tfrac{3}{8}) = \tfrac{7}{25}.$$

It follows that

$$vs(Cn(\neg a\&b\&c)) < vs(Cn(\neg a\&\neg b)),$$

but

$$vs'(Cn(\neg a\&b\&c)) > vs'(Cn(\neg a\&\neg b)).$$

But a correct normalization must not change the order between different hypotheses.[19]

The normalization of vs does not play any important role in the theory of verisimilitude. We can very well work with the non-normalized measure vs alone. It can also be asked if it really is desirable that the contradiction have a minimum value. From the contradiction follow all true and all false statements in the language in question. Supposing that they convey more or less the same true and false information, that is that $ct_T(Contrad) = ct_T(T) \approx ct_F(F) = ct_F(Contrad)$, we get

$$vs(Contrad) \approx 0.$$

Thus the non-normalized measure, according to which $vs(Contrad) = -p(t)$ seems adequate, since in richer languages $p(t) \approx 0$.

It can even be argued that it is desirable that we should normalize vs to vs^+ so that

$$vs^+(Contrad) = vs^+(Taut) = 0,$$

where $Taut$ is the set of all tautologically true statements. This would be achieved by the following normalization of ct_F:[20]

$$ct_F^+(A) = ct_F(A)ct(T).$$

If we define

$$vs^+(A) = ct_T(A) - ct_F^+(A)$$

we get

$$vs^+(Contrad) = vs^+(Taut) = 0.$$

2.2 The Verisimilitude of General Theories

What is the verisimilitude of general theories? According to ordinary interpretations of the calculus of probability, the probability of a general theory is zero. If we are interested only in the content of a general theory A with $A = Cn(a)$, we get

$$ct(A) = 1 - p(a) = 1.$$

Thus every general theory has the measured content 1, and ct cannot discriminate between different general theories.

Nevertheless it is obvious that different general theories can convey different degrees of information. Consider the case when a theory A unilaterally entails a theory B. Then

$$B \subset A$$

and A conveys more information than B, simply because A is logically stronger and has more deductive consequences than B. But nevertheless $ct(A) = ct(B)$. The measure ct is not fine enough to discriminate between A and B.

Since ct_T and ct_F ultimately are defined in terms of probabilities of general statements, it is not astonishing that we get similar problems when we want to compare general theories with respect to verisimilitude (vs).

Suppose as earlier that $A = Cn(a)$, $B = Cn(b)$ and $T = Cn(t)$, where T is the set of true statements. If

$$p(a) = p(t) = 0$$

as we have reasons to assume for a general a and for t in richer languages, we get that

$$p(a) = p(a \lor t) = 0$$

and that

$$ct_T(A) = ct(T) = 1$$

irrespective of whether A is true or false, and that $ct_F(A)$ is undefined if A is false.

Miller takes $p(A)$ as a measure over the set of models of A. Under the presupposition that the set of true statements T is unaxiomatizable, he has argued that we have to set $ct_T(A) = 1$ and $ct_F(A) = 0$ for every general theory A, irrespective of whether A is true or false.[21] Thus

$$vs(A) = ct_T(A) - ct_F(A) = 1 - 0 = 1$$

for every general true or false theory, and vs cannot even discriminate between true and false theories.

The reason for this problem is not the definition of vs, not even the definitions of ct_T and ct_F, but an interpretation of the calculus of probability such that the probability of every universal statement is zero.

Popper has given two suggestions for how the problem of comparing contents and probabilities of general theories can be solved. According to Popper's axiomatization of the calculus of probability we get when a unilaterally entails b[22]

$$p(a, b) = 0 \quad \text{and} \quad p(b, a) = 1.$$

In this case we can discriminate between general theories by contrasting $p(a, b)$ and $p(b, a)$. Using this idea Miller has shown that a sort of discrimination can be achieved for the verisimilitudes of different general theories, and that "to some extent it does seem possible to overcome the more distressing consequences of infinitude in this area".[23]

Popper's second idea for comparing contents and probabilities of general theories is an attempt to define a fine structure of content (CT) and probability (P).[24] For P, the fine structure of probability, we have the desiderata

(1) $A \subset B \rightarrow P(A) < P(B)$, and

(2) if $p(a) < p(b)$ for every sufficiently large finite universe, we put $P(A) < P(B)$ in the infinite case ($A = Cn(a)$ and $B = Cn(b)$), in spite of the fact that $p(a) = p(b)$.

The problem of giving a definition of the fine structure of content or probability is still unsolved. But there seem to be good reasons for thinking that it can be solved. In the simple case when a consequence class is a subset of another, it seems indisputable that such a fine structure exists. In the general case, when neither of the consequence classes A and B is a subset of the other, we notice that it is enough if we can compare $p(B)$ with some appropriate subset (or superset) of A. If $A' \subset A$ and if we know that $P(A') > P(B)$, it follows that $P(A) > P(B)$.

Assuming that the fine structure of probability (P) exists, we can define the fine structures of content (CT) and verisimilitude (VS):

$$VS(A) = CT_T(A) - CT_F(A),$$

where CT_T and CT_F are the fine structures of measured truth and falsity contents. Then the problem of comparing two general theories A and B would boil down to the problem of making conjectural estimations of $(CT_T(A) - CT_F(A))$ in relation to $(CT_T(A) - CT_F(B))$, since

$$VS(A) < VS(B) \leftrightarrow (CT_T(A) - CT_T(B)) < (CT_F(A) - CT_F(B)).$$

These considerations make the problem of the verisimilitude of general theories seem solvable, either by working with relative probabilities, or by working with the fine structures of probability, content and verisimilitude. So far we have had no reasons to suspect that the basic ideas behind the quantitative concept of verisimilitude are not adequate. The problem seems only to be to get fine enough measures for probability and content. In the next section we will however encounter more fundamental problems.

2.3. The Verisimilitudes of False Theories with the Same Measured Contents

A closer look at the definition of verisimilitude shows that all false theories with the same measured contents have the same degrees of verisimilitude. If for the sake of simplicity we assume axiomatizability of A and T, we get for a false theory A $vs(A) = ct_T(A) - ct_F(A) =$

$p(a)/(p(a) + p(t)) - (p(a) + p(t))$. But $p(t)$ is a constant. Thus $vs(A)$ is a function only of $p(a)$. This can also be expressed by saying that the verisimilitude of a false theory is a function only of its (measured) content.[25]

Intuitively false theories with the same measured contents can differ with respect to truthlikeness. Let us suppose that Newton's theory of gravitation is false. From the definition of vs it would follow that it has the same verisimilitude as an arbitrary false theory with the same measured content. This is startling, because we would think that the verisimilitude of Newton's theory is higher than that of any arbitrary false theory with the same measured content. We would also think that the verisimilitude of a false theory should be a function not only of *how much* we are saying, but also of *what* we are saying.

Intuitively we would think that (conjectural) estimation of the relative degrees of verisimilitude of two false hypotheses should be based not only on comparison of their (measured) contents, but also on the results of empirical tests. (The results of empirical tests would be one of the reasons for preferring Newton's theory of gravitation to any arbitrary theory with the same measured content.) But according to the definition of vs we do not have to consider the results of empirical tests at all once it is known that the hypotheses in question are false. The only information we need is that the falsity contents of the two theories to be compared are not empty. Then any further information about the elements of the falsity contents would be irrelevant for comparison with vs.

Why can't vs discriminate between different false hypotheses with the same measured contents? When we discussed the comparison of general hypotheses with respect to vs there was also a lack of discrimination, but there was some hope that it could be overcome by refined measures. Is a similar solution possible in this case?

Such a solution would be possible if there were significant differences between the truth and falsity contents of false hypotheses with the same measured content. Let us for a moment assume that consequence classes could be related as in Figure 4. In Figure 4 A and B have the same measured contents (since the circles representing them have the same diameters). Both have false elements. But B covers more of the true area T and less of the false area F than A does. Intuitively B would be closer to the truth than A. But according to our measure *vs*, all false hypotheses with the same measured contents have the same degrees of verisimilitude.

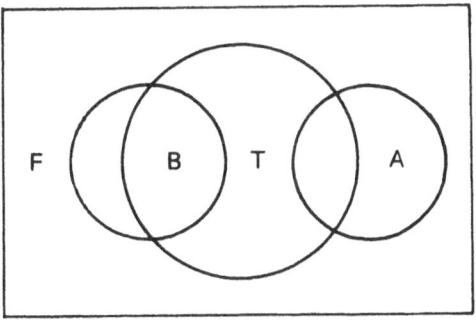

Fig. 4.

If the consequence classes of two false hypotheses with the same measured contents are related as in Figure 5 this result is to be expected. In Figure 5 the consequence classes of *A* and *B* are perfectly symmetrical with respect to *T* and *F*, and the reason for the lack of discrimination of *vs* between *A* and *B* would not be that our measures are not fine enough. The problem would be of a much more fundamental nature.

Are consequence classes of two false hypotheses related as in Figure 4 or 5? The measures ct_T and ct_F give us no answer to this question. If we look at the definition of ct_T we find (if $A = Cn(a)$ and $T = Cn(t)$)

$$ct_T(A) = 1 - p(a \vee t).$$

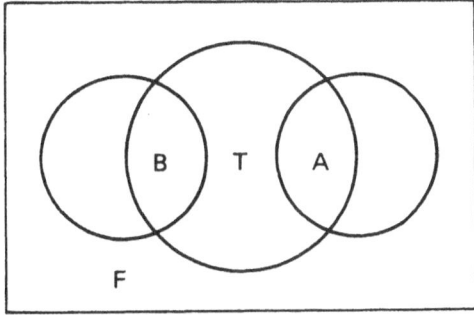

Fig. 5.

According to the calculus of probability

$$p(a \vee t) = p(a) + p(t),$$

if a is false and t is true. For the additivity of the probability measure it is enough that $A_F \neq \emptyset$. Thus if $A_F \neq \emptyset$, $B_F \neq \emptyset$, and $ct(A) = ct(B)$, $ct_T(A) = ct_T(B)$, irrespective of whether A and B are related as in Figure 4 or 5. Similar considerations apply for ct_F. If $A_F \neq \emptyset$, $B_F \neq \emptyset$, and $ct(A) = ct(B)$ then $ct_F(A) = ct_F(B)$. Thus $vs(A) = vs(B)$ under these conditions. Is this the explanation for why vs cannot discriminate between false hypotheses with the same measured contents? This would be the case if consequence classes could be related as in Figure 4.

But can they? In order to answer this question an investigation of the properties of consequence classes is necessary.

In section 2.1 above we discussed the language L_3 generated by the primitives a, b and c. Here we will discuss the simpler language L_2 generated by the primitives a and b. The statements $a \& b$, $a \& \neg b$, $\neg a \& b$ and $\neg a \& \neg b$ are the constituents of L_2, and every statement in L_2 can be written as a disjunction of one or more constituents. Since there are four constituents in L_2 and since they are mutually incompatible, jointly exhaustive and of equal logical strength, the probability of one constituent is $\frac{1}{4}$. We assume that $T = Cn(a \& b)$.

Now we want to compare $Cn(a \& \neg b)$ with $Cn(\neg a \& \neg b)$. Are there any significant differences between the true and false consequences of $a \& \neg b$ and $\neg a \& \neg b$? Tichý has argued that intuitively $a \& \neg b$ is closer to the truth than $\neg a \& \neg b$. The reason is that $a \& \neg b$ gets one primitive right, but $\neg a \& \neg b$ none.[26]

In order to investigate the consequence classes of $a \& \neg b$ and of $\neg a \& \neg b$ we construct the lattice ordered by the entailment relation of all statements in L_2.

The lattice in Figure 6 is constructed by combining (disjunctively) the constituents of L_2 in all possible ways. So, for example, $a \leftrightarrow b = a \& b \vee \neg a \& \neg b$, and $a \rightarrow b = a \& b \vee \neg a \& b \vee \neg a \& \neg b$. We suppose that $T = Cn(a \& b)$.

By investigating the lattice of L_2 we find that the consequence classes of $a \& \neg b$ and $\neg a \& \neg b$ are related as in the Figures 7 and 8. For every element in one of the consequence classes there corresponds an element in the other consequence class with the same truth value and measured content. To the true element a in

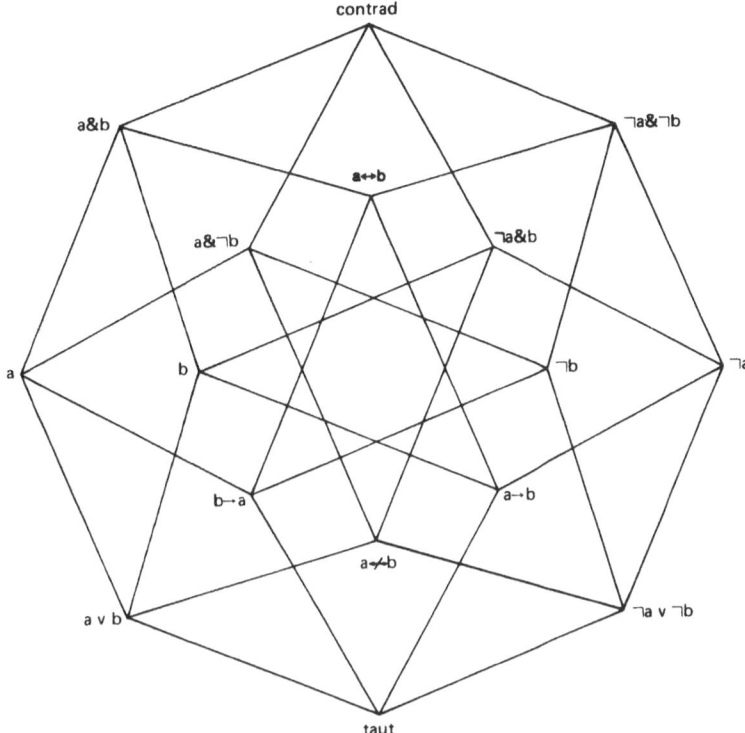

Fig. 6.

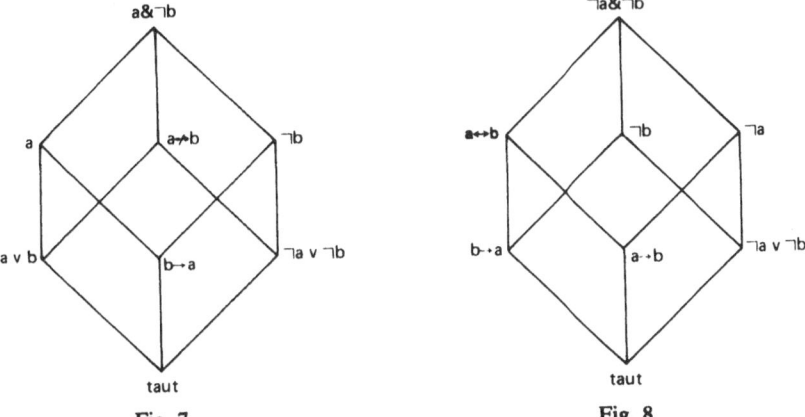

Fig. 7.   Fig. 8.

$Cn(a\&\neg b)$ there corresponds for example the true element $a\leftrightarrow b$ in $Cn(\neg a\&\neg b)$, and $ct(a) = ct(a\leftrightarrow b) = \frac{1}{2}$.

The truth contents of the hypotheses are:

$$Cn(a\&\neg b) \cap T = \{a, b \rightarrow a, a \vee b, taut\}$$
$$Cn(\neg a\&\neg b) \cap T = \{a\leftrightarrow b, b \rightarrow a, a \rightarrow b, taut\}.$$

The falsity contents of the hypotheses are:

$$Cn(a\&\neg b) \cap F = \{a\&\neg b, \neg b, a\nleftrightarrow b, \neg a \vee \neg b\}$$
$$Cn(\neg a\&\neg b) \cap F = \{\neg a\&\neg b, \neg b, \neg a, \neg a \vee \neg b\}.$$

The reasons for these symmetries are combinatorial. The falsity content of $a\&\neg b$ can be found by investigating all possible (disjunctive) combinations with the other two false constituents $\neg a\&b$ and $\neg a\&\neg b$. In order to find the falsity content of $\neg a\&\neg b$ we must investigate all possible (disjunctive) combinations with the false constituents $\neg a\&b$ and $a\&\neg b$. In a similar way the truth content of $a\&\neg b$ is found by investigating all possible combinations of $a\&\neg b \vee a\&b = b$ with the constituents $\neg a\&b$ and $\neg a\&\neg b$, and the truth content of $\neg a\&\neg b$ is found by investigating all possible combinations of $\neg a\&\neg b \vee a\&b = a\leftrightarrow b$ with $\neg a\&b$ and $a\&\neg b$.

There are no reasons to prefer for example the true element a in $Cn(a\&\neg b)$ to the true element $a\leftrightarrow b$ in $Cn(\neg a\&\neg b)$. Both elements are true, with the same measured content, and tell us different things. From the point of view of the information conveyed by them, none is more 'natural', relevant or interesting than the other. And if we are considering the true and false consequences of statements, there are no reasons to prefer $a\&\neg b$ to $\neg a\&\neg b$ with respect to truthlikeness.

All false statements with same measured content consists of the same number of constituents in the (disjunctive) normal form. Therefore, if we consider a language L_n generated by n primitives, the same symmetries will appear as in L_2. The language L_n has 2^n constituents. All statements in L_n can be constructed by combining (disjunctively) the constituents in all possible ways. For combinatorial reasons the truth and falsity contents of statements with the same measured content (and hence consisting of the same number of constituents) will be perfectly symmetrical, as in L_2.

The falsity content of a false statement in L_n is found by writing it in the (disjunctive) normal form and by combining it in all possible ways with the other false constituents (which do not appear in the normal form of the statement). Similarly, the truth content is found by first adding (disjunctively) t, and then investigating all possible combinations with the other constituents.

Thus if we are considering consequence classes, there are no significant differences between the truth and falsity contents of false statements with the same measured contents. We have no reasons to prefer one of them to another with respect to verisimilitude (vs).

What are the consequences of these results? Should we say that as a matter of fact and against our intuitions all false statements with the same measured contents have the same verisimilitude? Or should we say that we have not succeeded in giving an adequate definition of verisimilitude? I think that the latter conclusion is the correct one.

Popper has discussed the following example. We want to compare the truthlikeness of the two statements "it is now 9.45 p.m." and "it is now 9.40 p.m." when in fact it is 9.48 p.m.[27] Of course we would say that "it is now 9.45 p.m." is closer to the truth. But (assuming that both conjectures have the same measured contents) according to the definition of the quantitative concept of verisimilitude (vs), we must say that both statements have the same degree of verisimilitude. This result is so counterintuitive that we must conclude that the quantitative concept of verisimilitude is inadequate.

Popper has suggested that in order to overcome this difficulty we should consider the statements "it is now between 9.45 p.m. and 9.48 p.m." and "it is now between 9.40 p.m. and 9.48 p.m.". With these *interval statements* the quantitative verisimilitude (vs) of the first statement is greater than that of the second statement, since both statements are true, and since the first statement has more content than the second one.[28] This means that the original attempt to get a measure for truthlikeness by considering consequence classes alone has been given up.

If we return to the earlier example in the language L_2, we had $t = a \& b$, $h = a \& \neg b$ and $h' = \neg a \& \neg b$. $h \vee t = a \& \neg b \vee a \& b = a$, and $h' \vee t = \neg a \& \neg b \vee a \& b = a \leftrightarrow b$. Now we could argue that $\neg a \& \neg b$ is separated from the true statement $a \& b$ by $a \& \neg b$ (or by $\neg a \& b$), and that thus the interval statement corresponding to h' should be $h'' = \neg a \& \neg b \vee a \& \neg b \vee a \& b = b \rightarrow a$. If we look at the truth contents:

308   GUNNAR ANDERSSON

$$Cn(h') \cap T = \{a \leftrightarrow b, a \rightarrow b, b \rightarrow a, taut\}$$
$$Cn(h'') \cap T = \{ \qquad\qquad b \rightarrow a, taut\}$$
$$Cn(h) \cap T = \{a \quad , a \vee b, b \rightarrow a, taut)$$
$$ct(Cn(h'')) < ct(Cn(h')) = ct(Cn(h)), \text{ and } vs(Cn(h \vee t)) > vs(Cn(h'')).$$

Thus we can say that h is closer to the truth than h', since from h we can derive the true interval statement (or approximation theory) $a \& b \vee a \& \neg b = h \vee t$, while from h' we can only derive the true interval statement $a \& b \vee a \& \neg b \vee \neg a \& \neg b = h''$, and $vs(Cn(h \vee t)) > vs(Cn(h''))$.

But if we stick to the original idea of defining verisimilitude by considering the true and false consequences of statements, it is unclear why we should disregard some true consequences of h', as we do by investigating $Cn(h'')$ instead of $Cn(h')$. h' has the true consequences $a \leftrightarrow b$ and $a \rightarrow b$, which do not follow from the interval statement h''. A possible advantage of considering interval statements is that we can compare them directly, without using the idea of consequence classes, since one is logically stronger than the other.

For theories predicting a numerical constant the interval statements can be called *approximation theories*. So we could say that from the statement "it is now 9.45 p.m." follows the true approximation theory "it is now between 9.45 p.m. and 9.48 p.m.", which does not follow from the statement "it is now 9.40 p.m.". But it is still an open problem whether this approach can be generalized for theories predicting more than numerical constants.[29]

3. CONCLUSIONS

The conclusion of this paper is that the problem of giving an adequate definition of verisimilitude is still unsolved. For the time being we can only use the concept of truthlikeness or of better correspondence with the facts in an intuitive way. Perhaps those impressed by *l'esprit géométrique* will respond that one could defend the any triviality in this way. On the other hand the history of the development of for example the infinitesimal calculus teaches us that precise definitions are not always a necessary condition for progress.

The absence of a precise or formal definition alone should not prevent us from using the concept of truthlikeness in an intuitive way. This does not mean that the task of giving such a definition is unimportant. On the contrary, it is one of the most fascinating unsolved problems in the modern philosophy of science.

But it would be an exaggeration to regard a precise definition of truthlikeness as an necessary part of the philosophy of science of critical rationalism. We can have many other reasons for preferring a false hypothesis to another than the conjecture that one of them is closer to the truth. As Watkins has shown in this volume, we can for example compare their empirical contents.

Universität Trier

NOTES

*I am indepted to Barbara Hill for criticism of this article, and to David Miller for criticism of an earlier draft of it.

[1] Popper [1963], p. 234.

[2] Miller [1976], p. 380.

[3] Popper [1972], pp. 47ff and 329ff.

[4] Popper [1972], p. 54.

[5] Popper [1963], p. 233. Cf. Miller [1974], p. 170, and Tichý [1974], p. 156. The tautological consequences are unproblematic in Definition 1, since they belong to every truth content, and hence no two truth contents differ only with respect to them.

[6] (R1) and (R2) can also be proved with the theorems in Miller [1974]. In Tichý [1974] p. 156f and Harris [1974] p. 163 weaker forms of the relations are proved. For Miller's and Tichý's criticism of v, see Miller [1974] and Tichý [1974].

[7] Popper [1966] pp. 350f, Miller [1974], p. 174.

[8] Cf. Miller [1974], p. 173f.

[9] Miller [1974], p. 174.

[10] If T and the theories discussed are not axiomatizable, other considerations are brought to bear. See Popper [1972], Chapter 9. Popper indicates in his [1963], p. 392, that in expressions like $p(T)$ T should be understood as the (finite or infinite) conjunction of all the elements belonging to T. If $p(A)$ is understood in the same way, and if $A = Cn(a_1)$, then we get according to the calculus of probability $p(A) = p(a_1 \& a_2 \& a_3 \& ...) = p(a_1)p(a_2 \& a_3 \& ... , a_1) = p(a_1)$.

[11] Popper [1963], p. 392.

[12] Popper [1963], pp. 392f.

[13] Popper [1963], p. 393.

[14] Cf. Popper [1963], p. 393: "[A_F] is not a consequence class. For while T contains all the logical consequences of T – since the logical consequence of anything true must also be true – F does not contain all its logical consequences: while from a true statement only true statements follow, from a false statement follow not only false statements but always true statements also." In a private communication Miller has shown that if $A_F \neq \emptyset$, then $Cn(A_F) = A$. If $f \in A_F$ and $a \in A$, then $a \& f \in A_F$. Thus $a \in Cn(A_F)$, and $A = Cn(A_F)$.

[15] Popper [1963], pp. 393ff. Cf. Popper [1972], pp. 333f.

GUNNAR ANDERSSON

[16] Popper [1963], p. 234, and [1972], p. 334.
[17] Popper [1963], pp. 396f. Cf. Popper [1972], p. 334.
[18] Cf. Popper [1963], p. 396, desiderata (iv) and (v).
[19] Cf. Tichý [1974], pp. 158f, for a discussion of a similar example. The values for vs_1 $(Cn(\neg p \& \neg q))$ and $vs_2(Cn(\neg p \& \neg q))$ (corresponding to $vs(Cn(\neg a \& \neg b))$ and $vs'(Cn(\neg a \& \neg b)))$ in the table on p. 159 are not correct, and hence Tichý has not seen the inadequacy of the normalization of vs.
[20] The normalizations ct_F^+ and vs^+ have been suggested by Miller in a private communication.
[21] Miller [1972], p. 54.
[22] Popper [1959], p. 375.
[23] Miller [1972], p. 55.
[24] Popper [1959], pp. 375ff. For $P(A)$, cf. Note 10 above.
[25] Cf. Tichý [1974], pp. 157ff, and Miller [1975], p. 162.
[26] Tichý [1974], p. 159.
[27] Popper [1972], pp. 55f.
[28] Popper [1963], pp. 397f; and Popper [1972], p. 56.
[29] Popper [1976], pp. 155f.

BIBLIOGRAPHY

Harris, J. H.: 'Popper's Definitions of Verisimilitude', *British Journal for the Philosophy of Science* **25**, 160–66 (1974).
Miller, D.: 'The Truth-Likeness of Truthlikeness', *Analysis* **33**, 50–55 (1972).
Miller, D.: 'Popper's Qualitative Theory of Verisimilitude', *British Journal for the Philosophy of Science* **25**, 166–77 (1974).
Miller, D.: 'The Accuracy of Predictions', *Synthese* **30**, 159–91 (1975).
Miller, D.: 'Verisimilitude Redeflated', *British Journal for the Philosophy of Science* **27**, 373–81 (1976).
Popper, K. R.: *The Logic of Scientific Discovery*, Hutchinson, London, 1959.
Popper, K. R.: *Conjectures and Refutations*, Routledge & Kegan Paul, London, 1963.
Popper, K. R.: 'A Theorem on Truth-Content', in P. Feyerabend and G. Maxwell (eds.), *Mind, Matter, and Method. Essays in Philosophy and Science in Honor of Herbert Feigl*, University of Minnesota Press, Minneapolis, 1966.
Popper, K. R.: *Objective Knowledge*, Oxford University Press, London, 1972.
Popper, K. R.: 'A Note on Verisimilitude', *British Journal for the Philosophy of Science* **27**, 147–59 (1976).
Tichý, P.: 'On Popper's Definitions of Verisimilitude', *British Journal for the Philosophy of Science* **25**, 155–60 (1974).

HEINZ POST

OBJECTIVISM VS SOCIOLOGISM

1. INTRODUCTION

The aim of this paper is to bring out a major (philosophic) difference between the direction in which Lakatos and his followers are trying to develop Popper's methodology of appraisal, and what I might call the 'objective' methodology of appraisal to which I subscribe.[1]

I define a methodology of appraisal of scientific theories as 'objective' if all its criteria bar one are in principle 'internal', i.e., depend entirely on an analysis of the structure and content of the theories. The remaining criterion is the criterion 'agreement with experience'.

This last criterion, in common with all scientific criteria, is, of course, problematic. However, since that problem is irrelevant to the difference between 'MSRP' and 'objective' methodology, I will not pursue it here. I might even add that I am in general sympathy with Imre's softening up of 'refutationism' in favour of a more realistic description of scientific practice.

In the same spirit I shall make some points which I believe to be neutral on the question 'MSRP vs objectivity' but which are necessary as a basis for any theory of theories. Any polemic might well start, like a legal discussion, by specifying points of agreement. Unlike Kuhn, Agassi and others, I share the somewhat schizoid attitude of Popperians towards truth: We must have the notion of truth in any case at least at a meta-level, although we have no way of achieving certain truth in science. Since there is no ground for quarrel here, I will not discuss this question further. In any case I take it that our letter to Santa Claus should refer to actual problems of appraisal in science rather than be a rehearsal of general philosophy.

Since the difference in methodology I wish to discuss does not lie in a difference in approach to the problem of truth, or refutation, we might as well use the criterion of refutation (admitting this is to be a naive approximation) and say that our one external criterion will be whether a theory has been tested, and whether the tests did or did not result in refutation.

I am not primarily concerned with the substitution of 'program' for

G. Radnitzky and G. Andersson (eds.), Progress and Rationality in Science, 311–318.
All Rights Reserved.
Copyright © 1978 by D. Reidel Publishing Company, Dordrecht, Holland.

'theory' except in so far as it involves the importation of sociology. Again, I would say that I agree that for an *historian* of science references to research programmes may offer a more realistic description than references to individual theories.[2]

2. THE PROBLEM

The problem, then, narrows down to this: Given two theories T_1 and T_2, by what criteria can and should we decide to prefer T_2 over T_1?[3]

3. NEUTRAL BACKGROUND REMARKS

I shall list here some points which I take to be neutral with respect to the polemic between the sociological approach of LSE and the objective approach.

(1) *Background theory*: In comparing two theories it will, in general, be necessary to consider all other relevant theory for the sake of fair comparison and coherence. The relevant background theory may well include statements that have not appeared explicitly in print precisely because they were taken for granted. Thus, some 'ad hoc' theory such as attributing significance to the correlation between the value of the electronic charge obtained in experiments and the initials (or even IQ's) of the experimenters are, rightly or wrongly, rejected on the grounds that the theory offends against an implicit doctrine of the irrelevance of the names (or IQ's) of experimenters.

(2) *Truth*: The criterion 'agreement with experience' is formulated as 'survived all potentially refuting tests'.

(3) *Interpretation*: All actual theories are only partially interpreted empirically. We shall divide the consequence class of theorems, T, into a subset, E, containing all statements that are at least partially interpreted empirically, and the non-empty complement $T - E$, containing statements that are not empirically interpreted at all. This complement contains amongst other things logic, mathematics and what Popperians call 'metaphysics'. Empirical interpretation is taken to mean here refutability by empirical means.

(4) *Commensurability*: We shall side-step this and related problem by confining ourselves as far as possible to the relation of class inclusion between T_1 and T_2.[4]

(5) *Universality*: Since we are concerned with deductive systems we shall not refer to singular statements which can only be derived by the addition of 'initial conditions'. We shall test theories against agreed universal statements based on experiment, such as 'the charge of the electron is 4.8. e.s.u.' This is a realistic account of the reported results of experiments, and incidentally side-steps some of Imre's ceteris paribus problems.

4. THE AIM OF OBJECTIVITY

My letter to Santa Claus reads, in toto: "Please give me adequate acceptable objective criteria for preferring T_2 to T_1".

I will explain this message, not for the benefit of Santa Claus, who can be presumed all-wise, but for the benefit of those Popperians who have strayed. The function of philosophy of science is to explain the phenomenon of science – that science is possible, that it is successful, that there is progress in science. This function disappears if philosophy of science adopts a 'positivistic' sociological definition of what constitutes success. Again, our normative programme must not include sociological criteria, i.e., criteria that depend explicitly on sociological phenomena such as the behaviour of a majority or elite of scientists or 'psychological' considerations such as the state of awareness of the author or the theory, or 'historical' conditions such as the chronological ordering of tests and theories.[5]

This is not to say that we may not glance at history of science once we have formulated our criterion, to see how far it was observed in a particular period. This is the usual explicans-explicandum situation: The explicans should be sharply defined formally. This can only be done if a consistent demarcation is formally possible, a fact which is itself philosophically significant. It may however turn out that the explicans so defined does not overlap the vague explicandum in practice at all. In that case the labelling of the explicans should be changed.

The mere fact that a certain procedure is frequently followed in practice does not render that practice acceptable to an objective methodology of appraisal. Indeed even the fact that a practice following certain criteria generally leads to "success" does not justify those criteria in an objective programme: A practice of editors to decide on the acceptability of papers on the basis of what is known about the authors may well lead to 'success' in the sense of weeding out the worst and encouraging some of the best work. It may indeed be *reasonable* for editors to follow that practice. Nevertheless, the criterion of 'personal reputation of the author' is no solution to the problem of the philosopher to find criteria of evaluation. The matching to scientific practice is no guarantee of philosophic respectability: Scientists may be *well advised* to use subjective external criteria such as the reputation of an author, his schooling etc, in deciding whether to take a paper of his seriously. But philosophy of science has *failed* in *its* function if it merely follows this practice. So much for the LSE 'Case Studies' to 'justify' criteria.

Conversely, the mere historic fact that a 'research programme' stagnates during a certain period is not an objective ground to declare the programme 'degenerate'. Scientists may be doped, complacent, biassed, lazy, uninformed, lacking opportunity or equipment, or under political constraint causing the programme to be pursued less energetically than it might have been.

We must sharply distinguish between the *formulation* of a criterion and its application: The state of knowledge of the author of a theory ('psychologism') plays no part in the evaluation. We are not concerned with awarding medals for personal ingenuity of authors but with the objective worth of the theory. However, in *applying* an objective criterion such as 'agreement with experience' we can only use what experience is to hand, the knowledge whether the theory has survived or failed in tests.

A good deal of the LSE programme can be explained as an occasionally rather desperate attempt to eliminate 'ad-hocness'. It should be stated here that from the objective point of view it is quite irrelevant whether a theory has been created pragmatically 'ad-hoc'. We are solely concerned with the objective value of the result. A scientist may well have 'cheated' in using experimental results that were known privately to him, without declaring the fact; but we are not concerned with the worthiness of the scientist but with the usefulness of the theory to us. Indeed, it is clear that ad-hocness in the pragmatic sense is not only non-vicious from the objective point of view, but can of course not be eliminated by formal means.[6]

The main draw-back to appraisal by the 'methodology of research programmes' is the latitude it allows to subjectivity in the 'reconstruction' of the 'research programme', the 'core', and the sociological component in its measurement of 'success'.

5. SOLUTION OF THE PROBLEM

I claim that the following condition is sufficient[7] to meet the requirements stated in the letter to Santa Claus:

$$\Delta E > 0$$

ΔE stands for $E_2 - E_1$ where E_1, E_2 are respectively the empirical parts of T_1 and T_2.

A theory T is here taken to mean the consequence class of all theorems. E refers to that subset of T that contains only statements that are at least partially interpreted empirically, i.e., are at least partially refutable. Thus T is closed under deduction whilst E is not. $T - E$, the complement, is deductively closed being the set of all theorems that are not empirically refutable at all. This may be represented by the following diagram:

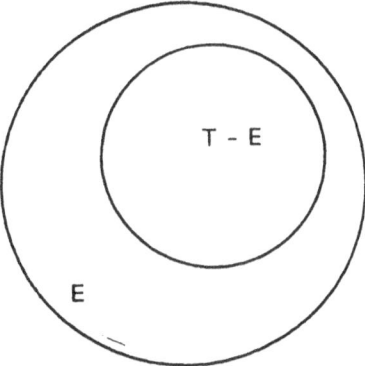

The convexity of $T - E$ and of T indicate their closure under deduction. It can be seen that E, not being convex, is not closed under deduction.

As explained above, we are side-stepping the independent problem of 'commensurability' by concentrating on the most trivial case, the relation of class inclusion between two theories.

Our condition for rational grounds of preference for T_2 as against T_1 is then that there should be an increase formally in the empirical content of the theory in going from T_1 to T_2: That this increase should have been tested to some extent; and that these tests should not have led to refutation.

The latter two pragmatic requirements have been explained above on pages 311 and 312. They are the implementation of the requirement 'agreement with experience'. For some purposes it is more convenient to revert to the Popperian 'class of potential falsifiers', R, as long as it is made clear that all the statements in R are universal statements as explained in Section 3(5).[8]

6. APPENDIX ON 'NORMAL SCIENCE'

This is to discuss the case of 'normal science' defined as scientific achievements *not* involving any change in the foundations.[9] (It will be seen that we do without reference to any more or less ineffable socio-cultural concept of 'paradigm' or 'core'.)

This common activity of scientists, much despised by Popperians and others, should perhaps receive an 'honourable mention' since, speaking informally, much progress in knowledge and understanding can be and is achieved by it, and it in fact covers the activity of most scientists, particularly in an experimental context. The question then arises whether normal science so defined stands condemned if we take our condition $\Delta E > 0$ to be necessary for progress. The answer is, I think, no, for at least two reasons:

(1) Normal Science consists in the application of already existing theories to a particular problem (which still covers a potential infinity of instances). There is still a ΔE in the hypothesis that a particular theory applies to the particular 'case'. Thus it was an advance in the theory of crystals to explain their actual behaviour by calculating the density of imperfections on the basis of standard ('given') statistical mechanics.

Indeed, the essential theoretical background to any advance in 'experimental' science consists in the 'ad-hoc' hypotheses as to the applicability of given theories to the case in hand.

(2) Normal Science frequently involves the non-trivial joining of two already 'given' theories. By non-trivial joining we mean that the union of T_1 and T_2 corresponds to an empirical part E_{12} which not only is larger than $E_1 \cup E_2$ under deductive closure of $T_1 \cup T_2$ but whose class of potential refutations R includes statements which refute neither T_1 nor T_2.

Laue's joining together of the wave-theory of X-rays and the lattice theory of crystals is an example of a non-trivial union.

Chelsea College, London

NOTES

[1] This discussion is complicated by the fact that even some quotations from Popper are open to the charge of advocating 'sociological', 'psychological', or 'historical' criteria of appraisal, whilst on the other hand certainly Lakatos himself and occasionally some followers, pay lip service to objectivism. However, I claim that in practice the so-called 'methodology of research programmes' is, amongst other things, 'sociological'.

[2] See my article 'Novel Predictions as a Criterion of Merit' in Lakatos Memorial Volume (Cohen, Feyerabend, Wartofsky, Eds.) to be published in Boston Studies in the Philosophy of Science.

[3] Pace Feyerabend, we are concerned with theories that at least 'overlap' (if necessary, after suitable translation of the language of T_2 into the language of T_1).

[4] Watkins treats the more general case (though not the most general case of overlap) when theory T_2 partially disagrees (corrects) theory T_1. In so far as T_2 corrects T_1, it is of course preferable on the grounds of the criterion 'agreement with experience'. As explained in the introduction I have simplified this criterion, by taking a primitive refutationist position, in order to bring out my objections to MSRP on other grounds.

[5] The psychological relevance and objective irrelevance of the chronological ordering of evidence is perhaps illustrated by the following example: A policeman is told to apprehend a criminal whose characteristics include a cauliflower ear and a tendency to wear loud dress. Consider now the following two situations:

(1) The policeman notices a man wearing loud dress and finds that this man has a cauliflower ear.

(2) The policeman notices a man with a cauliflower ear and finds that he is wearing loud dress.

I suggest that the policeman may well feel more confidence in the identification of the man in case 1 than in case 2. This would be due to the vagueness of the evidence relating to 'loud dress'. It would be reasonable for the policeman to distrust his own judgement as to the loudness of the dress in case 2 since he had already begun to suspect the man. On the other hand, the objective strength of evidence is of course the same in both cases.

[6] Touching references to 'design' (implying abandonment of objectivity) in attempts to eliminate 'ad-hocness' in theories are found throughout the series of discussion of this point (H. R. Paneth: *Science News* 24, 169 (1952); K. R. Popper: *Conjectures and Refutations*, 1963, p. 241; E. Zahar: *BJPS* 24, 103 (1973)). Private correspondence may be an amusing source for the *historian* but references to it, or indeed to any (notoriously misleading) autobiographical sources, do not raise the methodology of appraisal back to the level of objectivity.

[7] I do not hold that the condition is necessary since I believe that a theory may be preferred entirely on the grounds of *linguistic* simplicity. (See H. R. Post: 'Simplicity in Scientific Theories', *BJPS* **XI**, No. 41 (1960).)

[8] 'We set a theory to catch out a theory'.

[9] Thus any activity in physics since 1935 can *at best* be described as normal science; that is providing it has been progressive at all!

PART III

THE LSE REPLY

JOHN WORRALL

RESEARCH PROGRAMMES, EMPIRICAL SUPPORT, AND THE DUHEM PROBLEM: REPLIES TO CRITICISM*

Many of the papers delivered at the Kronberg conference did not directly criticize the LSE position. Rather they suggested alternatives. I shall not attempt to criticize these alternatives here, but shall simply try to show that the various specific criticisms which *were* directed at the methodology of scientific research programmes miss their mark. This will involve me in replying directly only to the papers by Professors Koertge, Musgrave and Post.[1] (I shall also make a few remarks on the paper by Professor Feyerabend.)

In my part of the position paper I argued that MSRP provides novel solutions of two connected problems: the problem of empirical support and the problem of appraising rival scientific theories. I shall consider criticisms of these two proposed solutions in Sections 1 and 2 respectively.

1. AD HOC EXPLANATIONS AND EMPIRICAL SUPPORT

There is a long standing debate in philosophy of science, sometimes called the 'weight of evidence' debate. The point at issue was recently given a sharp formulation by Lakatos [1968], and then by Musgrave [1974]. A large part of my position paper is also devoted to this debate. According to one side (the side usually associated with Mill[2] and with Keynes), whether or not some theory T is supported by a piece of evidence described by the statement e depends solely on the logical relation between T and e. I shall, following Musgrave, call this attractive position the 'purely logical' position. According to it, if two theories, T_1 and T_2, both imply a correct description, e, of some set of facts, then these facts support both T_1 and T_2.

The main argument I urged against this view in Chapter 3 is that it is generally possible to modify a theory in a trivial '*ad hoc*' way so as to entail a correct description of any fact successfully predicted by a rival theory. If, originally, T_1 entails e, but T_2 does not, then it is always possible to produce a T_2', sufficiently similar to T_2 to be

A unified bibliography can be found on pp. 379–383.

321

G. Radnitzky and G. Andersson (eds.), Progress and Rationality in Science, 321–338.
All Rights Reserved.
Copyright © 1978 by D. Reidel Publishing Company, Dordrecht, Holland.

'essentially the same' theory, which *does* entail *e*. On the 'purely logical' view *e* could then in general provide no basis for discriminating between T_1 and T_2'. I argued that on the contrary we should not regard *e* as supporting such a theory as T_2' if that theory was indeed modified in an *ad hoc* way precisely so as to entail *e*.

Clearly both Koertge and Post have a good deal of sympathy for the purely logical position on empirical support, because both regard the problem of *ad hoc* explanations as, in fact, a non-problem. Koertge claims (above, p. 269) that "there is no problem of *ad hoc*ness in the MSRP sense of the term and ... the whole business is really very simple". Similarly Post asserts (above, p. 314) that, while a "good deal of the LSE programme can be viewed as an occasionally desperate attempt to eliminate *ad hoc*ness", this attempt is misdirected, for, "from the objective point of view it is quite irrelevant whether a theory has been created pragmatically '*ad hoc*' ".

But neither Post nor Koertge counter my argument that the problem of *ad hoc*ness is fundamental since to ignore it is to give up the hope of discriminating between any two rival theories on the basis of the support they receive from the facts.

Let me state my argument more fully. As Duhem so forcefully demonstrated, those statements (or conjunctions of statements) which we usually regard as amounting to single scientific theories (such as Newton's theory or Maxwell's theory) do not on their own have consequences which are unequivocally testable against experience. In order for such observation statements to be derivable auxiliary assumptions must be added to these scientific theories. I shall therefore regard a testable scientific theory as consisting of two parts: the 'basic' theory (or 'hard core') T, and the auxiliary assumptions, A. Let one theory be $T_1 \wedge A_1$ and another be $T_2 \wedge A_2$, and suppose that $T_1 \wedge A_1$ does, but $T_2 \wedge A_2$ does not, entail *e*. Then, unless the 'basic' theory T_2 alone entails the negation of *e*, it is always possible to modify the auxiliary assumptions to A_2', so that $T_2 \wedge A_2'$ is both consistent and entails *e*. We should certainly speak of this new theory as a modified version of the old one since its 'basic' part remains unchanged. But it follows that we shall not, in general, be able to distinguish between two rival 'basic' theories (say the wave theory of light and the corpuscular theory of light, or classical mechanics and relativistic mechanics) on the basis of the strengths of their empirical support unless we distinguish between genuine and *ad hoc* explanations. A *particular* wave theory of light, say, may, of course, entail correct descriptions of facts not entailed by a particular corpuscular theory. The particular theories would then be distinguishable on the 'purely logical' view. However, it would then be open to the corpuscularist to modify his particular theory so that, as thus modified, it did entail correct descriptions of these facts. Post and Koertge must then say that the two theories receive equal support from the facts.

Moreover, as I showed in my paper *above*, all this is far from being merely a problem of armchair philosophy of science. It is a problem which has been faced by working scientists: the possibility of an apparently superseded theory 'catching up' with its rival by *ad hoc* modifications has often been exploited during science's development. Take the example of the wave and corpuscular theories of light. All the effects (such as those of diffraction and of the 'interference' of polarised light) which are generally taken as supporting the wave theory but not its corpuscular rival, *were* in fact given corpuscular theoretic explanations (many of which were provided by Biot). Scientists at the time were aware of this, and most of them were suitably unimpressed. For example, the experimental physicist Humphrey Lloyd wrote:

An unfruitful theory may ... be fertilised by the addition of new hypotheses. By such subsidiary principles it may be brought up to the level of experimental science, and appear to meet the accumulating weight of evidence furnished by new phenomena. But a theory thus overloaded does not merit the name. It is a union of unconnected principles ... The theory of emission [of light] in its present state exhibits all these symptoms of unsoundness ... (Lloyd [1833], p. 296).

Translated into MSRP terms, Lloyd is saying that such 'overloaded' theories are not given genuine support by the facts they were concocted to explain.

William Whewell too recognized the flexibility of the emission theory of light. Indeed this recognition may well have been one of the starting points of his philosophical position that a piece of evidence may be entailed by each of two theories and may yet weigh more heavily in favour of one than the other. Whewell writes ([1837] p. 340):

When we look at the history of the emission theory of light, we see exactly what we may consider as the natural course of things in the career of a false theory. Such a theory may, to a certain extent, explain the phenomena which it was at first contrived to meet; but every new class of facts requires a new supposition, – an addition to the machinery; and as observation goes on, these incoherent appendages accumulate, till they overwhelm and upset the original framework. Such was the history of the hypothesis of solid epicycles; such has been the history of the hypothesis of the material emission of light. In its simple form it explained reflection and refraction; but the colours of thin plates added to it the hypothesis of fits of easy transmission and reflection; the phenomena of diffraction further invested the particles with complex hypothetical laws of attraction and repulsion; polarization gave them sides; double refraction subjected them to peculiar forces emanating from the axes of crystals; finally, dipolarization loaded them with the complex and unconnected contrivance of moveable polarization ... There is no unexpected success, no happy coincidence, no convergence of principles from remote quarters: the philosopher builds the machine, but its parts do not fit; they hold together only while he presses them: this is not the character of truth.

Of course, Post and Koertge could give up their claim that the problem of *ad hoc*ness is a pseudoproblem without embracing the MSRP account of empirical support. They could, for example, say that if two theories entail correct descriptions of the same set of facts then they are indeed equally supported by the facts, but the two theories may nevertheless be distinguishable on other grounds, for example, on the ground of 'simplicity'. Indeed there are some clear indications that this is the position both Post and Koertge *would* adopt.[3] But so long as one aims to distinguish between empirically equivalent theories on *any* grounds, then one is taking the problem of *ad hoc* explanations seriously. Indeed, those theories which are intuitively regarded as complex (such as Ptolemy's planetary theory or Biot's corpuscular theory of light) are precisely those that have been modified *ad hoc* to explain certain facts which, in their original form, they had been unable to explain.

(It may seem that the heuristic approach I advocate differs only in name from an approach which, while making empirical support a purely logical affair, allows simplicity as an extra ground for rational preference of theories. These two approaches are indeed aimed at capturing the same intuitions about theory appraisal, but it seems to me that the heuristic approach is superior on two counts. First the notion of simplicity is notoriously difficult adequately to characterise in precise terms. The heuristic account of empirical support is certainly not free from sin in this respect, but it does seem to be sharper than any existing account of simplicity. Secondly, in regarding a theory's empirical support and its simplicity as two entirely separate questions the simplicity approach underplays the role of facts. For it is precisely through modifications of the original theory aimed at capturing certain hitherto unexplained facts that those theories intuitively regarded as complex or 'overloaded' have become so. It is originally recalcitrant facts which induce the complexity. In the heuristic approach, on the other hand, the role of facts is given full force: if a theory has been able to 'explain' a certain fact only through *ad hoc* modifications which have made the theory intuitively complex, then it receives no support from that fact.)

According to both Post and Koertge, the heuristic approach to empirical support is not only unnecessary (since directed at a non-problem) but also faulty. Post's criticism of it is that it is 'sociologistic'. (Indeed Post charges MSRP as a whole with 'sociologism'. No approach is acceptable unless it is 'objective'. Post includes as 'sociological' any consideration of the " 'historical' conditions such as the chronological ordering of tests and theories".[4] Now although the heuristic characterisation of empirical support is definitely not 'objective' in Post's sense, it is certainly objective.

I shall show this by restating the heuristic characterisation in rather more precise terms. According to this approach, facts support not simply a theory but rather a research programme, or a theory *together with* a heuristic. (These are equivalent formulations. A research programme at a particular stage of its development is characterised by a pair of entities: the latest theory it has produced, and its heuristic.) A fact described by the statement e supports (or corroborates) a research programme R at a particular stage of its development if (i) e is implied by T (the latest theory produced by R), and (ii) the programme's heuristic guided the construction of T, *independently of the fact described by e.*

There are certain standard cases in which clause (ii) is not satisfied:

(a) Some constant appears in T in a place where a free parameter had occurred in T's predecessor, and the value of this constant was not dictated by the heuristic of the programme but rather had to be 'read off' from the facts described by e;

(b) (i) e says that in conditions X, Y does not occur; (ii) T says, as did its predecessor, that Y occurs if, *and only if*, conditions Z hold, but (iii) T's characterisation of those cases in which conditions Z hold differs from its predecessor's precisely in excluding the condition X. (This is the general schema which underlies for instance, the Thomas Young example I discussed in my [1976], pp. 140–141.

This characterisation of empirical support remains less precise than one would like, but imprecision does not amount to the abandonment of objectivism. This characterisation makes no reference to, and is quite independent of, any subjective or 'sociological' consideration. No psyches or social structures need be inspected in determining whether the relation holds: one needs to look only at theories, facts and heuristics.

A more serious charge of 'sociologism' is hinted at by Noretta Koertge[5] and explicitly levelled against MSRP by Paul Feyerabend. Feyerabend would, no doubt, admit that MSRP's *criteria* are objective; what he denies is that they have been given an objective epistemological rationale. Feyerabend denies that anyone has shown that preferring that theory which is the better according to MSRP criteria is the most rational course of action, or the one most likely to lead to the truth, *etc*. He claims that MSRP advocates are nothing more than anthropologists with a special interest in the tribe of scientists. But Feyerabend's claim is, I take it, based not on any charge that MSRP's criteria are sociological but rather on the charge that the only *rationale* we have succeeded in giving these criteria is a sociological one: namely that they do seem to capture better than others the inutitive appraisals of competing theories which scientists have as a matter of fact made. But if MSRP is to be *more than* a simple descriptive generalisation of scientists' past preferences, it must give its methodological rules an at least tentative and conjectural underpinning of a *general epistemological kind.*

I think Feyerabend's claim is correct: such an underpinning has not so far been given. (Although I should stress that if MSRP's rules provide an accurate general characterisation of scientist's specific appraisals then this is already quite an achievement. And the tribe of scientists is, after all, a rather exceptional one.) This deficiency can certainly be made good by adopting some suitably vague metaphysical principle which states that God's universal blueprint was 'simple' or 'organically unified'. In that case only those explanations would be acceptable which stemmed from some unifying principle, since patched up *ad hoc* theories could not, given this metaphysical principle, possibly be true of the world. This 'solution' of the problem is analogous to the Popperian 'solution' of the problem of induction advocated by Lakatos.[6] Like the latter, it certainly has all 'the advantages of theft over honest toil'. But perhaps, in this area, theft is the only option.

However, I offer the following tentative proposal for an epistemological underpinning of the MSRP criterion which avoids the
simple postulation of a metaphysical principle and hence, whatever its
shortcomings, at least involves some toil.

Assume that scientists have in fact preferred those theories which
are better supported by the facts in the sense which I have specified.
Is this just a reflection of their quirks? Or are there good grounds of
an objective kind for this preference?

Presumably if we can agree that the scientific enterprise has certain
aims then we can also agree that there are objective grounds for
awarding higher marks to those theories which seem to have contributed to the achievement of one or more of those aims than to those
theories which have not so contributed. Presumably one of the aims
of science is to extend our factual knowledge. It follows that if our
view of empirical support were the 'strictly temporal' one (which of
course it isn't) we should have an immediate answer to the question
posed in the previous paragraph. Those theories which receive
genuine empirical support on this view have contributed to one of the
aims of science and are, therefore, on this ground objectively better
than those which have no empirical support (and so have not contributed to the achievement of this aim). (Of course whether these
theories are to be preferred as closer to the truth than others or as
more reliable for technological application or whatever are different
matters. My aim here remember is the rather modest one of establishing that there are *some* objective grounds for awarding high
marks to those theories pronounced best by MSRP.)

But the argument I have just given for the case of 'strictly
temporal' empirical support will not work when empirical support is
construed, as I have argued it should be, in the heuristic sense. For a
theory may be well supported in this latter sense without having
taught us any facts of which we were unaware prior to the theory's
proposal. Must we then fall back on simply reporting that scientists
have as a matter of fact generally preferred those theories which are
better supported in this sense?

Perhaps an epistemological underpinning can be derived from the
obvious one just given for the 'novel fact' criterion. Begin with the
premise that it is 'objectively' reasonable, other things being equal, to
prefer those theories which have taught us some new fact by correctly predicting the observable existence of some hitherto unobserved event. Notice, however, that whether or not a theory

predicts 'novel facts' depends not just on the theory itself but also on
the state of scientific knowledge at the time. Nevertheless there may
be a property of the theory itself which makes it *possible* for the
theory to contribute to the extension of our factual knowledge
(whether or not a theory which possesses this property actually does
contribute to the extension of our factual knowledge will then depend
simply on the historical conditions). Focussing attention on such a
property would mark out the merits of the theory itself and would
remove the unfairness of historical accidents. It is, in fact, easy to see
that the property of non *ad hoc*ness highlighted by MSRP is such a
property. If a theory entails some factual statement *e* and is non-*ad
hoc* relative to *e* then it is at least possible for the theory to have
contributed to the extension of our factual knowledge by predicting
the facts described by *e*. If, on the other hand, the theory is entirely
ad hoc then it could not, no matter what the historical circumstances
at the time of its proposal, have led to the discovery of any facts: for
such facts as the theory does entail had to be already known in order
for the theory to be constructed. *Ad hoc* theories of *necessity* lag
behind the facts; whether non *ad hoc* theories lag behind or anticipate
the facts is a matter of historical accident. It thus seems 'objectively'
reasonable for scientists to award higher marks to those theories
which are given some empirical support (in my sense) to those which
are not: for the former could, while the latter could not, have
contributed to the achievement of one of the aims of science – the
extension of our factual knowledge.

Noretta Koertge has an interesting specific criticism of the heuristic
notion of empirical support. It is concerned with the time dependence
of the heuristic criterion.[7] Her criticism (which is similar to an earlier
one of Alan Musgrave's[8]) is the following:

Suppose *e* was, as a matter of historical fact, used to construct *S*, but there existed at
the time a positive heuristic *P* which could have been used instead. In this case does *e*
support *S*? If [the Lakatosians] say 'No', I claim that their criterion is objectionable on the
grounds that the order of events enters into it. If they say 'Yes', then I claim they must give
up Lakatos's method of comparing research programmes ... [For] suppose I am working in
RP_1 and I construct *S* on the basis of *e* in an *ad hoc* way ... Surely it is the intent of MSRP
that *e* should not count as a success for RP_1. But suppose there is a competing research
programme RP_2 which contains a positive heuristic which could have been used. If one
does not care which problem situation in fact resulted in the production of *S*, one must now
say that *e* does support *S*. (See, p. 268.)

Now I do not accept Koertge's (unargued) assertion that introducing consideration of the time order of theory and experiment into the question of empirical support is undesirable. But my main answer is this: as I stressed in my original paper, the question "does e support S?" is not, for me, a complete question if S is simply a theory or conjunction of theories. The reason is that, like Popper and like Zahar, I make the relation of empirical support not a two-place, but rather a *three*-place relation.[9] Thus, just as for Popper e may support S given background knowledge b while e does *not* support S given different background knowledge b', so for me e may support S given one method of construction of S while e does *not* support S given a different method of construction. Thus my answer to Noretta Koertge's question "does e support S?" is "It depends". What *is* true, in the case she envisages, is that e supports RP_2 but not RP_1 and this despite the fact that the latest theories produced by both programmes are empirically equivalent. This sounds like an unacceptable consequence of my approach, but, as I stressed in my position paper, it is not. Indeed one of my starting points was the claim that it is usually possible for the proponents of a degenerating research programme to generate out of it a theory which, though not of course logically equivalent to the latest theory produced by its progressive rival, is *empirically* equivalent to it. This, as the passages from Lloyd and from Whewell quoted above suggest, is what happened, for example, in the case of the wave-corpuscle rivalry in optics. Nevertheless despite the empirical equivalence of the latest theories produced by such programmes, one programme may receive more support from the facts than the other. Or, if you like, the theory T as produced by research programme P is better supported by the facts than the *empirically equivalent* theory T' as produced by research programme P'.

(Koertge also asks (on p. 268) for clarification of the following problem. Assume that some theory S entails some factual statement e but e does not support S according to the heuristic criterion. Noretta then asks "Does e fail to support S in the same sense as $\neg e$ 'fails to support' S?" Presumably what she is driving at is that some distinction ought to be drawn between facts which refute some particular version of a theory and facts which have been encompassed by the theory even if only *via ad hoc* adjustments. I agree, and I drew such a distinction in my position paper (p. 66, note 32) – though I ought to have drawn it in the text rather than in a note. But, as I argued *above*, this distinction, except in a few rare cases,[10] will be a mobile one. Unless the refuting fact hits not just the whole theoretical system but the 'basic theory' itself there will always be modified versions of the theory within which the originally refuting fact is apparently explained. But such an 'explanation' is more like a consistency proof than a genuine explanation. Of course it is better for some theory to be consistent with a fact than inconsistent with it, but it is still better for a theory to *pre*dict the fact.)

I shall now comment on some of Alan Musgrave's remarks about empirical support. Musgrave in an earlier paper, advocated against both the 'strictly temporal' view (held, for example, by Popper) and the heuristic view (held, for example, by Zahar and me) a 'theoretical' view of empirical support (what he meant by this I shall explain in a moment). **Musgrave now complains that, in my position paper, I compared the heuristic view only with the temporal view, thus ignoring the theoretical one. Now I (largely) ignored the theoretical view because** my brief was to compare the MSRP position with Popper's and Popper holds a version of the 'strictly temporal' view. However I am very ready to compare the heuristic account of empirical support with the theoretical one.

The theoretical view of empirical support says that if a theory T_2 is proposed as a rival to some already established theory T_1 then T_2 is not given genuine support by any fact that is already explained by T_1. Thus on this account, for example, Einstein's theory is given genuine support only by those facts which it predicts but which were not already explained by Newton's theory. I shall argue that this account both runs into a general difficulty and also fails to square with our intuitions about empirical support in particular cases.

The general difficulty is the following. Suppose T_1 and T_2 are rival theories in some field and that both are independent of e which, however, describes some fact within the theories' field. Then, on the 'theoretical' view of empirical support, it is possible for the proponents of either T_1 or T_2 to produce slightly modified versions of their theories which *are* supported by e, *provided they beat their rivals to it*. If the proponents of T_1 happen to hear of e first and immediately produce a new theory T_1' which is in fact just the conjunction of T_1 and e, then e supports T_1', but cannot then support *any* modification of T_2. This is true even if T_2's proponents eventually provide an impressive and contentful modified theory which was not constructed simply to imply e but nevertheless does imply it. But this 'tackers' race' consequence of the theoretical view is surely absurd.

When it comes to particular cases, the theoretical view is both too narrow and too wide. It is too narrow because it rules, for example, that the result of the Michelson-Morley experiment cannot support the special theory of relativity proposed in 1905 since this result had already been explained on the basis of Lorentz's rival theory in 1904. The theoretical view is too wide because it says that this result *must* support Lorentz's theory if this was the first theory to explain it, even if this theory amounted to no more than an *ad hoc* postulation of just sufficient contraction of the arms of the interferometer. The theoretical view is too narrow because it rules that various polarisation effects which are taken to provide dramatic support for the theories Fresnel proposed to explain them in the period 1818 to 1823, cannot, in fact, genuinely support these theories, because Biot had by then already produced a corpuscular theory from which descriptions of these effects follow. And this view of empirical support is too wide because it rules that these polarisation effects *must* be regarded as supporting Biot's corpuscular theory because no previous theory had had these effects as consequences. (Biot's explanation of these effects was in fact condemned as entirely *ad hoc*.) Any theory no matter how 'cooked up' can, on this theoretical view, gain support from the facts so long as it is the first to explain them. None of these cases which destroy the 'theoretical view' provides any difficulty for the 'heuristic view'. The Michelson-Morley result supports both Einstein's 1905 theory *and* Lorentz's 1904 theory of corresponding states, since the result was not involved in the construction of either theory,[11] and yet follows from both of them. On the heuristic account, Biot's theory was *not* supported by the various polarisation effects despite no other explanation of them being then available, because its explanation of them was *ad hoc*: all the essential features of the explanation were 'read off' from these already known facts. But Fresnel's account of these effects *did* derive support from them, despite the pre-existence of Biot's account, because Fresnel did not need to use these effects in constructing his theories.[12]

Despite his complaint about my ignoring it Musgrave himself no longer claims that the theoretical view is superior to the heuristic

view, as he did in his earlier article. Rather he now thinks (above, p.
186) that these two views:

may be complementary rather than competing ... The heuristic view enables us to
determine the evidential support of a *single* theory. But when we wish to decide
whether some theory is an improvement over its predecessor, then the theoretical view
comes into its own. For then we will only count those facts which do not also support
the old theory.

Musgrave's new move seems to me a conventionalist strategem of
the worst kind. Those holding the various positions on empirical
support would, *of course*, all agree that those facts which give one
theory extra support over another are those which support the former
but not the latter! Thus the proponent of the 'purely logical' position
(against which Musgrave argued in his [1974]) would agree completely
with Musgrave's new claim: those facts which provide *extra* support
for Einstein's theory over Newton's are those which follow from
Einstein's theory but not from Newton's. But there is a genuine
dispute here which no talk about complementarity can mask. The
dispute is over when a *single* theory derives support from a given
fact. On the heuristic account a fact may follow from one theory and
not from its rival and yet may not provide support for the first theory.
But such a fact is bound to support the first theory according to the
theoretical view. For example, as I pointed out, Biot's 1816 corpuscular theory was not, on the heuristic view, provided with extra
support over its wave theoretical rival by its explanation of polarisation phenomena, despite the fact that no wave theoretical account of
these phenomena then existed. Thus the heuristic and theoretical
accounts of empirical support are inconsistent and not complementary:
Musgrave's policy of 'strategic withdrawal' does not succeed.

2. THE DUHEM PROBLEM

In this section I shall again consider a specific scientific theory as
consisting of a basic theory T conjoined with a set of less basic,
'auxiliary' assumptions A. Assume that some such scientific theory is
inconsistent with some factual statement. Noretta Koertge hopes to
find a methodology which will demarcate those situations in which
the better or more promising or more rational solution of the inconsistency is for scientists to look for replacements for T, from
those situations in which the better solution is for scientists to look

for replacements for A.[13] I wish her luck, but I do not think she will succeed. Certainly if one must produce such a demarcation in order to solve the Duhem problem, then MSRP does not solve this problem. More especially, MSRP does *not* provide the crazy solution of the Duhem problem attributed to it by Noretta Koertge. She says that MSRP implies that, if a theory of the form $T \wedge A$ is inconsistent with the evidence, then "if T is the hard core of a research programme, one should *always* keep T and replace A".[14] But MSRP would be mad to give this advice, since if it were consistently followed it would endow the first research programme in any field with an eternal monopoly. I shall explain what MSRP does say about "holding on to hard cores" in a moment. First, given that it doesn't provide a solution of the kind sought by Koertge, let us see whether MSRP provides *any* sort of solution of the Duhem problem.

As Lakatos often stressed,[15] MSRP consists of a set of criteria for appraising already-articulated programmes. Like Popper, Lakatos explicitly eschews the hope of providing an acceptable methodology of the old Bacon-Descartes kind – *i.e.* one which provides a *method* for finding the truth. MSRP therefore gives only the following solution of the Duhem problem: it places no restrictions on the way a theory may be modified in the event of a clash between theory and evidence; however, once the modified theory has been produced, MSRP's rules will tell whether or not the new theory constitutes progress over the old (and if various modified theories are produced MSRP's rules will be able to say if one constitutes scientific progress over all the others).

But can MSRP say nothing about which part of a theory is *likely* to be modified in the light of clashes with experiment? I think it can say a little. According to MSRP, the development of modern mature science has consisted of the rivalry, not of mere theories, but rather of research programmes. The most important distinguishing mark of a research programme is its heuristic. This will consist in part of mathematical techniques for formulating, and drawing consequences from, the theories produced by the research programme. Another part of the heuristic may consist of techniques for resolving anomalies. When scientists, as Lakatos puts it, declare some of their assumptions "irrefutable by methodological fiat", they do not do so arbitrarily. Restricting themselves in this way would obviously be wrong, unless it brought compensating benefits. It is clear from the nature of a

heuristic what these benefits are. If a theory produced by a programme is inconsistent with the result of some experiment, one can
modify either the hard core assumptions or the auxiliary assumptions.
The former course will involve abandoning the whole programme.
The programme's 'hard core' will (in part, at least) underpin its
heuristic: abandoning the 'hard core' will also involve abandoning the
heuristic. One will then have to build up a whole new programme
(unless a new programme is already to hand). This is an enormous
undertaking which may involve, amongst other things, the development of entirely new mathematical techniques. By taking this course
one puts oneself, at least at first, in a theoretical void. On the other
hand, if one sticks to the hard core, and to the programme, and to the
programme's heuristic, one's search for modified auxiliary assumptions will be guided in various ways by that heuristic. No wonder then
that some theories (those which form integral parts of powerful
research programmes) have had such long lives, surviving many
clashes with experiment, in which the "arrow of the *modus tollens*"
has been directed away from them and towards auxiliary theories.

Thus MSRP explains why it is usually the auxiliary A's and not the
'basic' T's which are modified because of the theory-experiment
clashes. The methodology does *not*, however, say that this is what
ought always to happen, nor even that this is what *ought* usually to
happen. The methodology points to the enormity of the step which
those who intend to abandon some 'basic' theory T must take, but it
cannot advise scientists against taking the step. It is precisely by
taking it that certain geniuses like Newton or Fresnel or Einstein have
advanced science.

Finally, I shall remark on some of Alan Musgrave's criticisms of
MSRP's solution of the Duhem problem and, more generally, its
account of when one scientific theory is better than another. I want
simply to take up two points, one of them rather minor.[16]

Musgrave agrees with large parts of the MSRP position, but thinks
it is overly anti-falsificationist in a few respects. One of his charges
rests on a partial misunderstanding of my position. I claimed that
MSRP appraisals, unlike Popper's corroboration appraisals, can distinguish between *some* pairs of theories which are experimentally
refuted. This is because one theory may derive genuine support from
more facts than the other theory. In other words, the set of facts
which genuinely support one system may properly include the set of

facts which genuinely support the other.[17] Thus, following Lakatos, I described MSRP as "transferring the methodological spotlight from refutations ... [to] verifications of excess content" (above, p. 52). Musgrave attacks this slogan without caring about the full description for which it was meant to act as shorthand. His attack (on p. 184) consists of considering two theoretical systems S_1 and S_2:

Suppose S_1 is corroborated in all tests except one, while S_2 is refuted wherever S_1 is corroborated but corroborated in the single test which refutes S_1. All the examined content of S_2 is excess content, and some of it is corroborated.

Now I agree with Musgrave that it would be 'totally unacceptable' to maintain that S_2 has more evidential support than S_1. But I cannot really understand why Musgrave thinks I would maintain this. As he says (*ibid.*) S_2 is not "supported by more facts than its rival". Since support by more facts is the criterion I advocated, I would have thought it obvious that S_2 is *not* better supported than S_1 on my criterion (nor is S_1 better supported than S_2). But, more importantly, Musgrave is wrong to think that this example shows that in assessing the evidential support of S_1 and S_2:

we must take into account the refutations as well as the corroborations of each (p. 184).

Musgrave may prefer a system in which refutations are taken into account, but then he must argue for it. Certainly it is not true that we 'must' take refutations into account here, for MSRP explains, without considering refutations, why neither is S_1 better supported than S_2, nor *vice versa*. The set of facts which support S_2 is not a subset of the set of facts which support S_1 and nor is the set of facts which support S_1 a subset of the set of those facts which support S_2.

But Musgrave has a much more interesting claim about over-zealous anti-falsificationism within the ranks of MSRP-supporters. This concerns the stress on the large degree of theoretical or heuristic autonomy which some programmes have. No doubt, Lakatos in some ways overreacted to the claim that "scientific theories are often overthrown by experiments, and the overthrow of theories is indeed the vehicle of scientific progress", and overstated the case for heuristic autonomy. In particular, I agree that it is a mistake to think that a 'positive heuristic' can *predict* that certain theories will face anomalies. (Although Musgrave here endows Lakatos with a more extreme position than his writings warrant; Lakatos never says that heuristics can predict the *precise* anomalies a theory will face!) Nevertheless, there is in Lakatos's position, an important kernel of truth, to which Musgrave, to judge from his comments, remains blind. I touched on this kernel when I claimed that a heuristic "will often point to shortcomings in existing theories in the programme ... *quite independently of any empirical* difficulties". Musgrave quotes this disapprovingly (p. 189). Let me try to defend this claim by means of a simple example.

By the time of Fresnel's work in optics, the law of the reflection of a light ray (that the angles which the incident and the reflected rays make with the normal are the same) had been an accepted experimental law for centuries. The law was considered as completely confirmed and entirely free from empirical difficulties. Huygens had, in the 17th Century, produced a wave theory of reflection. Since this theory implied the correct law of reflection, and nothing else, it can scarcely be said to have been in empirical difficulties. Yet Huygens's theory was unacceptable as Fresnel clearly saw. This was because a wave optics *programme* existed whose aim was the explanation of all optical phenomena on the basis of the straightforward mechanics of elastic media. Huygens's theory involved the assumption of a full, particulate ether in which the motion which a disturbed particle passed on to those in contact with it was 'infinitely feeble' in all directions, except one, *viz.*, the direction of the straight line joining the disturbed particle to the original source of the disturbance. This obviously had to be wrong, unless the ether had the most weird and wonderful mechanical properties. Fresnel saw this and produced instead a theory which allowed particles disturbed from equilibrium to pass on this disturbance in 'an infinity of senses'. Fresnel's theory (which also took interference into account) not only explained the success of the usual law of reflection (by yielding it as a 'limiting case'), it also *corrected* it at various points. That is, it yielded entirely different predictions for certain cases, principally where the reflecting surface was very narrow. (These hitherto unsuspected predictions were subsequently confirmed.) Here then it was the heuristic of the programme and not anomalous experimental results which pointed to defects in an existing theory. (*Of course*, once Fresnel had seen that Huygens's theory was unacceptable, and had constructed a new theory more in line with the heuristic of the wave optics programme, the new theory had to be tested against experience.)

Hence Musgrave's 'anti-empiricist' and 'anti-falsificationist' charges are valid only insofar as MSRP does not hold that empirical refutations are the only, or even the principal, driving force of progress in the developed sciences.

London School of Economics

NOTES

* My thanks for critical comments on a previous draft of this paper are due to Peter Clark, Greg Currie, Peter Urbach and John Watkins.

[1] Professor Scheibe also made some interesting points. I entirely agree with his main criticism (which was also expressed by many of the Kronberg symposiasts). This was that the MSRP idea of empirical support has not been made sufficiently precise. Although we have, I think, characterised quite clearly a few specific ways in which facts can be 'used' in the construction of theories ('parameter adjustment' and 'monster adjustment' – see particularly pp. 345–6), we have not yet given a *general* account of what it means for a fact to be involved in the construction of a theory.

[2] Not entirely correctly as Grünbaum persuasively argues above, p. 120 (cf. Watkins's comments below).

[3] Koertge introduces considerations of 'plausibility' (p. 262) and Post refers several times to his earlier attempt to characterise a theory's 'simplicity' (p. 318, note 7). Indeed I find it difficult to understand why Post is so much against our addressing the problem of *ad hoc* explanations when he confronted the *very same* problem in his earlier [1960] paper on simplicity. There (p. 32) he gives the problem the following clear formulation (his *solution* of the problem is, I fear, much less satisfactory):

"The merit of a scientific theory is judged not only by its logical consistency and its correspondence with experience ... (Indeed many crank theories, precisely the ones most difficult to eliminate, would qualify for acceptance under these two criteria!)"

[4] See above, p. 313.

[5] See above, p. 268.

[6] See Lakatos [1974].

[7] Koertge expresses some doubt about whether or not the criterion is, in fact, time-dependent. I had hoped that this was clear. I said, following Zahar, that if a hitherto unknown fact was first discovered as a result of its being predicted by a theory, then this is a sufficient condition for the fact to support the theory, but it is *not* a necessary condition. (See above, p. 49.)

[8] Musgrave [1974].

[9] Some confusion may have been caused by the fact that I sometimes speak of empirical support as a two place relation between a fact and a *research programme*. But as I pointed out above, these are just two equivalent ways of expressing the same thing. For a research programme at a particular stage of its development is characterised by a *pair* of entities: this research programme's heuristic and its latest theory. Thus again *three* entities are really involved in the empirical support relation.

[10] A case which *seemed* for a while to be such an exception is that of the wave theory

of light and rectilinear propagation. It seemed, *e.g.* to Newton, that *no possible* wave theory of light could explain rectilinear propagation. In such cases 'consistency proofs' may be very important.

[11] See Zahar [1973].

[12] See Worrall [1976].

[13] See her contribution above, (especially pp. 261–267). In a much earlier and famous paper, Adolf Grünbaum pursued the same hope. His solution was that the second of the two above courses was the correct one, whenever the (posterior) probability of A was extremely high. This solution seems to me to suffer from two defects. First there is no generally accepted inductive logic to provide us with values for the probability of A in the light of the evidence. Secondly, if there were such an inductive logic it would surely give high probabilities to 'well-entrenched', 'plausible', 'well-tested' theories. Yet many major scientific innovations have involved the overthrow of precisely such theories.

[14] Above, p. 260.

[15] See particularly his [1971], p. 92.

[16] There is one other point which Musgrave makes here, and on which I think he is quite right. I should have been more careful *always* to make explicit the distinction between a 'basic' theory as I called it above and a full theory or theoretical system, consisting of the 'basic' theory together with auxiliary theories. Only the latter can be directly inconsistent with experimental reports. But I did stress this point at least in my Section 3 (p. 52), and I certainly never followed Lakatos in talking about it being permissible to 'ignore' refutations of theories. This locution gives MSRP an unwarranted anti-falsificationist air. All refutations constitute problems. Lakatos did, however have an important point even if it was infelicitously expressed. Lakatos saw that the driving force for science was in large part provided not by refutations of existing theories but by much higher level 'heuristic' considerations. (See p. 336.)

[17] I assume throughout that the facts we are concerned with in each particular case are those 'atomic facts' with which scientists confronted by the two rival theories would be concerned. This blocks the kinds of construction which otherwise would show that the subset relation discussed in the text holds in no interesting case of theory comparison.

The notes refer to the unified bibliography for Parts I and III on pp. 379–383.

JOHN WATKINS

CORROBORATION AND THE PROBLEM OF CONTENT-COMPARISON

I will confine this reply to arguments that hit, or at least are aimed at, the position defended in my position paper. This means that I will concentrate mainly on the papers of Feyerabend, Hübner, and Grünbaum. Grünbaum's paper I regard as the most serious challenge in the present volume to the Popperian position. He criticises, among other things, an assumption, concerning the possibility of content-comparisons between rival theories, which is central to our position. I think that his criticism can be met. But in the meanwhile a more deadly objection to this assumption has been made by Miller, and the last part of my reply will be concerned with this crucial issue.

1. FEYERABEND'S EFFORTLESS ONE-UPMANSHIP

Feyerabend proceeds from a discussion of Aristotle to the conclusion that money should be taken away from critical rationalists. His method of 'arguing' against us is to be always one-up by taking an assumption that was not under debate, declaring *it* to be "the question at issue", and then giving it his usual Dadaist treatment. Our task in the position paper was to discuss criteria of scientific progress; it was not to discuss whether scientific progress is good or bad for mankind. This gave Feyerabend an easy opening: we dogmatically took it for granted that science is good and failed to examine the (meta-)question: "*What's so great about science?*" ([1976a], p. 310). But suppose we had dealt with this question and that our answer had boiled down to this, that science seems to have done better than Azande magic, theology, etc. in its search for truth. Would that have met his complaint? Not at all. He would have complained that we dogmatically took it "for granted that Truth is something quite excellent" (loc. cit.) and failed to examine the (meta-meta-)question: "What's so great about truth?" (this volume, p. 167).

Feyerabend has said that his 'favourite pastime is to confuse rationalists by inventing compelling reasons for unreasonable

A unified bibliography can be found on pp. 379–383.

G. Radnitzky and G. Andersson (eds.), Progress and Rationality in Science, 339–378.
All Rights Reserved.
Copyright © 1978 by D. Reidel Publishing Company, Dordrecht, Holland.

doctrines" ([1975], p. 189). The chief unreasonable doctrine of his present paper is that we should discard logic. Let us see how compelling are the reasons he has invented for this. He claims that "there are facts whose only adequate description is inconsistent" (p. 165), but the moving-train example (p. 157) on which he bases this claim is so pathetic that I will not bore the reader with an exposure of it. He accuses critical rationalists of an "astounding dogmatism" in failing to use illogical theories as criticisms of the laws of logic (p. 155). Well, I once proposed that for a statement C to constitute a criticism of a theory T, C must have some sort of adverse implications for T and must itself at least be a candidate for rational acceptance, a minimal condition for the latter being that C is not self-contradictory ([1971], p. 58). And I find no compelling reason here to change that. Feyerabend also says that "inconsistent theories are better to handle and lead to more discoveries than their decontaminated rivals" (p. 165). What he means by 'better to handle' I do not know (nor, I suspect, does he); but I do understand the second part of this claim: the idea that contradictions are fertile and lead to progress seems to be the least uncompelling reason he has been able to find (he did not invent it, of course) for this unreasonable doctrine.

Before examining further what he says in this connection I wish first to quote a passage, written some forty years ago, from Popper's "What is Dialectic?" It will prove most instructive to compare this with what Feyerabend now says. Popper wrote:

But the most important misunderstandings and muddles arise out of the loose way in which dialecticians speak about contradictions.

They observe, correctly, that contradictions are of the greatest importance in the history of thought ... without contradictions, without criticism, there would be no rational motive for changing our theories: there would be no intellectual progress.

Having thus correctly observed that contradictions ... are extremely fertile, and indeed the moving forces of any progress of thought, dialecticians conclude – wrongly as we shall see – that there is no need to avoid these fertile contradictions ...

Such an assertion amounts to an attack upon the so-called 'law of contradiction' ... ([1963] p. 316).

Now read carefully the following sentences from Feyerabend's paper:

If we can give reasons for the usefulness of contradictions in science as I think we can then we have also reasons against the selected system [i.e. a system of logic that asserts non-contradiction] (p. 155).

Assume somebody shows as I think I have shown (with the help of ... Hegel, Kuhn,

and Lakatos) that science often violates those laws of logic which our critical rationalists regard as a conditio sine qua non of rationality ... Now if one discovers that knowledge is beset by contradictions, progresses because of the contradictions *and right through them* [my italics], ... then the lesson is obvious: rationality as defined [by critical rationalists] has to be given up ...
But is it really so easy to get rid of fundamental principles of logic? (pp. 156–157).

The passage from Popper consists of categorical statements and its message is unequivocal: (a) contradictions cause progress because we have to work to eliminate them; (b) dialecticians wrongly conclude that contradictions can be retained because they cause progress. A hasty reading of the passages from Feyerabend might give the impression that they too contain an unequivocal message, namely that the fundamental principles of logic, including the law of contradiction, have to be given up. A close reading, however, reveals that *he is not actually asserting anything.* There is not one categorical statement in the passages I have quoted. He proceeds from a series of hypothetical statements ("If we can give reasons ...", "Assume somebody shows ...", "Now if one discovers ...") to an *insinuated* conclusion couched in the form of a question ("But is it really so easy ...?").

Suppose, however, that we replace those hypotheticals by categorical premises ('Contradictions in science *are* useful", "Science *does* often violate the laws of logic", "Knowledge *is* beset by contradictions, progresses because of the contradictions and right through them"), and the insinuated conclusion by the categorical conclusion: "Therefore, the fundamental principles of logic have to be got rid of". How compelling or otherwise would this argument be? I say that any persuasiveness it might have for the unwary reader would be the result of its *fudging* of Popper's sharp distinction between claim (a) (contradictions cause progress because we have to work to eliminate them) and claim (b) (contradictions can be retained because they cause progress). The phrase where the fudging occurs is the one I have italicised: "Knowledge ... progresses because of the contradictions *and right through them*"; this is nicely ambiguous between (a) and (b). If it meant that knowledge progresses beyond the contradictions by surmounting them it would provide no reason at all for the present 'unreasonable doctrine'. If it meant that knowledge progresses because of contradictions which it *retains*, it would be exposed to the obvious retort: but if we do not have to work for the elimination of contradictions, how do they cause progress? Is not the doctrine that we can retain contradictions which have come to light a recipe for, among other things, intellectual laziness?

Now the difference between (a) and (b) is one of which Feyerabend must have been aware. (If he were not, he ought to be raising questions about *his* salary.) For one thing, he knows that paper by Popper very well. (In the course of the passage from which I have quoted he refers to it (note 20): to p. 317; my quotation is from p. 316.) The conclusion seems inescapable. He *wished* to confuse us by inventing a compelling reason for this unreasonable doctrine; unfortunately, this wish was not father to any new thought. So he decided to concoct a sham reason by deliberately obfuscating the clear and obvious distinction between (a) and (b); and to be on the safe side he incorporated this sham reason into a passage in which he actually refrains from venturing any categorical assertion. In this way the unwary reader would be given the impression that an argument for discarding logic has been provided, while the wary reader would be presented with no statement that is both unequivocal and false.

Feyerabend often complains that he is not read properly. I say that he often writes so that he *cannot* be read properly. Since that is a serious accusation I shall now digress, briefly, in order to support it with an independent example. Feyerabend has never made this complaint so frequently and bitterly as in his reply [1976] to a scathing review of *Against Method* by Gellner [1975]. In that review Feyerabend was quoted (p. 340, from *AM* p. 187) as saying that violence is beneficial to the individual. To this he replied:

... don't you think you should have read the text a little more carefully or let somebody else explain it to you in case you can't read? The text says that violence is necessary *according to political anarchism* and adds that political anarchism is a doctrine I reject. The very first sentence of the book calls political anarchism 'not the most attractive political philosophy' (p. 17) and on p. 189 I again distinguish my views from political anarchism, just to be on the safe side. All in vain (p. 387, note 1).

That seems clear enough: Gellner has committed the reviewer's howler of attributing to an author a view which he had only reported, and moreover with disapproval. But let us investigate. To begin with, we find that on pp. 17 and 189 of *AM* he had been by no means as unequivocal as he here claims. On p. 17 he had written, not that anarchism is *not* the most attractive political philosophy but only: "... *anarchism*, while perhaps not the most attractive *political philosophy* ...". And on p. 189 he had given a characterisation of 'the epistemological anarchist' who is there said to resemble "the Dadaist ... much more than he resembles the political anarchist". There is in fact *no* clearcut rejection of political anarchism in *AM*. And in any case the question of political anarchism is irrelevant: it is *not true* that "The text says that violence is necessary *according to political anarchism*". The paragraph from which Gellner quoted does indeed *begin* with two sentences on political anarchism; but political anarchism is then left behind and the passage continues:

Occasionally one wishes to overcome not just some social circumstances but the entire physical world which is seen as being corrupt, unreal, transient, and of no importance. This *religious* or *esc[h]atological* anarchism denies not only social laws, but moral, physical and perceptual laws as well and it envisages a mode of existence that is no longer tied to the body, its reactions and its needs. *Violence, whether political or spiritual, plays an important role in almost all forms of anarchism. Violence is necessary to overcome the impediments erected by a well-organised society, or by one's own modes of behaviour (perception, thought, etc.), and it is beneficial for the individual, for it releases one's energies and makes one realize the powers at one's disposal (AM, p. 187).*

This says that violence plays an important role in almost all forms of anarchism, and no exception is made for the author's own form. So I think that Gellner was entirely justified in reading the passage he quoted as an expression of Feyerabend's own view. But I also think that Feyerabend may be technically right in denying that *he* was asserting that violence is beneficial for the individual; for I think that he was doing his best to *insinuate* this idea but in a way that would allow him to wriggle out of responsibility for it if challenged. End of digression.

I have counted at least seven references by Feyerabend to my "letter to Santa Claus" (two of them in his [1976a]). For instance, he writes (p. 166):

If one asks a Popperian why on earth one should accept his standards he will answer that this is how science proceeds (... Lakatos), or that the standards are fruitful in the sense that they make us understand science (... Popper ...), or that they lead to a science one "would like to have" (... Watkins in his "letter to Santa Claus").

He has "looked at every line of the position paper" (p. 153) but I wish that he had read them a little more carefully; for he makes a nonsense of what I was doing there. I did not, of course, propose as an aim for science what "we would like to have". What I did propose was that we should *begin* by erecting "a naively optimal ideal for science" and

then ask which extant methodology retrieves the most of that naive
ideal from the wreckage caused, chiefly, by Humean scepticism. I
suggested that a naively optimal ideal would include both *certainty* and
ever deeper (or even ultimate) *explanations*, and that these two aims
pull in opposite directions. And I claimed that the Popperian
methodology is able to capture more of that naive ideal by altogether
discarding the goal of certainty, than does any methodology which
hankers after something approaching certainty. (I now think that I
understated my case: I should have claimed that a methodology
which goes for certainty will capture *nothing* of that ideal, whereas
one which goes for ever deeper explanations can capture *everything*
except, of course, for the discarded certainty.)

It still seems to me that the right way to try to find an aim for
science which is both realistic and optimal, or neither over nor under
ambitious, is to begin with a most utopian and non-viable ideal and
then to consider in what ways it has to be cut back to make it viable
and non-utopian. But, of course, this way of proceeding is automatically exposed to Feyerabend's unfailing one-upmanship method. After
quoting my claim that the Popperian methodology preserves more of
the Bacon-Descartes ideal than does any rival methodology, he
comments: "The 'argument' assumes that the Bacon-Descartes ideal
is worth preserving – which is the question at issue" (p. 176, note 16).
I will not be so foolish as to argue for this assumption, for that ideal
involves, among other things, truth and consistency, and we know
what Feyerabend thinks of *these*.

Feyerabend's intellectual heroes change with bewildering rapidity.
Not long ago Cohn-Bendit was at the top. Now, the "much maligned
Bellarmine" (p. 172) is in the ascendant. For Feyerabend has
developed a solicitude for theology ("the theory of God"), and
considers it a criticism of science that "science disturbs theology" (p.
172). Whether this support from this unexpected quarter will enable
churchmen to sleep easier at night I do not know.

2. HÜBNER ON FALSIFICATION, 'GRUE', AND TRUTH

It is a relief to turn to Hübner's hard-hitting paper. In the following
key passage from it I will italicise the two points on which I disagree:

But every falsification ... has some premises, such as axioms of certain observational
theories. Now if these premises are conjectural, and this is not denied by LSE

philosophers, the falsification is conjectural too. This conjecture may be either purely arbitrary, *and consequently the falsification would be practically meaningless* [my italics]; or the scientist has some reasons [italics omitted] for his conjectures; but in that case *he cannot avoid using inductions* [my italics], either (p. 280).

I agree that a falsification will typically involve general statements as well as basis statements; but let us begin with the latter. On Popper's [1934] view, an accepted basis statement *is* 'purely arbitrary', logically considered: it is introduced by a conventional decision, if you like by *fiat*, and is not justified by other statements.

For my part, I now prefer to regard such a statement as a main premiss in a conjectural but testable explanation-sketch of the observers' perceptions; but I did not say that in my position-paper and in this reply I will go along with Popper's [1934] view. Does his conventionalist view of basis statements render falsifications 'practically meaningless'? Suppose that a low-level hypothesis predicts that if this knob is turned from 1 to 2 this pointer will move from 1 to 4. A group of experimenters agree to accept the following (conjunctive) basis statement: "This knob was turned from 1 to 2 and this pointer moved from 1 to $\frac{1}{2}$". Then this hypothesis has been falsified in a quite meaningful sense: a potential falsifier of it has been accepted.

Objections have often been voiced against Popper's conventionalist interpretation of the empirical basis. But *any* philosophy, including any inductivist philosophy, will have to halt the regress of criticisms or of justifications somewhere; and one which halts it at the bottom, with statements about pointer-readings, photographic plates, etc., and which makes the fate of higher level theories depend on decisions at this level, surely introduces much less arbitrariness than does a full-fledged conventionalism in which decisions directly control axioms, principles, frameworks, etc. Moreover, these bottom-level decisions are inter-subjectively arrived at and they are not irrevocable: if need be, a previously accepted basis-statement can be queried and perhaps rejected. Example: a photograph taken from the Mount Wilson Observatory in 1919 was interpreted as not showing the conjectured planet Pluto in its predicted position; it was discovered later that there was a flaw in the photographic emulsion just where its image should have fallen. (Holton and Roller [1958], p. 197.)

I now turn to the general statements typically involved in the falsification of a scientific theory. (Popper, it will be recalled, required for this the acceptance not just of "a few stray basic statements" but

of what he called a falsifying hypothesis which offers a general recipe
for *reproducing* effects prohibited by the theory.) I agree with Hübner
that the scientist should have *reasons* for accepting those general
statements which play a role in the falsification of a scientific theory.
But I do not agree that this reintroduces induction through the back
door. It is a main feature of Popper's theory of scientific rationality
that we may be rationally justified in accepting a statement which is
not itself justified by other statements (for instance, because it offers
the best available conjectural explanation for a range of accepted
basis statements). A reason for accepting a general statement which
figures in the falsification of a scientific theory may be that it is well
tested and that all known alternatives to it are less corroborated.

But, Hübner will no doubt reply, this means that a so-called
'falsification' of a scientific theory T is really a choice between two
theories: T itself and the interpretative theory (or theories) I required
to enrich and expand certain experimental reports (accepted basis
statements) into statements strong enough to be logically incompatible with T. (This is a point which Lakatos developed: see for
instance his [1970], pp. 128–131.) And why should we not instead treat
T as an interpretative theory and rely on it in a 'falsification' of I?
The first point that should be made in reply to this is that the
overthrow of a scientific theory always *is* a conjectural business. (As
I said in my position paper, there is no empirical certainty at any
scientific level.) Moreover, as Lakatos stressed, it does sometimes
happen in the history of science that such a falsification is later
reversed, T being rehabilitated and I being replaced by I', where I' no
longer interprets those experimental reports in a way that renders
them inconsistent with T. But the absence of an algorithm for
deciding, given certain accepted basis statements, whether T is or is
not falsified does not mean that a falsification must therefore be
'purely arbitrary' and 'practically meaningless'. Consider the following case: E_1 and E_2 are two very different experimental reports; when
interpreted respectively by I_1 and I_2, both are inconsistent with T; I_1
and I_2 are quite independent of one another; both have been well
tested and have run into no empirical difficulty (apart of course from
their clashes with T over E_1 and E_2). Then the decision to retain E_1
and I_1 and E_2 and I_2 and to reject T, though it *may* be wrong and *may*
be reversed later, looks more rational than the decision to retain T
and to reject E_1 or I_1 and E_2 or I_2. If moreover a new theory T' were

available which stood to T in roughly the relation depicted in Section 3 of my position paper, then the decision to reject T in favour of T', I_1, I_2, E_1 and E_2 would be eminently rational. There can be scientific rationality without induction.

Hübner rather overdoes the importance that we attach to Goodman's paradox. My point was this. On a verificationist theory of confirmation, if evidence E confirms a universal hypothesis H then E is supposed to raise the credibility (or probability or whatever) of H, and hence to lower that of rival alternatives to H. What Goodman showed is that any such H will be a member of an indefinite set of possible alternatives, H, H', H'' ... all standing in similar logical relations to E. Thus the confirming effect of E must be *rationed out* among H, H', H'' ... (It is not difficult to show, though I did not do this in my position paper, that this means that the confirming effect of E on any *one* of these hypotheses, such as H itself, will be zero – unless some principle of discrimination is introduced which allows us to concentrate it on some privileged member(s) of the set.)

Nothing analogous to this occurs under a falsificationist theory of corroboration. Let E report the result of an experiment which constituted a severe test on both H and H', where these are mutually incompatible hypotheses which, however, happen to agree in predicting E. Then H will be well corroborated by E and so will H'. The corroborating effect of E will *not* have to be rationed out between them: it is not being said that E raises the probability that either H or H' is true; all that is being said is that H has passed a severe test and so has H'. The test on H would have been neither more nor less severe if no other, or many other, hypotheses had also been tested by this test.

So my first point was that the existence of 'grue'-type alternatives to a well-corroborated hypothesis H does nothing to reduce the degree of corroboration enjoyed by H. But will not these alternatives be just as well corroborated, and hence just as worthy of rational acceptance, as H? In my position-paper I stressed that a consequence of Popper's non-verificationist philosophy of science, not shared by verificationist or inductivist philosophies, is that we can afford to ignore possible hypotheses which have not been 'laid on the table' for serious consideration by the scientific community. Musgrave objects that I lazily assumed that no scientist in his right mind would ever propose a 'grue'-type hypothesis for serious consideration; and he adds (pp. 181–182) that if a 'grue'-variant H' of an existing hypothesis H were seriously proposed, it would be less corroborated than H, since the early corroborations of H would have been absorbed into background knowledge when H' was proposed. But Hübner points out (p. 289) that if the author of H' got in first, the situation would be reversed: Musgrave, so to speak, lazily assumed that no scientist in his right mind would ever propose a 'grue'-type hypothesis *straight off*. What would we say if a 'catastrophe'-theory H', which postulated some systematic discontinuity at some specified future time t, were seriously proposed? One thing we could say is that a time will eventually come when we can decide between H' and a 'normal' or 'non-catastrophic' variant of it. (If in 1682 there had been a rival to Newtonian theory whose predictive implications were practically indistinguishable except that it predicted the non-return of Halley's comet, the scientific community would have had a long wait, until 1759, before it could decide between them.) We could also say that it is logically possible that H' will be stunningly corroborated after t, in which case our conjecture that such discontinuities never occur in nature will have been badly hit. In the meanwhile, however, there is no strong evidence against this conjecture: we are not obliged to scrap it on the say-so of the inventor of H'; unless evidence against it arrives in the interim, we may retain it until t and see what happens then.

We do not say that "Science *is* [my italics] constantly drawing nearer to absolute truth" (p. 279). We used to say, before the theory of verisimilitude got into serious trouble, that the following conjectural statement is *meaningful*: even if every theory in a historical sequence of scientific theories is false it may be that each later theory is closer to the truth than its predecessors. (I will revert to this important question in Section 10.) I was shocked that Hübner should appeal to Kant, of all people, in his attack on our claim that we can retain a regulative idea of truth even though there is no criterion for truth. No one can question Kant's commitment to the idea of truth (in the sense of the agreement, or conformity, or correspondence of a true judge-ment with its object). So he must have spurned any suggestion that the idea of truth would be empty without a criterion of truth since he flatly repudiated the possibility of such a criterion:

... it is quite impossible, and indeed absurd, to ask for a general test of the truth of such content. A sufficient and at the same time general criterion of truth cannot possibly be given ... Such a criterion would by its very nature be self-contradictory. (*Critique of Pure Reason*, A59 = B83.)

3. GRÜNBAUM'S CHALLENGE

Adolf Grünbaum's paper ends with this ringing challenge:

> In the light of the several groups of considerations presented in this paper, how can Popperians justify adhering to Popper's portrayal and indictment of inductivism while maintaining that he has given us a viable epistemological alternative on genuinely deductivist foundations?

Grünbaum is in the fortunate position of being a belligerent neutral in the battle between Popper and inductivism. He is against Popper without being for inductivism, his attitude to which is cool and often negative:

> ... I think that the inductivists have no more succeeded than Popper in stating *general* criteria for effecting a *neat* demarcation of science from non-science.
>
> Thus, the upshot of my comparison ... is *not* the claim that there is a viable inductivist counterpart to Popper's defective demarcation criterion ... (pp. 124–125).

Nor does Grünbaum point beyond both Popper and inductivism to some third alternative.

However, the force of Grünbaum's criticisms is not lessened by the fact that they do not stem from a rival viewpoint. I will begin by indicating which of his criticisms I shall not contest.

First, I agree with his rejection of Popper's proposal that we may use as a measure of a theory's content the complement of its (absolute) logical probability. I have never used this proposed measure myself. I will turn to the question of content-comparisons between competing theories in Sections 7–9 below.

Second, I concede that Bayesianism can account for the diminishing returns from repetitions of the same experiment in the confirmation of a hypothesis, and also that Popper is in some difficulty over this, as Musgrave showed [1975]. However, I will attempt a resolution of this difficulty, within a Popperian framework, in Section 6 below.

Third, I agree that there have been famous inductivists, such as Bacon and Mill, who repudiated what Grünbaum calls the 'Nicod criterion' or the 'instantiation condition' (whereby all positive instances of a hypothesis support it equally). Since Grünbaum chides me, as well as Popper, for misportraying inductivism 'as invariably *instantionist*' (p. 122) I would like to explain my own view here, not in order to ward off criticism, but as a way of entry into the main discussion.

I think that it is permissible to distil an idea out of a historical discussion and to examine it in a pure form, even though historically it was often combined with other ideas of a different tendency. The idea which I distilled out of inductivist writings was this: evidence E confirms hypothesis H if E entails part of the content of H. And *this* idea surely does involve the instantiation condition: if e_1 and e_2 report separate instantiations of H, then each of them contributes to the confirmation of H, irrespective of whether one of them was already well known while the other reports the verification of a risky prediction. But I should have added that this idea *had* to be supplemented by falsificationist ideas if unwanted consequences were to be cut off: for instance, the unwanted consequence that the evidence which confirms "All men are mortal" *also* confirms "All mortals are men" (see Popper [1974], pp. 991–2). So I am quite ready to admit that there were falsificationist tendencies in Bacon and Mill. Mill, for instance, allowed that a very large number of positive instances (and no counter-instances) could *fail* to support an inductive conclusion, whereas a single positive *could* support a universal law:

> That all swans are white, cannot have been a good induction, since the conclusion has turned out erroneous. The experience, however, on which the conclusion rested, was genuine. From the earliest records, the testimony of the inhabitants of the known world was unanimous on the point ...
>
> When a chemist announces the existence and properties of a newly-discovered substance, if we confide in his accuracy, we feel assured that the conclusion he has arrived at will hold universally, though the induction be founded but on a single instance ... (*Logic*, III, iii, Section 3.)

So whether an inference from evidence E to conclusion H is a good or a bad induction depends, according to Mill, *not* on the formal relation between E and H, but on whether H is going to withstand subsequent tests. I have claimed elsewhere ([1975a], p. 227) that the epistemology behind Mill's argument for free discussion in *On Liberty* was a kind of 'conjectures and refutations' view hardly compatible with the 'strict rules' to which he (erroneously) claimed to have reduced 'the inductive process' in his *Logic*.

Grünbaum is, again, right in saying that Bacon discriminated between instances of different kinds, regarding some as strongly supportive and others as hardly supportive at all. But here too my

reply is that there were strong falsificationist tendencies in his thinking and that these enabled him so to discriminate. Consider his
account of the opening stages of a 'true induction':

The *first* work therefore of true induction (as far as regards the discovery of Forms) is
the *rejection and exclusion* of the several natures which are not found in some instance
where the given nature is present. (*Novum Organum*, Part II, aphorism 16, my italics.)

In introducing his idea of prerogative instances Bacon writes: "such
instances make the way short, and accelerate and strengthen the
process of exclusion" (aphorism 22); again: "These instances therefore should be employed as a sort of preparative for ... purging the
understanding" (aphorism 32). The most famous of Bacon's types of
prerogative instance were his "Instances of the Fingerpost". These
were supposed to decide the question between two or more possible
causes, the erroneous possibilities being '*dismissed and rejected*',
(aphorism 36, my italics).

4. 'ALL MEN ARE MORTAL'

Grünbaum claims that Popper's demarcation criterion is at once too
strong and too weak: too strong because it excludes as unscientific
such hypotheses as "All men are mortal", and too weak because,
contrary to Popper's intention, it fails to exclude Freud's psycho-
analytical theory. About Grünbaum's methodological examination of
the scientific credentials, if any, of Freudian theory I will say two
things. First, I have read it with fascination: it brilliantly illuminates a
murky and tangled subject. But second, I do not see that it throws
much light on Popper's demarcation-criterion. Suppose (with Grünbaum) that psychoanalytic theory is testable; then Popper was wrong
about Freudian theory: it is better than he thought. On an earlier
occasion Grünbaum [1959] argued that the Lorentz-Fitzgerald
contraction hypothesis was not *ad hoc*, as Popper had claimed, but
yielded a prediction which conflicted with the result of the 1932
Kennedy-Thorndike experiment; and Popper, in reply, accepted this
correction: the contraction hypothesis is better than he had thought
('*it was an advance*'). In that earlier case there was no suggestion that
his demarcation-criterion was in trouble because it actually included
something which Popper himself had mistakenly excluded. But in
the present case Grünbaum seems to take the unscientific status of

Freudian theory as a datum against which that demarcation-criterion should be judged, rather than the other way round.

Does Popper's criterion wrongly exclude "All men are mortal"? It will exclude it if there could be no evidence against it. Could there be evidence against it? Suppose that Shadrach, Meshach and Abednego, after being cast into the burning fiery furnace, had continued to survive similar ordeals. We may imagine that Stalin, furious with them for refusing to fall down and worship him, had them tied to an atom-bomb; but after the explosion they landed on their feet and walked off; and they are alive and well today. (For the source of this story see my [1977], p. 76.) Would that be evidence against it? Not according to Grünbaum, who says that "we must produce at least one man who NEVER dies" (p. 123), which of course we could never do.

Now it seems to me that if this hypothesis is so construed that it would not be disturbed even by the kind of evidence I have imagined, then it is rightly excluded as non-scientific. After all, science does not need it, having at its disposal stronger, testable hypotheses about human mortality, that are well corroborated (e.g. "No man lives to be 200 years old"). So far as I can see, the case for including it would have to rely on some consequence condition or hereditarian principle whereby each of the non-analytic consequences of a well-confirmed scientific hypothesis is itself a well-confirmed scientific hypothesis even though it may be untested or actually untestable. But I object to this principle. For one thing, it 'solves' the problem of induction all too easily. Let H say "All ravens are black" and let E report that the individual objects $a_1, a_2, \ldots, a_n$ have been observed to be both ravens and black. Now split H into H_1 and H_2 where H_1 says

$$\forall x((x = a_1 \vee x = a_2 \vee \cdots \vee x = a_n) \rightarrow (Rx \rightarrow Bx))$$

and H_2 says

$$\forall x((x \neq a_1 \wedge x \neq a_2 \wedge \cdots \wedge x \neq a_n) \rightarrow (Rx \rightarrow Bx)).$$

In short, H_1 says of everything mentioned in E, while H_2 says of everything *not* mentioned in E, that if it is a raven then it is black. (In the language of Hempel's [1945], H_1 is the 'development' of H for E.) Then E verifies H_1 and confirms H; and H entails H_2; so by our hereditarian principle we have bridged the inductive gap between a report about observed instances and a prediction about unobserved instances. (I was pretty startled when Popper came dangerously close to endorsing such a principle in his [1972], pp. 19–20.) An inductivist might *welcome* a principle which achieves this. But easy 'solutions' of the problem of induction have a nasty tendency to backfire. We could play a Goodman-type trick here by replacing H by H' where H' is the conjunction of the previous H_1 and a new H'_2 which says:

$$\forall x((x \neq a_1 \wedge x \neq a_2 \wedge \cdots x \neq a_n) \rightarrow (Rx \rightarrow \sim Bx)).$$

Now we can proceed 'inductively' from E via H_1 and H' to the conclusion that all ravens not mentioned in E are non-black.

If "All men are mortal" is construed as an unfalsifiable all-some statement, then according to my old [1958] view it should be classed as metaphysical. However, I have suggested in my [1975b] that we take the metaphysical core, if any, of a scientific theory to consist of those consequences of it that are neither consequences of its Ramsey-sentence nor testable. I suggested calling any untestable consequences of its Ramsey-sentence 'quasi-empirical'. If "All men are mortal" is to be put into one or other of three boxes labelled respectively 'scientific', 'metaphysical', and 'quasi-empirical', I now consider the last the most appropriate.

5. DO CORROBORATIONS MATTER?

I turn now to Grünbaum on Popperian corroboration and the significance or otherwise of successful risky predictions. If I understand him, Grünbaum presents us with this dilemma: *either* we are inductivists in disguise, *or* we have no right to say that to pass a severe test counts in favour of a hypothesis.

Apropos the first horn of this dilemma Grünbaum quotes Popper's statement that "we try to select for our tests those crucial cases in which we should expect the theory to fail if it is not true" ("if it is not TRUE", Grünbaum repeats in a shocked tone). He reads this as sanctioning the inductivist inference that if the theory passes such a test, then this raises the likelihood that the theory is true: "Popper's reasoning here is of a piece with the reasoning of the Bayesian proponent of inductive probabilities ..." (p. 131). He must take a low view of Popper's concern for logical consistency, since he quotes several passages (on pp. 121 and 125–126) which flatly forbid such an interpretation. In the same connection he quotes (p. 132) my statement that "we want highly testable theories because we hope that, if such theories are severely tested they have a high chance of being weeded out if they are in fact false".

But nothing like Bayesian inductivism is lurking in these phrases. They reflect the following points: (1) we do not expect a theory to fail tests if it is true; (2) if T_1 is a false theory entailed by the more testable theory T_2, then T_2 has a higher chance of being weeded out by tests than T_1 (since T_2 will automatically fail any tests that T_1 fails and T_2 may fail tests that were not tests of T_1); (3) a false theory has a higher chance of failing a severe test than a non-severe test.

It seems to be our adherence to point (3) to which Grünbaum objects. He seems to be turning the tables on us by claiming that it is we anti-inductivists who are committed by our deductivism to a falsificationist analogue of the Nicod-condition. Are not all tests on a theory equally tests on it? If we do not resort to extralogical assumptions of an inductivist nature, have we any right to discriminate between tests of different kinds, regarding some as more and some as less severe? From a deductivist standpoint, does not the question whether an experiment constitutes a test for a theory call for a Yes-or-No answer, 'Yes' if it is logically possible that the result will tell against the theory, otherwise 'No'?

Well, one test on a theory in mathematical physics, say, may be more severe than another just because it is made with a more precise kind of measuring instrument; and if the theory under test is false it has a higher chance of failing the more severe test.

True, we also say that a test on a theory is more severe (other things, such as experimental error, being equal) the more unlikely is the tested prediction relative to current background knowledge minus the theory in question. Grünbaum objects that we here make inductivist use of the past experience incorporated in background knowledge. Well, past experience is being used, but to what use is it being put? Not to establish anything but to suggest promising lines of attack on the theory. True, we will regard the theory as better corroborated if it passes a test which is severe in this sense; but a high degree of corroboration carries no inductivist implications as to how the theory will stand up to future tests. No *inductivist* use is made of background knowledge when it is used to assess the relative severity of tests.

But perhaps we are caught on the other horn of the alleged dilemma? Grünbaum writes:

the inductivist can justly complain that Popper's pure deductivism has no non-trivial answer to the question: "What does it mean to say that a successful risky prediction COUNTS in favor of the theory that made it?" (p. 125).

Again:

What does the word '*count*' mean ... when Popper declares that "Confirmations should count only if they are the result of *risky predictions*"? Count toward or for what?? (p. 130).

Behind these questions there is obviously the assumption that if a successful risky prediction does not count in favour of the theory that made it *in some inductivist sense*, then it does not count at all: either it contributes towards the verification (probabilification, credibilification) of the theory or it contributes nothing to it.

There are passages with a similar tendency in Ernan McMullin's paper; for instance:

Popper's preoccupation with the Humean problem of induction led him to an exaggerated rejection of any suggestion of verificationist elements in proper scientific methodology. But unless there is *some* sense in which a theory can be 'confirmed' by progressive successes in accounting for new sorts of data, ... Hume's critique has been allowed to triumph, and scepticism must be the result (*above*, p. 235).

Some ten years earlier Wesley Salmon had posed the (alleged) dilemma very sharply: Popperian corroboration is either *empty* or *inductive* ([1968], pp. 28f.).

My position paper was intended as an answer to this kind of objection. Of course, someone who is drawn to what I there called the certainty pole of the Bacon-Descartes ideal will regard Popper's kind of non-inductive or non-verificationist corroboration as pointless; but not those of us who are drawn to the other pole. What we ask from science is good conjectural explanations, or rather, ever better conjectural explanations. And we regard T' as a better conjectural explanation than T, other things being equal, if T' is a deeper and more unified and empirically more powerful theory than T. We could summarise this in Popperian shorthand by saying that, other things being equal, T' is better than T if T' is more corroborable, that is, more testable, than T. But other things are not equal if T', in virtue of its greater testability, has been refuted whereas T has withstood all tests on it. If T' is to be unambiguously better than T it should be both more corroborable and better corroborated. In particular, it should prove its superiority to T by gaining corroboration in regions where it revises or goes beyond T.

The main message of my position paper was that the comparative goodness of scientific theories both can and should be assessed in terms of explanatory power and depth *without* recourse to anything like degree of partial verification, which would pull us towards hypotheses that are *ad hoc* and merely phenomenal. The answer provided there to Grünbaum's question is that successful risky predictions, especially predictions which diverge from, or go beyond, those of a rival theory, *count* in that they raise the theory's degree of corroboration which is an index of its comparative goodness as a conjectural explanation.

6. THE PROBLEM OF DIMINISHING RETURNS FROM REPEATED TESTS

I turn now to a problem, first raised by Musgrave in [1975], which Grünbaum takes up.

Let C_1, C_2, C_3, ... be a series of singular predictions, each got from a hypothesis H by conjoining the same general statement in H with different initial conditions. Let C_1 be the first of these predictions to be put to the test, C_2 the second, and so on. Let E_1 be a report of the outcome of the first test, E_2 of the second and so on, and suppose that E_1, E_2, E_3 ... are all favourable to H. Let B_0 be background knowledge as it was prior to the first test, B_1 as it became after the incorporation of E_1, and so on. Popper claimed that, for a case like this where later tests are mere repetitions of the first test, his account yields 'diminishing returns' of corroboration. We might express this claim by:

$$c(H, E_1) > c(H, E_2) > c(H, E_3) ... ,$$

that is, the additional corroboration separately afforded to H by each further test of the same type is less than that afforded by the previous test. For Popper, degree of corroboration, when the outcome of the test is favourable, varies with the severity of the test, and severity is a function of $r - q$, where r is the probability of the prediction given the hypothesis (plus initial conditions and auxiliary assumptions drawn from background knowledge) and q is its probability given all our background knowledge minus the hypothesis. Thus it would seem that, if we are to get diminishing corroboration returns, the value of q must steadily rise as the series of tests continues; and Musgrave declared in [1975] that "this clearly involves a straightforward *inductive argument*" (p. 250). Let us look into this.

It will make things simpler if we have the first test of maximum severity by putting

$$p(C_1, B_0) = 0,$$
$$p(C_1, H) = 1.$$

(For simplicity's sake I omit the initial conditions and auxiliary assumptions.) Then it would seem that, in order to get diminishing returns from repetitions of the test, we need:

(1) $0 = p(C_1, B_0) < p(C_2, B_1) < \cdots < p(C_n, B_{n-1}).$

But what has B_1 got that B_0 has not got? Answer: E_1. So (1) appears to rely on the assumption that

$$p(C_2, E_1) > p(C_2);$$

and this is surely an inductivist assumption. It would seem that in place of (1) a strict non-inductivist should put:

(2) $0 = p(C_1, B_0) = p(C_2, B_1) = \cdots = p(C_n, B_{n-1}).$

But (2) would not yield diminishing returns. There is another alternative which Musgrave considered, namely that after 'sufficiently many' tests we incorporate into background knowledge either H itself, or else a low-level law-statement in H which yields the predictions $C_1, C_2, C_3 \ldots$ This would give:

(3) $0 = p(C_1, B_0) = \cdots = p(C_m, B_{m-1}) < p(C_{m+1}, B_m) = \cdots$
          $= p(C_n, B_{n-1}) = 1,$
          $(0 < m < n).$

After m tests, each of maximum severity, the severity of the next test (and of all subsequent tests) suddenly drops to zero.

Thus we seem to be faced by a trilemma: either (i) steadily diminishing returns at the price of inductivism, or (ii) constant high returns forever, or (iii) constant high returns for some unspecified time, abruptly followed by constant zero returns.

I will try to find a way out of this trilemma. I begin by considering what kinds of modification to background knowledge, in response to $E_1, E_2, E_3, \ldots$, are permissible from a falsificationist point of view.

Background knowledge will be strengthened in one way, just

because E_1, E_2, E_2, ... will be successively added to it. But that, I hold, should be the only way in which it is strengthened. When B_0 is modified into B_1 in response to E_1, E_1 should have only a falsifying or negative effect: it should *reduce* (eliminate or at least damp down) the component in B_0 relative to which C_1 has zero probability without introducing some positive new component. If we subtract E_1 from B_1 what remains should be merely a weakened version of B_0; likewise when we proceed from B_1 to B_2 in response to E_2.

We are operating here with the notional idea that each $p(C_i, B_{i-1})$ has a definite numerical value. But, except in the two extreme cases where B_{i-1} either entails or denies C_i, we may safely assume that we cannot precisely determine this value. However, I shall now suppose that a given B_{i-1} does allow us to make a *best estimate* of the value, or that there is a smallest interval within which we can confine the value. And I further suppose that as background knowledge gets successively weakened when the evidence E_1, E_2, E_3 ... mounts up against that part of B_0 which gave C_1 zero probability, so this smallest interval widens. In other words, as background knowledge weakens we become increasingly agnostic about the probability, relative to it, of the next prediction. On this supposition we will not be entitled to assert (1), or for that matter (2) or (3). The most that we are here entitled to assert is the following statement which is weak enough to be consistent with each of those three:

(4) $0 = p(C_1, B_0)$

$0 \leqslant p(C_2, B_1) \leqslant \delta$

$0 \leq p(C_3, B_2) \leq \delta'$

...

where $0 < \delta < \delta' < \cdots 1$.

Here, the lower bound of the smallest interval within which the value of each $p(C_i, B_{i-1})$ can be located is assumed to remain fixed at zero, while the lowest upper bound is progressively revised upwards.

Suppose that the initial B_0 can be split into two parts, say b_N and b_0, where b_N is neutral with respect to the predictions $C_1, C_2, C_3, ...$ and remains unchanged throughout the series of tests, while b_0 actually excludes every C_i. We are supposing that as the reports $E_1, E_2, E_3, ...$ of verifications of $C_1, C_2, C_3, ...$ come in, background knowledge is progressively weakened except for just the addition of

$E_1, E_2, E_3, \ldots$ Thus if b_0 is replaced by b_1 after the arrival of E_1, and b_1 by b_2 after $E_2, \ldots$ the changing background knowledge may be represented by

$$B_0 = b_N \wedge b_0$$
$$B_1 = b_N \wedge E_1 \wedge b_1$$
$$B_2 = b_N \wedge E_1 \wedge E_2 \wedge b_2$$
$$\ldots$$

where b_0 entails (without being entailed by) b_1, b_1 entails (without being entailed by) b_2, and so on. Thus the best estimate that can be made of the probability of any C_i on the basis of b_0 must entail the more tolerant best estimate that can be made on the basis of b_1, and so on. The sequence of best estimates available on the basis of $b_0, b_1, b_2 \ldots$ might be of this form:

$$b_0: (\forall i)[p(C_i) = 0]$$
$$b_1: (\forall i, i > 1)[0 \leq p(C_i) \leq \delta]$$
$$b_2: (\forall i, i > 2)[0 \leq p(C_i) \leq \delta']$$
$$\ldots$$
where $0 < \delta < \delta' < \cdots 1$.

It is easy, now, to amend Popper's measure for degree of corroboration in a way which yields diminishing returns from successful repetitions of a test. Let r be (as before) the probability of a singular prediction given our hypothesis H. But instead of q (the probability of the prediction given background knowledge minus H) let us take an interval with a and b as its greatest lower bound and lowest upper bound, respectively. Now, instead of making degree of corroboration vary with $r - q$ we may make it vary with, first, $r - (a + b)/2$, i.e. with the difference between r and the mid-point of the interval, and second, $1 + a - b$, i.e. with the smallness of the interval. Thus we might take as our measure of the degree to which H is corroborated by a particular test-result E_i:

$$c(H, E_i) = \left(r - \frac{a + b}{2}\right) \times (1 + a - b).$$

If (as we have been supposing) we have $r = 1$ and $a = 0$ throughout, this simplifies to

$$c(H, E_i) = (1 - b/2) \times (1 - b).$$

On the above construal of the effect of each E_i on the best estimate of $p(C_{i+1}, B_i)$, the value of b will, in the situation we have been considering, increase as i increases, giving diminishing degrees of corroboration. We will get

$$c(H, E_1) > c(H, E_2) > c(H, E_3) \dots .$$

7. CONTENT-COMPARISONS BETWEEN RIVAL THEORIES

No idea is more central to the Popperian conception of scientific progress, or growth, than that of *content*, especially empirical content. This conjecturalist and falsificationist philosophy allows us to contemplate the soaring superstructure of theoretical science with an admiration unmarred by verificationist fears that it may have grown too high for inductive support and need to be cut down to a less unverifiable size. The idea of content involved here is an essentially *comparative* one: if a new theory T' is to constitute a clearcut advance over T, one requirement is that T' has *more* empirical content than T. Another requirement, as we saw, is that T' revises T. This latter requirement means that T' will be strictly incompatible with T (though perhaps yielding T as a limiting case). But what relation must one theory bear to another, in order to have more empirical content than it, when the two theories are incompatible? One answer which Popper gave to this question, and the answer which I adopted in my position paper, was this: the *questions* answered by T must be *a proper subclass* of the questions answered by T'.

This idea was attacked briefly by Miller in his [1975] and at length by Grünbaum in his [1976]. I think that Grünbaum's criticism, which I will consider first, can be met; but Miller's is fatal.

A favourite example, for Popper, of content-dominance by one theory over another is that of Einstein's over Newton's. According to him, "to every question to which Newton's theory has an answer, Einstein's theory has an answer which is at least as precise"; moreover, there are questions which Einstein's theory can answer but Newton's cannot ([1972], pp. 52–3). (I now think that the qualification 'at least as precise' is redundant: if the best answer which T' can give to a certain question is less precise than an answer which T can give, then there will be a more exacting question which T can answer and

T' cannot.) In opposition to this claim of Popper's, Grünbaum (who does not do things by halves) has constructed no less than five sets of questions which, he claims, Newton's theory can answer but which Einstein's theory cannot answer but only 'obviate'. These questions have a common pattern: they involve a presupposition which is endorsed by Newton but rejected by Einstein. One such question is: How long would it take for a force F to accelerate a mass m to a velocity of $2c$? (p. 36).

In order to simplify the discussion I am going to switch to the following surrogates for, respectively, Newton's and Einstein's theories:

> A: The French monarch at t_1 was male and bald.
> B: The French president at t_1 was male and not bald and tall.

I think that the relation between A and B replicates that between Newton's and Einstein's theories well enough for present purposes: B repudiates A's royalist ontology concerning the French head of state at t_1 in favour of a republican ontology; and B partly endorses, partly revises, and partly goes beyond what A says about the physical characteristics of the person in question. Transposed to this example, Grünbaum's objection would be that B cannot answer, but only obviate, the question "Was the French monarch at t_1 bald?"

This question presupposes that there was a French monarch at t_1 which I will express by

$$\exists x (Mx \wedge \forall y (My \rightarrow y = x)).$$

The answer *yes* to this question says, in effect:

$$\exists x (Mx \wedge \forall y (My \rightarrow y = x) \wedge Bx);$$

that is, it endorses and adds to the presupposition of the question. How should we construe the answer *no*? There are two alternatives. The simpler is to construe it as just the negation of the *yes*-answer, i.e. as

$$\sim [\exists x (Mx \wedge \forall y (My \rightarrow y = x) \wedge Bx)].$$

Elsewhere (in my [1978]) I have called this the either-*yes*-or-*no* construal. The other alternative is to construe it as expressing the statement: "The French monarch at t_1 was not bald" or

$$\exists x(Mx \land \forall y(My \rightarrow y = x) \land \sim Bx).$$

On this construal the answer *no* endorses the presupposition of the question just as *yes* does, and adds to it the opposite of what *yes* adds. But if we adopt this construal we must allow for a third answer besides *yes* and *no*. For let P and R be, respectively, the presupposition of some *yes/no* question and the information conveyed by the answer *yes* so that question. Since R entails P there are three truth-table possibilities for P and R, namely TT, TF and FF. On the former construal, *no* did duty for both TF and FF; but on the present construal *no* is tied just to TF. I propose to use *n/a* ('not applicable') as the answer which singles out just *FF*. On this construal *yes*, *no*, and *n/a* are respectively equivalent to:

$$P \land R$$
$$P \land \sim R$$
$$\sim P \land \sim R$$

I call this the *yes-no-n/a* construal. Notice that it involves no departure from a two-valued logic. Notice also that *n/a* is quite unlike "Don't know": it assigns truth-values as unequivocally as *yes* does (and less equivocally than *no* does on the either-*yes*-or-*no* construal).

Let us now consider how well A and B fare, on either of these two construals of the answer *no* to a *yes/no* question, in answering the following questions (the words 'French' and 'at t_1' are omitted).

		Either-*yes* -or-*no*		Yes-no-n/a	
		A	B	A	B
Was the head of state	male	Yes	Yes	Yes	Yes
	bald	Yes	No	Yes	No
	tall	Don't know	Yes	Don't know	Yes
Was the monarch	male	Yes	No	Yes	N/A
	bald	Yes	No	Yes	N/A
	tall	Don't know	No	Don't know	N/A
Was the president	male	No	Yes	N/A	Yes
	bald	No	No	N/A	No
	tall	No	Yes	N/A	Yes

On the face of it, it seems that B maintains its question-answering dominance over A on either construal. I think that Grünbaum's mistake was to equate what he called *obviating* a question, which I am representing by n/a, with the reply "Don't know".

8. MILLER'S OBJECTION

If only *yes/no* questions are admitted, then to say that the questions which T answers are a proper subclass of those which T' answers becomes equivalent to saying that the propositions to which T assigns truth-values are a proper subclass of those to which T' assigns truth-values. The claim implicit in my diagram on p. 33 of my

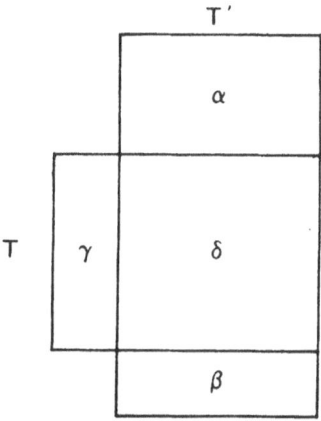

position paper was that the T-determinate propositions are indeed a proper subclass of the T'-determinate propositions; for every consequence c of T in the γ-area is paired with a consequence c' of T' in the β-area which is its negation; and there are consequences of T' in the α-area to which T assigns no truth-value. So from that diagram, if it represents the situation correctly, we can proceed to the diagram on p. 364 where the smaller circle represents the class of propositions to which T assigns truth-values and the larger one those to which T' assigns truth-values.

But is it true that T' may assign a truth-value to every proposition to which T assigns one even though T and T' are incompatible? Alas, it is not (at least on the assumption, here taken for granted, that T' is not a complete theory which assigns a truth-value to *every* proposi-

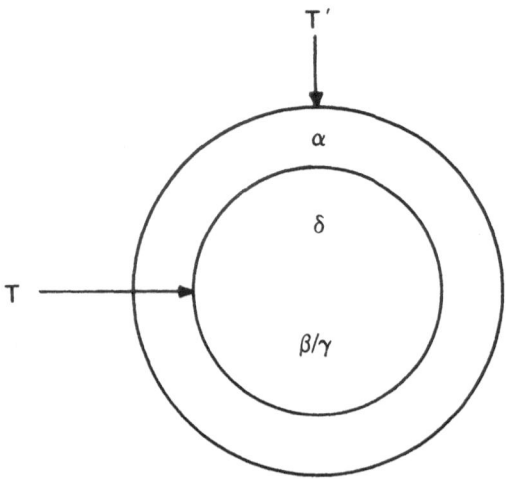

tion expressible in the language in which it is formulated). The reason
was given by Miller in [1975]: those consequences of T in the γ-area,
which are denied by T', will have consequences which are neither
denied nor affirmed by T'. For instance, let p be any proposition that
is T'-indeterminate and let c be a consequence of T that is denied
by T'; then $c \vee p$ will be a consequence of T which is T'-indeter-
minate. Thus my diagram was wrong. In addition to the α-area
(consisting of T-indeterminate consequences of T') we must introduce an ε-area consisting of T'-indeterminate consequences of T. The
diagram should rather look like this:

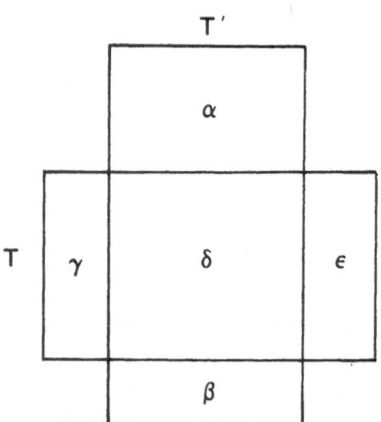

And from this we must now proceed to the diagram below where the protuberance on the right represents the T'-indeterminate consequences of T.

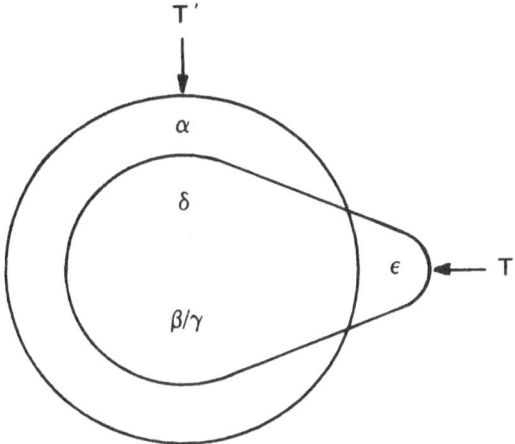

Actually, I ought to have realised without this prod from David Miller that even if the T-determinate propositions *could* be a proper subclass of the T'-determinate propositions, that would by no means ensure that T' has more content than T; for let U' be a class of universal propositions of which U is a proper subclass; and suppose that T' assigns the truth-value *false* to every proposition in U' while T assigns *true* to every proposition in U; then T might be highly falsifiable whereas T' would be altogether unfalsifiable.

To have to drop content-comparisons between competing theories would be a shattering blow to the whole Popperian view of scientific progress. That view has already been dealt a rather severe blow by the discovery (Miller [1974], Tichý [1974]) of major flaws in Popper's definitions of verisimilitude. I hope that the efforts currently being made by Miller, Tichý, Niiniluoto, Tuomela, Hilpinen, Oddie and others, not to mention Popper himself, to repair the damage will succeed. But suppose they do not. Well, Popper got along well enough without the idea of verisimilitude for a quarter of a century after 1934; and while he would no longer be able to give a clear meaning to the claim that T', though false, is closer to the truth than T, he could still claim, for suitable T and T' and provided always that

they can be compared for empirical content, that T' is a better
conjectural explanation than T because it has more content and has
gained only corroborations in areas where it revises or goes beyond
T. Indeed, verisimilitude comparisons first came in on the coat-tails of
content-comparisons. Here are some extracts from near the opening
of the paper in which Popper introduced the idea of verisimilitude:

... we can know of a theory, even before it has been tested, that *if* it passes certain tests
it will be better than some other theory.
   ... we have a criterion of ... *potential* progressiveness ...
   This criterion ... is extremely simple and intuitive. It characterizes as preferable the
theory which tells us more; that is to say, the theory which contains the greater amount
of empirical information or content; ... and which can therefore be more severely tested
...
   All these properties which ... we desire in a theory can be shown to amount to one
and the same thing: to a higher degree of empirical *content* ... ([1963], p. 217, all italics
in the original.)

Without the possibility of content-comparisons between logically incompatible theories, this philosophy of science would be in total
disarray.

9. COUNTERPART CONSEQUENCES AND CORRECTION-RULES

For years I assumed that only by using the subclass relation in some
way or another could one ever compare the contents of two universal
theories; for their consequence-classes will both be infinite, and this
surely means that there is no possibility of *counting* how many 'units'
of content each of them has. However, Graham Oddie has shown me
that we may compare two universal theories T and T' with respect to
the *finite* number of *yes/no* questions which each of them can answer
at some depth d (in the sense of Hintikka's [1973]); and if, as seems
reasonable, all depth-d constituents, that is, complete or maximal
sentences at depth-d, are given the same weight, it will be possible
actually to count and compare the number of depth-d constituents
excluded respectively by T and T'; and it may further be possible to
prove that if T' dominates T in this way at depth-d, then T' will
continue to dominate T as $d \to \infty$. I believe that his forthcoming paper
provides a powerful solution for our problem.
   However, since we need not fear having too many solutions for this
problem, I am going to try for a solution which relies once more upon

the subclass relation. Briefly, my idea is to define a notion of (congruent and incongruent) *counterparts* so that we may say that every T'-indeterminate consequence of T in the ε-area has an incongruent counterpart in the α-area, but not vice-versa, so that the situation looks like this:

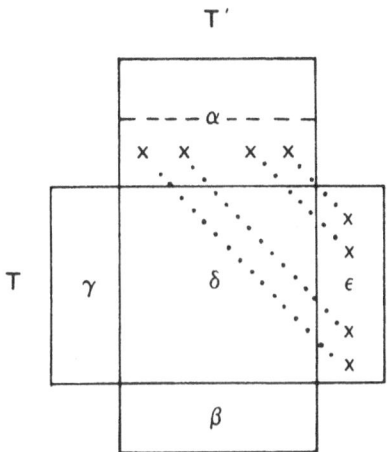

More generally, I will try to show, for suitable T and T', that every consequence of T has a counterpart, whether congruent or incongruent, among those of T', while some consequences of T' have no counterpart among those of T; so that a proper subclass relation is re-established between T and T' with respect, not to the questions they answer, but to the consequences of one and the counterpart consequences of the other.

It is a theory's propositional content, more specifically its empirical content, that I want to render comparable with that of a competing theory. But the method I shall propose will effect such comparisons by a consideration of the relation between certain sentential expressions of the theories being compared. So the language in which the theories are assumed to be formulated becomes important. I will begin with a very restricted language: a first-order language, with identity, whose primitive predicates are all (a) monadic, (b) observational, and (c) atomic in Carnap's sense. I will later drop restriction (c) and restriction (b).

To begin with, then, I assume that theories up for pair-wise content-comparison are expressible in a first-order monadic language L_1 whose primitive predicates $P_1, \ldots, P_k$ are atomic (so that, for any P_i and P_j, where $i \neq j$ and $1 \leq i \leq k$ and $1 \leq j \leq k$, if P_i is asserted of an individual then either P_j or $\sim P_j$ can be asserted of that individual without inconsistency); observational (so that, for a given individual and for any P_i it can be determined observationally whether that individual is P_i or $\sim P_i$); and lexicographically ordered. I shall assume that the theories we want to compare for content are expressible in general sentences without individual constants. I shall also assume that there is a canonical form, say prenex normal form, for sentences in L_1 such that, for any sentence i, there is just one sentence j such that, of all the sentences logically equivalent to i, j is *the* canonical one: in the sentence j all quantifiers will be in front, with existential quantifiers (if any) coming first; the predicates which occur in j will all be primitive, and they will occur in a uniquely determined order. My reason for introducing this last assumption is that we need to have the consequences of theories in some standardised form before we proceed to compare them.

I now introduce the idea of *counterpart consequences.* Let c and c' be canonical sentences in L_1 which respectively express a consequence of the theory T and of the theory T'. I say, first and trivially, that c and c' are congruent counterparts if they are identical; and I say that they are incongruent counterparts if: (1) they are identical except that there occurs in them at least one disputed predicate, i.e. a predicate which is negated in one of them at a place where it is not negated in the other; (2) c could be turned into c' by a *systematic* reversal of the sign of each disputed predicate wherever it occurs in c, i.e. by inserting the negation sign wherever the predicate occurs unnegated and deleting it where it occurs negated. Thus if c were

$$\forall x((\sim P_1 x \wedge P_2 x) \vee (\sim P_1 x \wedge P_3 x) \vee (P_1 x \wedge P_4 x))$$

and c' were

$$\forall x((P_1 x \wedge P_2 x) \vee (P_1 x \wedge P_3 x) \vee (\sim P_1 x \wedge P_4 x)),$$

then c and c' would be incongruent counterparts: c could be turned into c' by reversing the sign of the disputed predicate P_1 wherever it occurs. If we allow an equal strength to each of the sixteen Q-

predicates (in Carnap's sense) generated by $P_1, \dots, P_4$, then c and c' are equal in content in the sense that each says that everything falls under a disjunction of the same number (in this case, ten) of Q-predicates.

When any two canonical sentences c and c' are incongruent counterparts, every consequence of c will have a counterpart among those of c' and vice versa. Both sentences will be in disjunctive or conjunctive normal form and any consequence of c can be obtained by dropping conjuncts, or adding disjuncts, from or to c. If a conjunct is dropped, dropping the corresponding conjunct from c' will yield a counterpart consequence of c'. If a disjunct is added to c, then at the same place the same disjunct should be added to c' except that, if a disputed predicate is involved, its sign must always be reversed.

I now say that T' has more content than T if there is a consequence $c_{T'}$ of T' (where $c_{T'}$ does not entail T') such that $c_{T'}$ and T are counterparts (whether congruent or incongruent). For in that case every consequence of T will have a counterpart among those of T' but not vice versa.

Let us try this out on those two specimen 'theories' A and B which I substituted for Newton's and Einstein's theories in Section 7. We cannot immediately apply the above to the two sentences "The French monarch at t_1 was male and bald" and "The French president at t_1 was male and not bald and tall". However, it is permissible to tinker with the linguistic formulations of theories in order to render them comparable provided, of course, that one does not thereby change their propositional content. It is sometimes said that Newton's and Einstein's theories are incomparable because Newton just had mass whereas Einstein distinguished mass (m) from rest-mass (m_0) and related them by the equation $m = m_0/\sqrt{1 - v^2/c^2}$. But of course we can easily get round this difficulty by slightly reformulating Newton's theory so that it first distinguishes m from m_0 and then asserts that $m = m_0$. This tinkering would of course leave its propositional content unchanged.

In the present case I propose to reformulate A and B as: "The French head of state at t_1 was a monarch and not a president and male and bald" and "The French head of state at t_1 was not a monarch and a president and male and not bald and tall". And I shall suppose that in L_1 the canonical sentences for these two statements are:

A: $\exists x \forall y (P_1 x \wedge (\sim P_1 y \vee y = x) \wedge P_2 x$

      $\wedge \sim P_3 x \wedge P_4 x \wedge P_5 x)$,

B: $\exists x \forall y (P_1 x \wedge (\sim P_1 y \vee y = x) \wedge \sim P_2 x \wedge P_3 x$

      $\wedge P_4 x \wedge \sim P_5 x \wedge P_6 x)$,

where $P_1 =$ "French head of state at t_1", $P_2 =$ 'monarch', $P_3 =$ 'president', $P_4 =$ 'male', $P_5 =$ 'bald', $P_6 =$ 'tall'. Now let c_B be that consequence of B that is obtained by dropping its last conjunct ('$\wedge P_6 x$'). Then A and c_B are incongruent counterparts: one could be converted into the other by reversing the signs of the disputed predicates P_2, P_3 and P_5. Hence B has more content than A.

When every consequence of a theory T has a counterpart among the consequences of theory T', some of them incongruent, there exists a $T' - T$ correction-rule, as I will call it, whereby T can be 'corrected' by T', or converted into a theorem of T', simply by the systematic reversal of the signs of the disputed predicates in T. We may regard the canonical sentence for T as a certain function of the primitive predicates that occur in T. If T' is in the above relation to T, then there will be a consequence of T', say $c_{T'}$, such that the canonical sentence for $c_{T'}$ is *the very same function* of those same predicates except that in at least one instance a predicate is now negated. (I owe this way of looking at the matter to Elie Zahar.) Thus if we have

$$T = f \langle P_1, P_2, P_3 \rangle,$$

we will have

$$c_{T'} = f \langle \pm P_1, \pm P_2, \pm P_3 \rangle,$$

where '$\pm$' is resolved one way or the other in each case and negatively in at least one case. Suppose that we in fact have

$$c_{T'} = f \langle P_1, P_2, \sim P_3 \rangle.$$

Then the $T' - T$ correction-rule for converting T into a theorem of T' will be: "Replace P_3 wherever it occurs in T by $\sim P_3$, and $\sim P_3$ by P_3"; or more briefly: "Reverse the sign of P_3 wherever it occurs in T".

We can now say, rather crisply, that T' has more content than T if there is a $T' - T$ correction-rule but not a $T - T'$ correction-rule. (I will somewhat modify this idea of a $T' - T$ correction-rule when quantitative predicates have been introduced.)

Before I turn to a quantitative language L_2 I will consider an objection raised by David Miller, in a personal communication, to the foregoing method of content-comparison. Like his [1974] objection to Tichý's [1974] definition of verisimilitude, his objection here is that my method is *language-dependent*. However, the charge is *not* that orderings may be *reversed* when T and T' are reformulated in a new language L_1', but only that they may be incomparable for content in L_1', or be comparable in L_1' if incomparable in L_1. His argument goes like this. Let L_1 contain the atomic predicates P_1, P_2, P_3, and let the molecular predicates M_1 and M_2 be defined as:

$$M_1 = (P_1 \wedge P_2) \vee (\sim P_1 \wedge P_3),$$
$$M_2 = (\sim P_1 \wedge P_2) \vee (P_1 \wedge P_3).$$

Now let L_1' be a language whose primitive predicates are P_1, M_1, M_2, and let the molecular predicates $\mathbf{M}_1$ and $\mathbf{M}_2$ be defined as:

$$\mathbf{M}_1 = (P_1 \wedge M_1) \vee (\sim P_1 \wedge M_2),$$
$$\mathbf{M}_2 = (\sim P_1 \wedge M_1) \vee (P_1 \wedge M_2).$$

Then the molecular predicates $\mathbf{M}_1$ and $\mathbf{M}_2$ in L_1' are respectively equivalent to the primitive predicates P_2 and P_3 in L_1. Suppose that c and c', when formulated canonically in L_1, are

c: $\forall x((P_1 x \wedge P_2 x) \vee (\sim P_1 x \wedge P_3 x))$,
c': $\forall x((\sim P_1 x \wedge P_2 x) \vee (P_1 x \wedge P_2 x))$.

Then c and c' are incongruent counterparts: one could be converted into the other by reversing the sign of the disputed predicate P_1.

Now reformulate c and c' in L_1'. In view of the equivalence of P_2 and P_3 with $\mathbf{M}_1$ and $\mathbf{M}_2$ we could first write:

c: $\forall x((P_1 x \wedge \mathbf{M}_1 x) \vee (\sim P_1 x \wedge \mathbf{M}_2 x))$,
c': $\forall x((\sim P_1 x \wedge \mathbf{M}_1 x) \vee (P_1 x \wedge \mathbf{M}_2 x))$.

However, these are not in canonical form, since $\mathbf{M}_1$ and $\mathbf{M}_2$ are not primitive predicates. Canonical sentences for them in L_1' would be:

c: $\forall x M_1 x$,
c': $\forall x M_2 x$.

And these are no longer incongruent counterparts.

Now consider a case where c and c' are not incongruent counterparts in L_1:

c: $\exists x P_2 x$,
c': $\exists x P_3 x$,

Reformulated in L_1' these become:

c: $\exists x \mathbf{M}_1 x$,
c': $\exists x \mathbf{M}_2 x$.

And when these are put into canonical form in L_1' they become:

c: $\exists x ((P_1 x \wedge M_1 x) \vee (\sim P_1 x \wedge M_2 x))$,
c': $\exists x ((\sim P_1 x \wedge M_1 x) \vee (P_1 x \wedge M_2 x))$,

and are now incongruent counterparts.

If the equivalences being relied on here were achieved by means of explicit definitions within L_1 no problem would arise, since M_1 and M_2 would still be molecular predicates requiring to be decomposed into P_1, P_2 and P_3. Miller's point relies on our somehow *knowing* that $\mathbf{M}_1$ and $\mathbf{M}_2$ in L_1' are in fact equivalent to P_2 and P_3 in L_1 although there is no way of *telling* this within either L_1 or L_1'. I suppose that it might just *happen* that certain primitive predicates in one language were exactly equivalent to certain molecular predicates in the other and vice versa. If it did, and if we could carry out a desired content comparison only in one, why then, we should of course adopt that language. What about the other kind of case where an unsought-for content-comparison becomes possible in another language, as '$\exists x P_2 x$' and '$\exists x P_3 x$' became comparable in L_1'? Well, one should never suppress even unwanted information. The discovery that these two propositions can be considered of equal content when expressed in L_1' tells us *something* significant about them (roughly, that each of them locates something under a disjunction of the same number of Q-predicates). So my conclusion, here, is that if two languages, related in the above way, should ever be available to us we should in each case prefer the one that allows a content-comparison to be made, whether that comparison was sought-for or not.

I will now proceed to a language L_2 which will allow us to make content-comparisons we could not make in L_1. I will now drop condition (c), which required the primitive predicates to be atomic, and borrow from Carnap ([1950], pp. 76f) the idea of a *family* of predicates, for instance of colour predicates or of weather predicates.

A family of predicates effects a partition that is collectively exhaustive and mutually exclusive: everything within a certain domain must fall under one, and only one, of the predicates in that family. The partition may be as fine as we like provided the number of predicates in the family remains finite. If the predicates have a natural order (as do, for instance, the 'gentle breeze', 'high wind', 'strong gale' etc. of the Beaufort scale) such a partition determines a (noncontinuous) *scale*. We would like the intervals of such a scale to be, in some sense, of roughly equal 'width' so that the scale is 'fair'. Since we are, at present, still retaining requirement (b) that the predicates be observational, I propose as a way of securing this equality that each interval be associated with one j.n.d. ('just noticeable difference') or with some multiple thereof.

I now suppose that our present language L_2 is like the previous L_1 except that instead of containing k atomic predicates, it contains k predicate-families, $F_1, \ldots, F_k$, many of which have many more than two members. I will assume that the predicates belonging to a given family are ordered in some way and numbered accordingly. A predicate in this language will be denoted by the letter P with a subscript to indicate which predicate-family it belongs to and a superscript to indicate its position in that family. Thus P_i^j is the j-th predicate in family F_i (or interval j on scale F_i).

I now need to adjust our previous two conditions for incongruent counterparts. It will help if before we do this we consider what a canonical sentence in L_2 might look like. Let proposition G say: "All middle-aged Pygmies are 3–5 feet tall". Assume that in L_2 the predicate-families F_1, F_2 and F_3 cover respectively: ethnic groups, age, and height; that P_1^1 in F_1 denotes Pygmies; that F_2 contains just three predicates, P_2^1 (= 'young'), P_2^2 (= 'middle-aged') and P_2^3 (= 'old'); and that F_3 contains the predicates P_3^1, P_3^2, P_3^3, ... which denote respectively the intervals [0, 1], [1, 2], [2, 3] ... on a height-scale divided into feet. (Thus G says that all middle-aged Pygmies fall under P_3^4 or P_3^5.) Then we might express G by:

$$\forall x((P_1^1 x \wedge P_2^2 x) \rightarrow (P_3^4 x \vee P_3^5 x)).$$

But I shall suppose that the canonical sentence for G is in disjunctive normal form and is:

$$G: \forall x(\sim P_1^1 x \vee \sim P_2^2 x \vee P_3^4 x \vee P_3^5 x).$$

When are sentences in L_2 incongruent counterparts? For L_1 our conditions for c and c' to be incongruent counterparts were that they should be identical except that (1) the sign of at least one predicate in c is reversed in c' and (2) such sign-reversals are systematic. Obviously, mere sign-reversal will not do in L_2 where P_i^j will typically be more determinate than $\sim P_i^j$. Now we might have said that L_1 with its k atomic predicates has k predicate-families, $F_1, \ldots, F_k$, where each F_i has just two members P_i^1 and P_i^2 (where P_i^2 is $\sim P_i^1$); and we could then have formulated condition (1) as: at least one predicate in c is replaced in c' by the other predicate from the same family. And we can now generalise this for L_2 with its many-membered predicate-families by saying: (1′) at least one predicate in c is replaced in c' by some other predicate from the same family. Quarrels between incongruent counterparts must be family quarrels.

What should be the analogue of condition (2) for L_2? We introduced it for L_1 because of the need to exclude partial or selective replacements of disputed predicates which would, typically, lead to significant variations in strength. And this need persists into L_2. Consider these two statements:

$$\forall x((\sim P_1{}^1 x \lor P_2{}^1 x) \land (\sim P_1{}^2 x \lor P_2{}^2 x)),$$
$$\forall x((\sim P_1{}^1 x \lor P_2{}^1 x) \land (\sim P_1{}^1 x \lor P_2{}^1 x)).$$

At first glance these may look like incongruent counterparts; in fact, the first is stronger than the second, which it entails. For L_1 we required that for c and c' to be incongruent counterparts they should be the same function of the same ordered set or predicates except that in c' one or more of the predicates is negated. Let us try to extend this requirement to L_2. Let c be a function of the predicates $P_i^k \ldots P_i^m$ from F_i and $P_j^l \ldots P_j^n$ from F_j, or:

$$c = f\langle P_i{}^k \ldots P_i{}^m, P_j{}^l \ldots P_j{}^n \rangle.$$

Clearly, for c' to be an incongruent counterpart of c it must be the same function f of the same numbers of predicates from F_i, and F_j. What is not so clear is what displacements within families should be permitted. One feels that the displacements should be of a regular or uniform or systematic nature. Consider the following two candidates for being incongruent counterparts of our previous hypothesis G:

$$G': \ \forall x(\sim P_1 x \lor \sim P_2{}^2 x \lor P_3{}^5 x \lor P_3{}^6 x),$$
$$G^+: \ \forall x(\sim P_1 x \lor \sim P_2{}^2 x \lor P_3{}^3 x \lor P_3{}^8 x).$$

G' 'corrects' G in a systematic way, raising the height of middle-aged Pygmies from 3–5 to 4–6 feet; and it seems a proper incongruent counterpart of G. But G^+ 'corrects' G in two opposed ways, making their height either 2–3 or 7–8 feet. It hardly seems to be an incongruent counterpart. The statement that all Pygmies are either very short or very tall is hardly on a par with the statement that they are all rather short.

So for L_2 I am going to replace condition (2), provisionally, by the following condition (2'): if a predicate P_i^j in c is displaced in c', then every predicate from that family F_i which occurs in c must be displaced in the same direction and by the same amount in c'. Thus for c' to be counterpart of our c we must have

$$c' = f(P_i^{k+a} \dots P_i^{m+a}, P_j^{l+b} \dots P_j^{n+b}),$$

where a and b may be positive, negative, or zero.

I shall suggest later that this condition (2') is too rigid, and I will replace it by one that is more flexible and which will yield (2') only as a limiting case. But there are certain advantages in using (2') provisionally. For one thing it makes possible an easy extension to L_2 of the idea of a $T' - T$ correction-rule. Let c and c' be incongruent counterparts where c and T entail each other and c' is entailed by but does not entail T'. Assume that the disputed predicates in c and c' are from predicate-families F_i and F_j. Then in L_2 the $T' - T$ correction-rule will have the form: "Displace each F_i-predicate by a and each F_j-predicate by b". (I will also replace this by something more flexible later.)

I now relax condition (b), which required the predicates of our language to be observational. Let L_3 be like L_2 except that it contains, in addition, theoretical predicates $\Pi_1, \Pi_2, \dots$ And assume that T and T' are theories one or both of which contains theoretical predicates. Since we are trying to rehabilitate the basic Popperian idea that one theory is better than another, other things being equal, if it is more testable and hence more corroborable, we are here interested in comparing only their empirical contents. So for the purpose of this kind of comparison we may replace T and T' by T_E and T'_E, where T_E is the strongest consequence of T that contains only observational terms, and ditto for T'_E. (If T contains no theoretical terms then T_E is T.) We then proceed as before. Suppose that H says: "All middle-aged Pygmies having gene α are 2–3 feet tall and brown-eyed, while all those not having gene α are 3–4 feet tall and green-eyed". We might express this in L_3 by:

$$H: \forall x(\sim P_1^1 x \vee \sim P_2^2 x \vee (\Pi_1 x \wedge P_3^3 x \wedge P_4^1 x)$$
$$\vee (\sim \Pi_1 x \wedge P_3^4 x \wedge P_4^2 x)),$$

where P_1^1, P_2^2, P_3^3, P_4^4 mean the same as before, P_4^1 = 'brown-eyed', P_4^2 = 'green-eyed', and Π_1 = 'having gene α'. To show that H has more empirical content than the previous G we first replace it by H_E:

$$H_E: \forall x(\sim P_1^1 x \vee \sim P_2^2 x \vee (P_3^3 x \wedge P_4^1 x) \vee (P_3^4 x \wedge P_4^2 x)).$$

From H_E we proceed to a consequence that is an incongruent counterpart of G by dropping from the third and fourth disjuncts their last conjunct, leaving

$$\forall x(\sim P_1^1 x \vee \sim P_2^2 x \vee P_3^3 x \vee P_3^4 x).$$

Comparing this with G:

$$\forall x(\sim P_1^1 x \vee \sim P_2^2 x \vee P_3^4 x \vee P_3^5 x)$$

we can at once read off the $H - G$ correction-rule: "Displace each F_3-predicate in G by -1". This rule converts G into a theorem of H.

10. THE CORRESPONDENCE PRINCIPLE AND A GLANCE AT THE VERISIMILITUDE-PROBLEM

Following customary usage I say that scientific theory T' is in the relation of correspondence to T if these conditions hold: First, T' assigns values to every parameter Φ_i to which T assigns values. Second, they do not everywhere assign the same values. Third, where T' and T disagree over some Φ_i, the values assigned to it by T' approach those assigned to it by T *asymptotically.* Typically, T will treat some Φ_i as invariant while T' treats it as a variable whose value tends to the value assigned to it by T as some other parameter Φ_j tends to some limiting value, usually 0 or ∞. I have already mentioned one much cited example: for Newton $m = m_0$, while for Einstein $m = m_0/\sqrt{1 - v^2/c^2}$ so that $m \to m_0$ as $v \to 0$.

I agree with the many recent authors (e.g. Koertge [1969], Krajewski [1977], Post [1971], Szumilewicz [1977], Yoshida [1977]) who hold that it typically happens in the history of science that when some hitherto dominant theory T is superseded by T', T' is in the relation of correspondence to T. And although Popper has referred to the Correspondence Principle only in passing ([1969], p. 101; [1972], p. 202), it is clear that the relation of correspondence is an exemplary case of his conception of the relation which ought ideally to hold between a superseding and a superseded theory. So it is important that our account of content-dominance by T' of T should extend to cases where T' and T are in this relation.

It will not do so as it stands because condition (2') for incongruent counterparts is too rigid. It would require T' always to 'correct' the values assigned by T to Φ_i by some *fixed* quantity, displacing them uniformly sideways, whereas we want to allow T' to distort them in some continuous way.

Hitherto I have taken the idea of a correction-rule as derivative: we first find that consequence (if there is one) of T' which (in canonical form) is an incongruent counterpart of T (in canonical form) and then read off the $T' - T$ correction-rule from a comparison of these two parallel sentences. But when T' is in the relation of correspondence to T, a $T' - T$ correction-rule is laid down in advance: for any parameter Φ_i occurring in T, there is a function which sends every T-value of Φ_i into the corresponding T'-value. We already know how T should be 'corrected' for it to be converted into a theorem of T'.

Suppose that particular values of the parameters Φ_i and Φ_j, when translated into our coarse-grained observational language, take positions on, respectively, scale F_i and scale F_j whose intervals are designated by $P_i^1, P_i^2, P_i^3, \dots$ and $P_j^1, P_j^2, P_j^3, \dots$ And suppose that there are functions g and h which send T-values of Φ_i and Φ_j respectively into T'-values of Φ_i and Φ_j. I will now revise my previous conditions for incongruent counterparts. I retain (1') and also this much of (2'): if we have

$$c = f\langle P_i^k \dots P_i^m, P_j^l \dots P_j^n \rangle,$$

then for c' to be a counterpart of c, c' must be the same function f of the same number of predicates from F_i and from F_j. But instead of the uniform displacement of the F_i-predicates by a and of the F_j-predicates by b which I had provisionally required, I now say that these displacements are to be determined by the functions g and h, so that c' will have this character:

$$c' = f\langle P_i^{g(k)} \dots P_i^{g(m)}, P_j^{h(l)} \dots P_j^{h(n)} \rangle.$$

And I now say that if T' is in the relation of correspondence to T, then T' has strictly more content than T if there is a $T' - T$ correction-function which would convert T into a theorem of T' having less content than T' itself.

In conclusion I will offer one diffident comment on the vexed question of verisimilitude. Suppose we have a sequence of theories T, T', T'', ... where each later theory is in the relation of correspondence to its predecessors. (As Krajewski points out ([1977], p. 52), this relation is transitive.) Popper's intention, when he introduced the notion of verisimilitude, was not to justify any claims to *know* that the theories in such a sequence are getting progressively closer to the truth, but only to show that it is *meaningful* to say that they are doing so even if we regard the latest theory in the sequence as false. He introduced it as a semantic notion analogous to truth, not as an epistemological notion analogous to degree of confirmation. So let us now make the conjecture that there is an as yet undiscovered true 'target theory' T^* such that T^* is in the relation of correspondence to ... T'', T', T. This conjecture is, no doubt, quite unfounded and unwarranted; but it is, I think, perfectly meaningful and either true or false. Suppose it is true: what would that imply concerning the merits, truth-wise, of T, T', T'' ...? One thing it would imply is that they are all *false*: in T'' there is still at least one parameter which is given a constant value instead of being treated as a variable which tends to that value in the limit. But it would also imply that T', in unfreezing parameters that had been frozen in T, was *right* so far as it went, though it did not go far enough; and again that T'' was *right* in going further than T' in the unfreezing of parameters. In short, our conjecture would imply that we have here a sequence of false theories which are getting closer to the truth at least with respect to those parameters which ought to be treated as variables which approach a fixed value asymptotically. That may not rescue much but it does rescue something of the idea of verisimilitude.

The London School of Economics and Political Science

UNIFIED BIBLIOGRAPHY FOR PARTS I AND III

Adler, R., Bazin, M., and Schiffer, M.: 1965, *Introduction to General Relativity*, McGraw-Hill, New York.

Agassi, J.: 1966, 'Sensationalism', *Mind* **75**, 1–24.

Bacon, F.: 1620, *Novum Organon*, in John M. Robertson (ed.), *The Philosophical Works of Francis Bacon*, George Routledge and Sons Ltd., New York, 1905, pp. 212–387.

Boltzmann, L.: 1897, 'On the Question of the Objective Existence of Processes in Inanimate Nature', in B. McGuiness (ed.), *Ludwig Boltzmann: Theoretical Physics and Philosophical Problems*, Vienna Circle Collection, Vol. 5, D. Reidel Publ. Co., Dordrecht, Holland, 1974, pp. 57–76.

Boltzmann, L.: 1905, *Populäre Schriften*, J. A. Barth, Leipzig.

Bridgman, P. W.: 1927, *The Logic of Modern Physics*, The Macmillan Co., New York.

Bucherer, A. H.: 1909, 'Die Experimentelle Bestätigung des Relativitätsprinzips', *Annalen der Physik* **28**, 513–536.

Carnap, R., 1928, *Der Logische Aufbau der Welt*, English translation by R. A. George, *The Logical Structure of the World*, Routledge and Kegan Paul, London, 1967.

Carnap, R.: 1950, *Logical Foundations of Probability*, Routledge and Kegan Paul, London.

Clark, P. J.: 1976, 'Atomism Versus Thermodynamics: A Case Study in the Methodology of Scientific Research Programmes', in Howson (ed.), 1976.

Clifford, W. K.: 1886, *Lectures and Essays*, Macmillan and Co., London.

Descartes, R.: 1628, 'Rules for the Direction of the Mind', in *The Philosophical Works of Descartes*, translated by E. S. Haldane and G. R. T. Ross, Volume 1, Cambridge University Press, Cambridge, pp. 1–77.

Duhem, P.: 1906, *La Theorie Physique: Son Objet, Sa Structure*, English translation by P. P. Wiener, *The Aim and Structure of Physical Theory*, Princeton University Press, Princeton, 1954.

Feyerabend, P. K.: 1963, 'How to Be a Good Epistemologist – A Plea for Tolerance in Matters Epistemological' in B. Baumrin (ed.), *Philosophy of Science: the Delaware Seminar*, Volume 2, Interscience Publishers, New York, pp. 3–39.

Feyerabend, P. K.: 1964, 'Realism and Instrumentalism: Comments on the Logic of Factual Support', in M. Bunge (ed.), *The Critical Approach to Science and Philosophy*, The Free Press, Glencoe, pp. 280–308.

Feyerabend, P. K.: 1972, 'Von der Beschränkten Gültigkeit Methodologischer Rageln', in R. Bubner (ed.), *Neue Hefte für Philosophie, Volume 3/3: Dialog als Methode*, pp. 124–171.

Feyerabend, P. K.: 1976a, 'On the Critique of Scientific Reason', in C. Howson (ed.), 1976.

G. Radnitzky and G. Andersson (eds.), Progress and Rationality in Science, 379–383.
All Rights Reserved.
Copyright © 1978 by D. Reidel Publishing Company. Dordrecht, Holland.

Feyerabend, P. K.: 1976b, 'Logic, Literacy and Professor Gellner', *BJPS* **27**, 381–391.

Gellner, E. A.: 1975, 'Beyond Truth and Falsehood', *BJPS* **26**, 331–342.

Gödel, K.: 1964, 'What is Cantor's Continuum Problem?', in P. Benacerraf and H. Putnam (eds.), *Philosophy of Mathematics: Selected Readings*, Basil Blackwell, Oxford, pp. 258–273.

Goodman, N.: 1947, 'The Problem of Counterfactual Conditionals', *Journal of Philosophy* **44**, 113–128. Reprinted, with minor changes, in N. Goodman, *Fact, Fiction and Forecast*, The Athlone Press, London, 1954.

Goodman, N.: 1961, 'Safety, Strength, Simplicity', *Philosophy of Science* **28**, 150–151.

Grünbaum, A.: 1959, 'The Falsifiability of the Lorentz-Fitzgerald Contraction Hypothesis', *BJPS* **10**, 48–50.

Grünbaum, A.: 1976, 'Can a Theory Answer More Questions than one of its Rivals?', *BJPS* **27**, 1–23.

Hempel, C. G.: 1945, 'Studies in the Logic of Confirmation', *Mind* **54**, 1–26 and 97–121. Reprinted in Hempel [1965], pp. 3–51.

Hempel, C. G.: 1965, *Aspects of Scientific Explanation*, The Free Press, New York.

Herschel, J. F. W.: 1831, *A Preliminary Discourse on the Study of Natural Philosophy*, Longmans, London.

Hintikka, J.: 1973, *Logic, Language, Games and Information*, Oxford.

Holton, G. and Roller, D. H. D.: 1958, *Foundations of Modern Physical Science*, Addison-Wesley.

Howson, C.: 1973, 'Must the Logical Probability of Laws be Zero?', *The British Journal for the Philosophy of Science* **24**, 153–163.

Howson, C. (ed.): 1976, *Method and Appraisal in the Physical Sciences*, Cambridge University Press.

Kaufmann, W.: 1905, 'Über die Konstitution des Elektrons', *Sitzungsberichte der Königlich Prüssichen Akademie der Wissenschaften* **2**, 949–956.

Kaufmann, W.: 1906, 'Über die Konstitution des Elektrons', *Annalen der Physik* **19**, 487–553.

Koertge, N.: 1971, 'Inter-Theoretic Criticism and the Growth of Science', in R. C. Buck and R. S. Cohen (eds.), *Boston Studies in the Philosophy of Science*, Vol. VIII, 160–173.

Kolmogorov, A. N.: 1933, *Grundbegriffe der Wahrscheinlichkeitsrechnung*. English translation by N. Morrison, *The Foundations of the Theory of Probability*, Chelsea Publishing Company, New York, 1956.

Krajewski, W.: 1977, *Correspondence Principle and Growth of Science*, D. Reidel Publ. Co., Dordrecht, Holland.

Kuhn, T. S.: 1962, *The Structure of Scientific Revolutions*, The University of Chicago Press, Chicago.

*Lakatos, I.: 1962, 'Infinite Regress and the Foundations of Mathematics', *Aristotelian Society*, Supplementary Volume XXXVI, pp. 155–184.

Lakatos, I.: 1963–64, 'Proofs and Refutations', *The British Journal for the Philosophy of Science* **14**, 1–25, 120–139, 221–245, 296–342, reprinted as Chapter 1 of *Proofs and Refutations*, Cambridge University Press, 1976.

*Lakatos, I.: 1967, 'A Renaissance of Empiricism in the Recent Philosophy of Mathematics?', *British Journal for the Philosophy of Science* **27**, 201–223.

*Lakatos, I.: 1968a, 'Changes in the Problem of Inductive Logic', in I. Lakatos (ed.), *The Problem of Inductive Logic*, North-Holland, Amsterdam, pp. 315–417.

*Lakatos, I.: 1970, 'Falsification and the Methodology of Scientific Research Programmes', in I. Lakatos and A. Musgrave (eds.), *Criticism and the Growth of Knowledge*, Cambridge University Press, Cambridge, pp. 91–196.

*Lakatos, I.: 1971a, 'History of Science and its Rational Reconstructions', in R. C. Buck and R. S. Cohen (eds.), *Boston Studies in the Philosophy of Science*, Vol. VIII, pp. 91–136.

Lakatos, I.: 1971b, 'Replies to Critics', in R. C. Buck and R. S. Cohen (eds.), *Boston Studies in the Philosophy of Science*, Vol. VIII, pp. 174–181.

Lakatos, I.: 1971c, 'Popper zum Abgrenzungs- und Induktionsproblem', in H. Lenk (ed.), *Neue Aspekte der Wissenschaftstheorie*, Vieweg, Braunschweig, pp. 75–110.

*Lakatos, I.: 1974, 'Popper on Demarcation and Induction', in P. A. Schilpp (ed.), *The Philosophy of Karl Popper*, Open Court, La Salle, Illinois, pp. 241–274.

*Lakatos, I and Zahar, E.: 1975, 'Why Did Copernicus's Programme Supersede Ptolemy's?', in R. Westerman (ed.), *The Copernican Achievement*, The University of California Press, Los Angeles.

Lakatos, I.: 1978, *Philosophical Papers*, 2 volumes, Worrall and Currie (eds.), Cambridge University Press.

Laplace, P. S.: 1820, *A Philosophical Essay on Probabilities*. English translation from the sixth French edition by F. W. Truscott and F. L..Emory, Dover Publications, New York, 1951.

Lloyd, H.: 1833, 'Report on the Progress and Present State of Physical Optics', *British Association for the Advancement of Science Reports* 4, 295–413.

Lorentz, H. A.: 1906, *The Theory of Electrons and its Applications to the Phenomena of Light and Radiant Heat: A Course of Lectures Delivered in Columbia University, New York, in March and April 1906*, Dover Publications, New York, 1952.

Lorentz, H. A.: 1914, 'La Gravitation', *Scientia* 16, 28–59.

Mach, E.: 1912, *The Science of Mechanics*, seventh edition. English translation by T. J. McCormack, Open Court, La Salle, Illinois, 1942.

Medawar, P.: 1969, *Induction and Intuition in Scientific Thought*, American Philosophical Society, Philadelphia.

Merton, R.: 1957, 'Priorities in Scientific Discovery', *American Sociological Review* 22, 635–59.

Merton, R.: 1961, 'Singletons and Multiples in Scientific Discovery', *Proceedings of the American Philosophical Society* 105, 470–86.

Merton, R.: 1963, 'Resistance to the Systematic Study of Multiple Discoveries in Science', *European Journal of Sociology* 4, 237–49.

Miller, D.: 1974, 'Popper's Qualitative Theory of Verisimilitude', *The British Journal for the Philosophy of Science* 25, 166–177.

Miller, D.: 1975, 'The Accuracy of Predictions', *Synthese* 30, 159.

Musgrave, A.: 1974, 'Logical versus Historical Theories of Confirmation', *The British Journal for the Philosophy of Science* 25, 1–23.

Neurath, O.: 1931, 'Empirical Sociology', in M. Neurath and R. S. Cohen (eds.), *Empiricism and Sociology*, D. Reidel Publ. Co., Dordrecht, Holland, 1973, pp. 319–421.

Ostwald, W.: 1895, 'Emancipation from Scientific Materialism', *Science Progress* **4**, 430–436.

Ostwald, W.: 1927, *Lebenslinien*, Volume 2, Klasing and Co., Berlin.

Planck, M.: 1897, *Vorlesungen über Thermodynamik*, Veit, Leipzig. English translation by A. Ogg, *Thermodynamics*, Dover Publications, New York, 1945.

Planck, M.: 1906a, 'Das Prinzip der Relativität und die Grundgleichungen der Mechanik', *Verhandlungen der Deutschen Physikalischen Gesellschaft* **8**, 136–141.

Planck, M.: 1906b, 'Die Kaufmannschen Messungen der Ablenkbarkeit der β-Strahlen in ihren Bedeutung für die Dynamik der Electronen', *Verhandlungen der Deutschen Physikalischen Gesellschaft* **8**, 418–432.

Planck, M.: 1907, 'Nachtrag zu der Besprechung der Kaufmannschen Ablenkungsmessungen', *Verhandlungen der Deutschen Physikalischen Gesellschaft* **9**, 301–305.

Poincaré, M.: 1902, *Science and Hypothesis*.

Polanyi, M.: 1972, 'Genius in Science', *Encounter*, pp. 43–50.

Popper, K. R.: 1934, *Logik der Forschung*, Expanded English edition: Popper, K. R. [1959].

Popper, K. R.: 1945, *The Open Society and its Enemies*, Volumes 1 and 2, Fifth edition, Routledge and Kegan Paul, London, 1966.

Popper, K. R.: 1957, 'The Aim of Science', *Ratio* **1**, 24–35. Reprinted in Popper [1972], pp. 191–205.

Popper, K. R.: 1959, *The Logic of Scientific Discovery*, Hutchinson, London.

Popper, K. R.: 1962, 'Addendum: Facts, Standards, and Truth', to the fourth, and subsequent editions of Popper [1945], pp. 369–396.

Popper, K. R.: 1963, *Conjectures and Refutations*, Routledge and Kegan Paul, London.

Popper, K. R.: 1972, *Objective Knowledge*, The Clarendon Press, Oxford.

Post, H.: 1968, 'Atomism', *Physics Education* **3**, 1–13.

Post, H.: 1971, 'Correspondence, Invariance and Heuristics', *Studies in the History and Philosophy of Science* **2**, 213–255.

Reichenbach, H.: 1951, *The Rise of Scientific Philosophy*.

Robinson; R.: 1964, *An Atheist's Values*, The Clarendon Press, Oxford.

Schaffner, K.: 1972, *Nineteenth Century Aether Theories*, Pergamon Press, Oxford.

Scheffler, I.: 1963, *The Anatomy of Inquiry*, Alfred A. Knopf, New York.

Scheffler, I.: 1974, *Four Pragmatists*, Routledge and Kegan Paul, London.

Szumilewicz, I.: 1977, 'Incommensurability and the Rationality of the Development of Science', *BJPS* **28**.

Tichý, P.: 1974, 'On Popper's Definitions of Verisimilitude', *The British Journal for the Philosophy of Science* **25**, 155–160.

Toulmin, S.: 1972, *Human Understanding*, Vol. 1.

Urbach, P.: 1974, 'Progress and Degeneration in the "I.Q. Debate"', *The British Journal for the Philosophy of Science* **25**, 99–135 and 235–259.

Watkins, J. W. N.: 1958, 'Confirmable and Influential Metaphysics', *Mind* **67**, 344–365.

Watkins, J. W. N.: 1964, 'Confirmation, the Paradoxes and Positivism', in M. Bunge (ed.), *The Critical Approach to Science and Philosophy*, The Free Press, Glencoe, pp. 92–115.

Watkins, J. W. N.: 1968a, 'Non-Inductive Corroboration', in I. Lakatos (ed.), *The Problem of Inductive Logic*, North-Holland, Amsterdam, pp. 61–66.

Watkins, J. W. N.: 1968b, 'Hume, Carnap and Popper', in I. Lakatos (ed.), *The Problem of Inductive Logic*, North-Holland, Amsterdam, pp. 271–282.

Watkins, J. W. N.: 1970a, 'Against "Normal Science" ', in I. Lakatos and A. Musgrave (eds.), *Criticism and the Growth of Knowledge*, Cambridge University Press, Cambridge, pp. 25–37.

Watkins, J. W. N.: 1970b, 'Imperfect Rationality', in R. Borger and F. Cioffi (eds.), *Explanation in the Behavioural Sciences*, Cambridge University Press, Cambridge.

Watkins, J. W. N.: 1971, 'CCR: A Refutation', *Philosophy* **46**, 56–61.

Watkins, J. W. N.: 1974, 'The Unity of Popper's Thought', in P. A. Schilpp (ed.), *The Philosophy of Karl Popper*, Open Court, La Salle, Illinois, pp. 371–412.

Watkins, J. W. N.: 1975a, 'Three Views Concerning Human Freedom', in R. S. Peters (ed.), *Nature and Conduct*, Macmillan.

Watkins, J. W. N.: 1975, 'Metaphysics and the Advancement of Science', *The British Journal for the Philosophy of Science* **26**, 91–121.

Watkins, J. W. N.: 1977, Chapter in Joan Abse (ed.), *My LSE*, Robson Books, London.

Watkins, J. W. N.: 1978, 'Minimal Presuppositions and Maximal Metaphysics', *Mind* **87**, 195–209.

Whewell, W.: 1837, *History of the Inductive Sciences*.

Whewell, W.: 1858, *Novum Organon Renovatum: Being the Second Part of the Philosophy of the Inductive Sciences*, third edition, Parker, London, 1858.

Whittaker, E. T.: 1910, *A History of the Theories of the Aether and Electricity, from the Age of Descartes to the Close of the Nineteenth Century*, Longmans, Green, London.

Wood, R. W.: 1905, *Physical Optics*, Macmillan, London.

Worrall, J.: 1975, *The 19th Century Revolution in Optics: A Case Study in the Interaction between Philosophy of Science and History and Sociology of Science*, University of London, Ph.D. Thesis, Unpublished.

Worrall, J.: 1976, 'Thomas Young and the "Refutation" of Newtonian Optics', in Howson (ed.), 1976.

Zahar, E. G.: 1973, 'Why Did Einstein's Programme Supersede Lorentz's?', *The British Journal for the Philosophy of Science* **24**, 95–123 and 223–262.

* These papers are also included in Lakatos (1978).

PART IV

TWO BRIEF REJOINDERS

PAUL FEYERABEND

THE GONG SHOW–POPPERIAN STYLE

John Watkins writes: "The chief unreasonable doctrine of his (i.e. my) present paper is that we should discard logic" (340). What I actually say is that "there is no one 'logic' ... there is a whole spectrum from formally rigorous systems with non-contradiction and excluded middle via more informal systems, intuitionistic systems without excluded middle, systems in which contradictions do not entail every statement to Hegel's logic and beyond to logics that are not expressed in explicitly formulated laws" (154f); that critical rationalists arbitrarily choose "one particular and rather simpleminded .... system" from among this manifold (155), and demand that science must conform to it; and that science actually conforms to a complex 'practical logic' where "contradictions do not lead to the unwanted assertion of every statement but to specific and highly useful results" (158).

Quoting a series of if-then statements from (156f) John Watkins complains that I am "not actually asserting anything. There is not one categorical statement in the passage I have quoted" (341). But two pages later I provide the antecedents and with them the categorical statements he is looking for.

John Watkins makes a great fuss about a distinction found in Popper's *What is Dialectic?* viz. the distinction between (a) contradictions cause progress because we have to work to eliminate them and (b) contradictions can be retained because they cause progress (341). He accuses me of either not being aware of the distinction (in which case, he adds, one "ought to be raising questions about (my) salary" – 342) or of trying to 'confuse' the reader "by deliberately obfuscating the clear and obvious distinction" (342). But when presenting my examples (157ff) I do not argue that theories are fertile, lead to discoveries, widen our horizon *because* they contain contradictions, I argue that some theories *that happen to contain contradictions* are fertile, lead to discoveries, widen our horizon and that the attempt to make them contradiction-free interferes with their discovery-producing potential. Again, I do not say that such an

387

G. Radnitzky and G. Andersson (eds.), *Progress and Rationality in Science*, 387–392.
All Rights Reserved.
Copyright © 1978 by D. Reidel Publishing Company, Dordrecht, Holland.

interference is a *necessary consequence* of the increase of rigour (or the elimination of contradictions), or that it is *brought about* by it, I say that there are many cases where removal of contradictions has sterility as an undesired *side effect*. Logicians are not bothered by this side effect for all they want is a theory that fits into their dream world (cf. the italicised clause on p. 177, note 26). Scientists, on the other hand, are interested in fruitfulness and so they will occasionally *try* to live with contradictions as well as they can (cf. the Bohr-quotation in the same footnote) and they *can* live with them as science has "ways of handling contradictions that do not lead to the ... assertion of every statement but to specific and highly useful results" (158). Popper's distinction had no place in *this* argument. It is true that I go a little further and assert that "inconsistent theories are better to handle, lead to more discoveries than their decontaminated rivals" (165) but I do so not on general grounds (contradictions being *necessary* for fruitfulness, for example) but because the examples I have considered and mention (154f) have this feature and because I conjecture that the feature will occur in other cases as well (I don't argue abstractly but from an analysis of concrete historical examples; and by 'fruitfulness' I don't mean what critical rationalists understand by the term, but what scientists mean by it). It is also true that on page 156 I write that knowledge progresses *because* of contradictions (it seems that it was this lonely statement that started John Watkins on his rampage) but note that the statement is an aside, it is neither a main thesis nor the summing up of an argument dealing with a main thesis. Secondly the statement is part of one of those hypothetical statements Watkins censures for 'not actually asserting anything' (341). Now, if it doesn't say anything, how can Watkins criticize it for saying the wrong thing? Thirdly, there is no need to import Popper even when asserting that contradictions cause progress. Popper's distinction is a distinction with a difference only if contradictions are fatal to knowledge and can be removed – which is the point at issue: for Hegel (and Lenin – cf. his notes on Hegel's *Logic*) contradictions create progress, the attempt to eliminate them is futile and it would be fatal if it succeeded.[1] But, as I said, this was an aside and not the point I wished to make.

I write (cf. the quotation in the last paragraph) that "inconsistent theories are better to handle ... than their decontaminated rivals". Watkins' comment: "What he means by 'better to handle' I do not know" and he adds "nor, I suspect, does he" (340). He could have obtained the missing information by following the hint at the end of the passage that puzzles him ('see above, Section iv') for in Section iv, note 26 I give a variety of references and examples to illustrate the phrase 'difficult to handle'. It is quite possible that these examples which are somewhat technical are too difficult to handle for Watkins but any physics undergraduate will be able to explain the matter in terms even a Popperian can understand: the limits of Popperian understanding are not the limits of the world.

John Watkins complains that my "intellectual heroes change with bewildering rapidity" (344) and he mentions Cohn-Bendit and Bellarmine as examples. But the fact that I quote both of them with approval does not mean that I have chosen them as my heroes (I also quoted John Watkins with approval in some earlier papers of mine) and the fact that John Watkins has never left the Popperian Church does not mean that he is more critical than I. He also mockingly remarks that I have "developed a solicitude for theology" (344) as if excursions into fields outside Popperian competence were a sign of mental aberration. Bellarmine realised that scientific progress might make it impossible for man to take care of his soul and he feared a loss of humanity. In this he was wiser, more critical and more humane than Popperians can ever aspire to be.

Considering one of my arguments John Watkins remarks that it is too 'pathetic' to deserve refutation (340). But I have discussed the argument with Profs. Ayer, Craig, Owen, Mates and others and I did not find their objections insurmountable. What obvious, earth-shattering, knock-down reason does Watkins have to remain silent? Or is he afraid that his reason may be found wanting and so prefers authoritarian gestures?

John Watkins accuses me of 'effortless oneupmanship'. By this he means that while he, John Watkins, tries to solve one problem, I take him to task for not solving another. "Our task in the position paper" he intones (339) "was to discuss criteria of scientific progress; it was not to discuss whether scientific progress is good or bad for mankind". What evidence does he produce to show that in my *present paper* I don't address myself to this problem? The fact that I don't deal with it in *another paper* which was written earlier and had an entirely different purpose (339). I agree with Watkins that in looking for criteria one should begin "with a most utopian and non-viable ideal" and work one's way down from there (344). This is why I chose Aristotle: Aristotle gives us certainty, progress, but no content increase. I claim that critical rationalists have not a shred of an argument against his 'criteria of progress'. Watkins does not say a word about this charge and about the arguments I use to support it, though they fill about three-quarters of my paper. It seems that he either has not read it (cf. the next paragraph) or not understood it, or has no argument against it and prefers harping on irrelevancies to a direct attack.

The impression one gets from Watkins' reply to my paper is that instead of reading it he threw it up in the air, stared balefully at the pages as they slowly returned to earth, managed to catch a sentence here, part of a sentence there, got upset by what he saw and gave vent to his anger without regard to context and without having ever read the paper in its entirety. He seems to be dimly aware of the need for an apology. The one he provides is the strangest part of an already rather strange performance. To prepare my reply, let me fill in some details. Some time ago Lakatos and I decided to write a book on rationalism; I was to attack the rationalist position while Imre Lakatos would defend it. When my part was finished Lakatos, having been held up by other business, was not ready to reply. He told me to go ahead and publish, he asked Prof. Koertge to review the book for BJPS and Prof. Koertge accepted. When Watkins succeeded Lakatos as editor he replaced Koertge by Gellner. This was a rather puzzling move. BJPS is a journal on the philosophy of science, my book is 4/5 pure history and philosophy of science, Koertge is an expert historian and philosopher of science; Gellner, on the other hand has no competence in these fields (as he freely admits in his review of my book) and one of his earlier books was rejected for review by *Mind* because of incompetence in other fields as well. Why did Watkins choose him instead of Koertge? The only reason I can imagine is that he wanted to do me in for if there is one thing in which Gellner excels it is the arts of verbal (*not* intellectual) assassination. The attempt misfired, it turned out that Gellner could not even read plain English and Watkins who prides himself on fairness, rationality, objectivity, good judgement and other virtues of this kind was seriously embarrassed. I conjecture that he now introduces Gellner in the hope to remove at one stroke both this old embarrassment and the uneasiness about his present reply. For he tries to show that Gellner not only *misread* my book but *could not possibly have read it properly* (342) the implication being that any error I might now find in Watkins is my fault and not his. How does he proceed? Quite simply: in *Against Method* I talk about political anarchism. On some occasions, such as in note 12 of the introduction I say explicitly that I reject it; on other occasions I describe some of its features without repeating my explicit rejection. John Watkins quotes one such passage (343) though incompletely: he omits a remark that contains the qualification he says is missing and so gives a somewhat biased account of my text. I am prepared to overlook this slip on his part.[2] I even admit that the passage does not reject political anarchism as decisively as does note 12 of the introduction. For Watkins who seems to know a lot of devious people the difference of emphasis is an indication of sinister machinations on my part: I do my best to insinuate political anarchism "but in a way that would allow (me) to wriggle out of responsibility for it when challenged" (343). But when you have said a thing clearly once you don't have to say it clearly a second time – unless you are dealing with morons.

John Watkins is quite upset about my suggestion of a financial criticism of intellectual matters. But as a taxpayer I have the right to inquire what happens with my money, to insist that it be not wasted and to suggest terminating certain ways of using it. Using this right I say that there are more important things than adding epicycles to content increase and verisimilitude, that millions of people have not enough food to live a decent life, that others, desirous to learn about the traditions of their forefathers have no teachers, that the welfare of the soul is more important than mental acrobatics and that a proper balance must be found. In some civilised countries there exist institutions that give political force to such considerations and so enable ordinary citizens to control the intellectuals' attempt to shape society in their image. I regard this as a great step forward but I also understand why Watkins prefers to be judged by people less critical of the enterprise in which he participates.

I am sorry that my reply to John Watkins had to be so completely negative. I have known the gentleman for now almost 30 years, I have participated with him in many amusing events, we fought some interesting battles together, I have often enjoyed talking to him, both on technical matters and on matters of 'life' and I have always profited from his observations. I do not understand why his present comments on my paper should be so far below his usual level of competence but I guess that debates at the boundary of religions have this character especially when one of the religions has lost its original vigour and simplicity and has become a cumbersome, well established and yet slowly crumbling dogma.

NOTES

All page numbers in parentheses refer to this volume.

[1] Popper's 'proof' that a contradiction entails everything is not an objection to Hegel for it uses rules of derivation no Hegelian would accept. It is as naive as Carnap's 'disproof' of 'Das Nichts Nichtet' which assumes that Heidegger, when writing the statement, was trying to express a property of the negation sign in some simple forms of propositional logic.

[2] There is another interesting slip. Watkins says that "there is in fact *no* clearcut rejection of political anarchism in AM" (342). This suggests that he has overlooked passages such as note 12 of the introduction. But when quoting from my reply to Gellner he omits parts that not only refer to that note but explicitly mention the objection to violence that occurs there and replaces them by dots (342 – dots at the beginning of the quotation) which means that oversight was generously aided by the wish to deceive.

KURT HÜBNER

REPLY TO WATKINS':
'HÜBNER ON FALSIFICATION, "GRUE" AND TRUTH'

Watkins starts with a question: "Does Popper's purely conventionalist view of basic statements render falsifications 'practically'
meaningless?" And he answers: No, this is not the case. *If* e.g. a
hypothesis predicts a special position of a pointer, which, according
to the opinion of a group of experimenters, did not occur, *then* a
potential falsifier of the hypothesis has been accepted and
consequently its falsification is meaningful (345). But the question is:
Why have these experimenters accepted the falsification? If they did it
purely arbitrarily then it cannot be taken seriously; if on the other
hand there are *reasons* for it, then, I think, there will also be some
inductions among them.

Watkins is convinced that it "surely introduces much less arbitrariness" (345) to stick conventionally to basic statements than to
axioms and principles. But why is he so convinced? Does he not
admit himself on the other hand that we cannot have basic statements without axioms and principles? (345). Exactly for that reason
not only basic statements "are intersubjectively arrived at" (345);
because the measure of their intersubjectivity depends on the
measure of intersubjectivity of those axioms and principles, from
which they follow. Only this way can we explain that in the history of
the sciences depending on the particular context scientists sometimes
used axioms and principles as arguments against basic statements and
sometimes basic statements against axioms and principles, without
necessarily being accused in the one or in the other case of having
acted arbitrarily. It is the basic error of falsificationism not to recognize
that the *same* kind of a priori presuppositions used for the setting up
of a theory and its axioms will be used also for its test.

Nevertheless Watkins agrees with me that the scientists should
have *reasons* for accepting those general statements which play a role
in a falsification. The induction, however, will not be reintroduced
through the back door because one of the reasons in question may be
that these statements have been well tested (346). But here I ask:
Does a test – it may be as good as possible – absolutely exclude

393

G. Radnitzky and G. Andersson (eds.), Progress and Rationality in Science, 393–396.
All Rights Reserved.
Copyright © 1978 by D. Reidel Publishing Company, Dordrecht, Holland.

induction? It was exactly this that I have denied by referring to special inductive expectations involved in the use of measuring instruments!

Watkins gives an example. Let us suppose that a theory T has been falsified by experiments. Let us suppose further that these experiments presuppose other different theories T_1 and T_2 which both are well tested. Then, he thinks, it looks more 'rational' (346) to reject T and to accept T_1 and T_2 than the other way round. This would be especially the case, if a theory T' existed competing with T which could be corroborated by the same experiments. And this 'scientific rationality' (347) too does not include any induction.

Is that certain? How about somebody who would not rest content with Watkins' assertion and would ask him *why exactly* it is more rational in the mentioned case to keep T_1 and T_2 and to reject T? And suppose he offers Watkins the following answer: It is more rational because the good experiences we had with T_1 and T_2 (but not with T) in the *past* justify our confidence in them also for the future. Now, to be sure this would be an inductive justification of Watkins scientific rationality – which he thinks is not inductive – but does he know a better one?

But this is not all. It is not even certain that in the case mentioned by Watkins it will always be rational to reject T. Because it is contrary to Watkins' opinion, absolutely not cogent to stick in any case to basic statements rather than to axioms scientists have often ignored his so plausible sounding rule and have sometimes made the biggest progress by doing exactly that. I refer to those parts of my contribution in which I have mentioned cases of that kind in the history of the sciences and where I also tried with system-theoretical means to show why it was there more rational to progress quite differently than recommended by Watkins. (I have treated extensively questions of that kind among others in my book *Kritik der wissenschaftlichen Vernunft*, Freiburg 1978).

Once more to Goodman's paradox. Watkins does not think the falsificationists are concerned with this paradox because contrary to the inductivists it is insignificant to them whether by the same set of facts any sum of hypotheses you please can be as well corroborated as that one you are just interested in (347). But Goodman's paradox does not consist in the fact that *any* hypothesis can be as well corroborated as that one you are just interested in but that any sum of *absurd* hypotheses you please can be as well corroborated as *reasonable* ones. And for that reason also the falsificationists are concerned with this problem because their postulate that a hypothesis has to be falsifiable guarantees as little the rationality of a hypothesis as the postulate that it has to be inductively verifiable.

What does Watkins hope to prove by his example of a 'catastrophe'-theory? (348) Goodman's point is that the absurd grue-hypotheses can be very well tested, but exactly this will never happen to Watkins' 'catastrophic'-hypothesis!

Now, Watkins says that I have overdone the importance of Goodman's paradox for the falsificationists (347) although I did nothing other than to reject his *expressis verbis* reference to it. But however that may be: Even if that would be true that the advantage of the falsificationists regarding the inductivists consists in the fact that they have to compare only those hypotheses which a scientific community laid on the table for serious consideration (347) (which means that these hypotheses are not absurd) even then the falsificationists have won nothing. Because now the question arises of how to choose among different hypotheses and it is exactly this question which the falsificationist not less than the inductivist cannot answer. The reason is this: If the empirical basis is exhausted then we have to rely on something that he calls a scientifically undiscussible metaphysical belief and unclear 'ideas' (Popper, *Logik der Forschung*, Tübingen 1966, S. 6) and that means we have to rely on *a priori* reasons in favour of one or the other hypothesis. To be sure Popper does not deny that something of that kind is unavoidable and important for progress; but people who think falsifiability to be the essence of wisdom underestimate easily that *rationality* which alone can help to make decisions in cases mentioned by Watkins; they underestimate too the rationality in those important realms of scientific research which serve less the testing of theories than the setting up of theory-contents. (To show also this rationality was one of the intentions of my system-theoretical interpretation of the history of the sciences.)

Watkins finally quotes Kant. In the quotation of the 'Kritik der reinen Vernunft' mentioned by him Kant says indeed that there can be no *general* criterion of truth, even if there can be no doubt that truth is recognizable on the occasion of a special given 'content' (B85). But Kant is speaking there neither of an *absolute* truth nor of the truth as an *idea* as Watkins insinuates (ideas are for Kant only world, soul, God). Now, Watkins by not noticing that misses here the decisive difference: Because in the case of the absolute truth there is according to the Popperians not only no general criterion but it also can never be recognized at all, not even in a special case and on the occasion of a 'given content'. For exactly that reason Watkins thinks absolute truth to be only a regulative idea which is never realizable. Now, as to myself, I have nowhere said that the concept of "truth would be empty without a criterion of truth" as Watkins insinuates (cf. especially Chapter XI of my book *Kritik der wissenschaftlichen Vernunft*, Freiburg 1978); my point was that it is absurd to speak of a measure of approximation or of distance to it if it would be, like absolute truth, only an idea and not recognizable at all.

BIOGRAPHICAL NOTES

HANS ALBERT is Professor of Sociology and Philosophy of Science, University of Mannheim. Among his published books are *Traktat über kritische Vernunft*, J. C. B. Mohr, Tübingen, 1968, 3. rev. ed. 1975; *Plädoyer für kritischen Rationalismus*, Piper, München, 1971, 4. ed. 1975; *Konstruktion und Kritik*, Hoffmann & Campe, Hamburg, 1972; *Aufklärung und Steuerung*, Hoffmann & Campe, Hamburg, 1976.

GUNNAR ANDERSSON is Associate Professor of Philosophy of Science at the University of Trier; author of *Vetenskapens nytta och frihet* (Göteborg, Sweden: Göteborg Uni., 1975); he has contributed to journals and collections in the Philosophy of Science.

PAUL FEYERABEND is Professor of Philosophy at the University of California, Berkeley. He has received the Austrian President's award for Science and Fine Arts and D.Litt. h.c. from Loyola University, Chicago. Among his publications are 'How to be a Good Empiricist – A Plea for Tolerance in Matters Epistemological', in B. Baumrin (ed.), *Philosophy of Science: The Delaware Seminar*, Vol. 2, Wiley, New York, 1963. 'Realism and Instrumentalism: Comments on the Logic of Factual Support', in M. Bunge (ed.), *The Critical Approach to Science and Philosophy*, The Free Press, Glencoe, 1964. *Against Method*, New Left Books, London, 1975.

ADOLF GRÜNBAUM has been William Wilson Selfridge Professor of Philosophy at Lehigh University (1956–1960); Visiting Research Professor, University of Minnesota (Center for Philosophy of Science), 1956 and 1959; since 1960 Andrew Mellon Professor of Philosophy and Director of the Center for Philosophy of Science, University of Pittsburgh. In addition to other offices he has been President of the Philosophy of Science Association 1965–67, 1968–70; Vice-president of the American Association for the Advancement of Science (1963, section L). He has received among others the following awards: J. Walker Tomb Prize from Princeton University (1958) and Alumni Honor Citation by Wesleyan University, Connecticut, 1959. He is a member of the Editorial Board of *Philosophy of Science*, *The Encyclopedia of Philosophy*, *The Philosopher's Index*, *The American Philosophical Quarterly*, *Studies in History and Philosophy of Science*, *Erkenntnis*. Among his publications are: *Philosophical Problems of Space and Time*, D. Reidel Publ. Co., Dordrecht, Holland, 1963, 2. enl. ed. 1973; *Geometry and Chronometry in Philosophical Perspective*, University of Minnesota Press, 1968. He has published over 100 articles.

KURT HÜBNER is Professor of Philosophy at the University of Kiel, The Federal Republic of Germany. President, Center for Cybernetics, Berlin, 1965–69; President, German Association for Philosophy (Deutsche Gesellschaft für Philosophie), 1969–75; Chairman of the colloquy and member of the scientific council of the 14th World Conference of Philosophy, Vienna, 1968; advisor to the planning commission of the 16th World Conference of Philosophy, Düsseldorf, 1978. He is a member of the Editorial Board of *Erkenntnis*, *Zeitschrift für allgemeine Wissenschaftstheorie*, *Philosophia Naturalis*. Among his publications are: *Beiträge zur Philosophie der Physik*, Beiheft 4 der Philosophischen Rundschau, J. C. B. Mohr, Tübingen, 1963; 'Philosophische Probleme der Technik', in H. Lenk and S. Moser (eds.), *Philosophische Probleme der Technik*, Dokumentation, Pullach/München, 1973; 'Grundlagen einer Theorie der Geschichtswissenschaften', in R. S. Schaefer and W. Zimmerli (eds.), *Grundlagen einer Theorie der Geisteswissenschaften*, Hoffmann & Campe, Hamburg, 1975.

NORETTA KOERTGE received her Ph.D. in philosophy of science at London University, 1969. She is Professor, Department of History and Philosophy of Science, Indiana University, Bloomington, Indiana. Among her articles are 'Inter-Theoretic Criticism and the Growth of Science', in Roger C. Buck and Robert S. Cohen (eds.), *Boston Studies in the Philosophy of Science*, Vol. VIII, 1971; 'Theory Change in Science', in Glenn Pearce (ed.), *Conceptual Change*, 1971; 'Popper's Metaphysical Research Program for the Human Sciences', *Inquiry* **18**, 437–62 (1975); 'Rational Reconstructions', in R. Cohen, P. Feyerabend, and M. Wartofsky (eds.), *Essays in Memory of Imre*

Lakatos. Boston Studies in the Philosophy of Science, Vol. XXXIX, D. Reidel Publ. Co., Dordrecht, Holland, 1976.

ERNAN McMULLIN received his Ph.D. from the University of Louvain, 1954. He is Professor of Philosophy at the University of Notre Dame, U.S.A. President, American Catholic Philosophical Association, 1966–7; President, Metaphysical Society of America, 1973–74; Chairman, Section L, American Association for the Advancement of Science, 1976–7; Chairman, Philosophy of Science Division, International Congresses of Philosophy, Vienna 1968 and Varna 1973. He has edited and written introductions to *The Concept of Matter*, Notre Dame University Press, 1963, and *Galileo, Man of Science*, New York: Basic Books, 1967. Among his other publications are: 'Two Faces of Science', *Review of Metaphysics* **27**, 655–676 (1974); 'What Difference Does Mind Make?', in A. Karczmar and J. C. Eccles (eds.), *Brain and Human Behaviour*, Springer, Berlin, 1972; 'History and Philosophy of Science: A Taxonomy', in R. Stuewer (ed.), *Historical and Philosophical Perspectives of Science*, University Press, Minneapolis, 1970.

ALAN MUSGRAVE is Professor of Philosophy, University of Otago, New Zealand. He was President (1975) of New Zealand Division of the Australasian Association for Philosophy. He has edited (together with Imre Lakatos) *Criticism and the Growth of Knowledge*, Cambridge University Press, London, 1970. Among his articles are: 'The Objectivism of Popper's Epistemology', in P. A. Schilpp (ed.), *The Philosophy of Karl Popper*, Open Court Publishing Company, La Salle, Illinois, 1974. 'Falsification and Its Critics', in P. Suppes *et al.* (eds.), *Logic, Methodology and Philosophy of Science*, Vol. 4, North-Holland Publishing Company, Amsterdam, 1973. 'Logical Versus Historical Theories of Confirmation', *British Journal for the Philosophy of Science* **25**, 1–23 (1974).

H. R. POST studied chemistry at Trinity College, Oxford and physics at the University of Chicago (Ph.D. 1950). He also studied philosophy of science under Carnap. Experimental work in chemical reaction kinetics, neutron diffusion, and neutrino recoil. Theoretical work in solid state physics and the mathematics of the several-identical-particle-problem in first quantization. Now teaching in the Department of History and Philosophy of Science at Chelsea College, University of London.

GERARD RADNITZKY is Professor of Philosophy of Science at the University of Trier, corresponding member of the Académie Internationale de Philosophie des Sciences; author of *Contemporary Schools of Metascience* (New York: Humanities, 1968, 1970), *Preconceptions in Research* (London: Literary Services and Production, 1974), *Epistemologia e Politica di Ricerca* (Rome: Armando 1978) and more than 60 papers.

PETER URBACH obtained a Ph.D. in Physical Chemistry from the University of Manchester 1969, teacher in philosophy at the London School of Economics and Political Sciences since 1973. Published paper: 'Progress and Degeneration in the "I.Q. Debate" ', Parts I and II, *British Journal for the Philosophy of Science* **25**, 99–135, 235–259 (1974).

J. W. N. WATKINS, Professor of Philosophy at the London School of Economics. President, British Society for The Philosophy of Science, 1972–75. Co-editor, *British Journal for the Philosophy of Science*, since 1974. Author of many papers in philosophical journals, and of *Hobbes's System of Ideas*, Hutchinson, London, 1965 (second ed. 1973), *Entscheidung und Freiheit*, J. C. B. Mohr (Paul Siebeck), Tübingen, forthcoming.

JOHN WORRALL is lecturer at the London School of Economics. Co-Editor of the *British Journal for the Philosophy of Science* since 1974. Author of 'Thomas Young and the "Refutation" of Newtonian Optics', in C. Howson (ed.), *Method and Appraisal in Physical Sciences*, Cambridge at the University Press, 1976, Joint editor of I. Lakatos's *Proofs and Refutations*, senior editor of Lakatos's *Collected Papers*.

AUTHOR INDEX

Compiled by Sieglinde Kordel
n = note, b = bibliography

SUBJECT INDEX

Compiled by Klaus Pähler

SYNTHESE LIBRARY

Studies in Epistemology, Logic, Methodology,
and Philosophy of Science

Managing Editor:
JAAKKO HINTIKKA, (Academy of Finland, Stanford University
and Florida State University)

Editors:
ROBERT S. COHEN (Boston University)
DONALD DAVIDSON (University of Chicago)
GABRIËL NUCHELMANS (University of Leyden)
WESLEY C. SALMON (University of Arizona)

1. J. M. Bocheński, *A Precis of Mathematical Logic.* 1959, X + 100 pp.
2. P. L. Guiraud, *Problèmes et méthodes de la statistique linguistique.* 1960, VI + 146 pp.
3. Hans Freudenthal (ed.), *The Concept and the Role of the Model in Mathematics and Natural and Social Sciences. Proceedings of a Colloquium held at Utrecht, The Netherlands, January 1960.* 1961, VI + 194 pp.
4. Evert W. Beth, *Formal Methods. An Introduction to Symbolic Logic and the Study of Effective Operations in Arithmetic and Logic.* 1962, XIV + 170 pp.
5. B. H. Kazemier and D. Vuysje (eds.), *Logic and Language. Studies Dedicated to Professor Rudolf Carnap on the Occasion of His Seventieth Birthday.* 1962, VI + 256 pp.
6. Marx W. Wartofsky (ed.), *Proceedings of the Boston Colloquium for the Philosophy of Science 1961-1962,* Boston Studies in the Philosophy of Science (ed. by Robert S. Cohen and Marx W. Wartofsky), Volume I. 1963, VIII + 212 pp.
7. A. A. Zinov'ev, *Philosophical Problems of Many-Valued Logic.* 1963, XIV + 155 pp.
8. Georges Gurvitch, *The Spectrum of Social Time.* 1964, XXVI + 152 pp.
9. Paul Lorenzen, *Formal Logic.* 1965, VIII + 123 pp.
10. Robert S. Cohen and Marx W. Wartofsky (eds.), *In Honor of Philipp Frank,* Boston Studies in the Philosophy of Science (ed. by Robert S. Cohen and Marx W. Wartofsky), Volume II. 1965, XXXIV + 475 pp.
11. Evert W. Beth, *Mathematical Thought. An Introduction to the Philosophy of Mathematics.* 1965, XII + 208 pp.
12. Evert W. Beth and Jean Piaget, *Mathematical Epistemology and Psychology.* 1966, XII + 326 pp.
13. Guido Küng, *Ontology and the Logistic Analysis of Language. An Enquiry into the Contemporary Views on Universals.* 1967, XI + 210 pp.
14. Robert S. Cohen and Marx W. Wartofsky (eds.), *Proceedings of the Boston Colloquium for the Philosophy of Science 1964-1966, in Memory of Norwood Russell Hanson,* Boston Studies in the Philosophy of Science (ed. by Robert S. Cohen and Marx W. Wartofsky), Volume III. 1967, XLIX + 489 pp.

15. C. D. Broad, *Induction, Probability, and Causation. Selected Papers*. 1968, XI + 296 pp.
16. Günther Patzig, *Aristotle's Theory of the Syllogism. A Logical-Philosophical Study of Book A of the Prior Analytics*. 1968, XVII + 215 pp.
17. Nicholas Rescher, *Topics in Philosophical Logic*. 1968, XIV + 347 pp.
18. Robert S. Cohen and Marx W. Wartofsky (eds.), *Proceedings of the Boston Colloquium for the Philosophy of Science 1966-1968*, Boston Studies in the Philosophy of Science (ed. by Robert S. Cohen and Marx W. Wartofsky), Volume IV. 1969, VIII + 537 pp.
19. Robert S. Cohen and Marx W. Wartofsky (eds.), *Proceedings of the Boston Colloquium for the Philosophy of Science 1966-1968*, Boston Studies in the Philosophy of Science (ed. by Robert S. Cohen and Marx W. Wartofsky), Volume V. 1969, VIII + 482 pp.
20. J.W. Davis, D. J. Hockney, and W. K. Wilson (eds.), *Philosophical Logic*. 1969, VIII + 277 pp.
21. D. Davidson and J. Hintikka (eds.), *Words and Objections: Essays on the Work of W.V. Quine*. 1969, VIII + 366 pp.
22. Patrick Suppes, *Studies in the Methodology and Foundations of Science. Selected Papers from 1911 to 1969*. 1969, XII + 473 pp.
23. Jaakko Hintikka, *Models for Modalities. Selected Essays*. 1969, IX + 220 pp.
24. Nicholas Rescher *et al.* (eds.), *Essays in Honor of Carl G. Hempel. A Tribute on the Occasion of His Sixty-Fifth Birthday*. 1969, VII + 272 pp.
25. P. V. Tavanec (ed.), *Problems of the Logic of Scientific Knowledge*. 1969, XII + 429 pp.
26. Marshall Swain (ed.), *Induction, Acceptance, and Rational Belief*. 1970, VII + 232 pp.
27. Robert S. Cohen and Raymond J. Seeger (eds.), *Ernst Mach: Physicist and Philosopher*, Boston Studies in the Philosophy of Science (ed. by Robert S. Cohen and Marx W. Wartofsky), Volume VI. 1970, VIII + 295 pp.
28. Jaakko Hintikka and Patrick Suppes, *Information and Inference*. 1970, X + 336 pp.
29. Karel Lambert, *Philosophical Problems in Logic. Some Recent Developments*. 1970, VII + 176 pp.
30. Rolf A. Eberle, *Nominalistic Systems*. 1970, IX + 217 pp.
31. Paul Weingartner and Gerhard Zecha (eds.), *Induction, Physics, and Ethics: Proceedings and Discussions of the 1968 Salzburg Colloquium in the Philosophy of Science*. 1970, X + 382 pp.
32. Evert W. Beth, *Aspects of Modern Logic*. 1970, XI + 176 pp.
33. Risto Hilpinen (ed.), *Deontic Logic: Introductory and Systematic Readings*. 1971, VII + 182 pp.
34. Jean-Louis Krivine, *Introduction to Axiomatic Set Theory*. 1971, VII + 98 pp.
35. Joseph D. Sneed, *The Logical Structure of Mathematical Physics*. 1971, XV + 311 pp.
36. Carl R. Kordig, *The Justification of Scientific Change*. 1971, XIV + 119 pp.
37. Milič Čapek, *Bergson and Modern Physics*, Boston Studies in the Philosophy of Science (ed. by Robert S. Cohen and Marx W. Wartofsky), Volume VII. 1971, XV + 414 pp.

38. Norwood Russell Hanson, *What I Do Not Believe, and Other Essays* (ed. by Stephen Toulmin and Harry Woolf), 1971, XII + 390 pp.
39. Roger C. Buck and Robert S. Cohen (eds.), *PSA 1970. In Memory of Rudolf Carnap*, Boston Studies in the Philosophy of Science (ed. by Robert S. Cohen and Marx W. Wartofsky), Volume VIII. 1971, LXVI + 615 pp. Also available as paperback.
40. Donald Davidson and Gilbert Harman (eds.), *Semantics of Natural Language*. 1972, X + 769 pp. Also available as paperback.
41. Yehoshua Bar-Hillel (ed.), *Pragmatics of Natural Languages*. 1971, VII + 231 pp.
42. Sören Stenlund, *Combinators, λ-Terms and Proof Theory*. 1972, 184 pp.
43. Martin Strauss, *Modern Physics and Its Philosophy. Selected Papers in the Logic, History, and Philosophy of Science*. 1972, X + 297 pp.
44. Mario Bunge, *Method, Model and Matter*. 1973, VII + 196 pp.
45. Mario Bunge, *Philosophy of Physics*. 1973, IX + 248 pp.
46. A. A. Zinov'ev, *Foundations of the Logical Theory of Scientific Knowledge (Complex Logic)*, Boston Studies in the Philosophy of Science (ed. by Robert S. Cohen and Marx W. Wartofsky), Volume IX. Revised and enlarged English edition with an appendix, by G. A. Smirnov, E. A. Sidorenka, A. M. Fedina, and L. A. Bobrova. 1973, XXII + 301 pp. Also available as paperback.
47. Ladislav Tondl, *Scientific Procedures*, Boston Studies in the Philosophy of Science (ed. by Robert S. Cohen and Marx W. Wartofsky), Volume X. 1973, XII + 268 pp. Also available as paperback.
48. Norwood Russell Hanson, *Constellations and Conjectures* (ed. by Willard C. Humphreys, Jr.). 1973, X + 282 pp.
49. K. J. J. Hintikka, J. M. E. Moravcsik, and P. Suppes (eds.), *Approaches to Natural Language. Proceedings of the 1970 Stanford Workshop on Grammar and Semantics*. 1973, VIII + 526 pp. Also available as paperback.
50. Mario Bunge (ed.), *Exact Philosophy – Problems, Tools, and Goals*. 1973, X + 214 pp.
51. Radu J. Bogdan and Ilkka Niiniluoto (eds.), *Logic, Language, and Probability. A Selection of Papers Contributed to Sections IV, VI, and XI of the Fourth International Congress for Logic, Methodology, and Philosophy of Science, Bucharest, September 1971*. 1973, X + 323 pp.
52. Glenn Pearce and Patrick Maynard (eds.), *Conceptual Change*. 1973, XII + 282 pp.
53. Ilkka Niiniluoto and Raimo Tuomela, *Theoretical Concepts and Hypothetico-Inductive Inference*. 1973, VII + 264 pp.
54. Roland Fraissé, *Course of Mathematical Logic – Volume 1: Relation and Logical Formula*. 1973, XVI + 186 pp. Also available as paperback.
55. Adolf Grünbaum, *Philosophical Problems of Space and Time*. Second, enlarged edition, Boston Studies in the Philosophy of Science (ed. by Robert S. Cohen and Marx W. Wartofsky), Volume XII. 1973, XXIII + 884 pp. Also available as paperback.
56. Patrick Suppes (ed.), *Space, Time, and Geometry*. 1973, XI + 424 pp.
57. Hans Kelsen, *Essays in Legal and Moral Philosophy*, selected and introduced by Ota Weinberger. 1973, XXVIII + 300 pp.
58. R. J. Seeger and Robert S. Cohen (eds.), *Philosophical Foundations of Science. Proceedings of an AAAS Program, 1969*, Boston Studies in the Philosophy of

Science (ed. by Robert S. Cohen and Marx W. Wartofsky), Volume XI. 1974, X + 545 pp. Also available as paperback.

59. Robert S. Cohen and Marx W. Wartofsky (eds.), *Logical and Epistemological Studies in Contemporary Physics*, Boston Studies in the Philosophy of Science (ed. by Robert S. Cohen and Marx W. Wartofsky), Volume XIII. 1973, VIII + 462 pp. Also available as paperback.

60. Robert S. Cohen and Marx W. Wartofsky (eds.), *Methodological and Historical Essays in the Natural and Social Sciences. Proceedings of the Boston Colloquium for the Philosophy of Science 1969-1972*, Boston Studies in the Philosophy of Science (ed. by Robert S. Cohen and Marx W. Wartofsky), Volume XIV. 1974, VIII + 405 pp. Also available as paperback.

61. Robert S. Cohen, J. J. Stachel and Marx W. Wartofsky (eds.), *For Dirk Struik. Scientific, Historical and Political Essays in Honor of Dirk J. Struik*, Boston Studies in the Philosophy of Science (ed. by Robert S. Cohen and Marx W. Wartofsky), Volume XV. 1974, XXVII + 652 pp. Also available as paperback.

62. Kazimierz Ajdukiewicz, *Pragmatic Logic*, transl. from the Polish by Olgierd Wojtasiewicz. 1974, XV + 460 pp.

63. Sören Stenlund (ed.), *Logical Theory and Semantic Analysis. Essays Dedicated to Stig Kanger on His Fiftieth Birthday*. 1974, V + 217 pp.

64. Kenneth F. Schaffner and Robert S. Cohen (eds.), *Proceedings of the 1972 Biennial Meeting, Philosophy of Science Association*, Boston Studies in the Philosophy of Science (ed. by Robert S. Cohen and Marx W. Wartofsky), Volume XX. 1974, IX + 444 pp. Also available as paperback.

65. Henry E. Kyburg, Jr., *The Logical Foundations of Statistical Inference*. 1974, IX + 421 pp.

66. Marjorie Grene, *The Understanding of Nature: Essays in the Philosophy of Biology*, Boston Studies in the Philosophy of Science (ed. by Robert S. Cohen and Marx W. Wartofsky), Volume XXIII. 1974, XII + 360 pp. Also available as paperback.

67. Jan M. Broekman, *Structuralism: Moscow, Prague, Paris*. 1974, IX + 117 pp.

68. Norman Geschwind, *Selected Papers on Language and the Brain*, Boston Studies in the Philosophy of Science (ed. by Robert S. Cohen and Marx W. Wartofsky), Volume XVI. 1974, XII + 549 pp. Also available as paperback.

69. Roland Fraïssé, *Course of Mathematical Logic – Volume 2: Model Theory*. 1974, XIX + 192 pp.

70. Andrzej Grzegorczyk, *An Outline of Mathematical Logic. Fundamental Results and Notions Explained with All Details*. 1974, X + 596 pp.

71. Franz von Kutschera, *Philosophy of Language*. 1975, VII + 305 pp.

72. Juha Manninen and Raimo Tuomela (eds.), *Essays on Explanation and Understanding. Studies in the Foundations of Humanities and Social Sciences*. 1976, VII + 440 pp.

73. Jaakko Hintikka (ed.), *Rudolf Carnap, Logical Empiricist. Materials and Perspectives*. 1975, LXVIII + 400 pp.

74. Milič Čapek (ed.), *The Concepts of Space and Time. Their Structure and Their Development*, Boston Studies in the Philosophy of Science (ed. by Robert S. Cohen and Marx W. Wartofsky), Volume XXII. 1976, LVI + 570 pp. Also available as paperback.

75. Jaakko Hintikka and Unto Remes, *The Method of Analysis. Its Geometrical Origin and Its General Significance*, Boston Studies in the Philosophy of Science (ed. by Robert S. Cohen and Marx W. Wartofsky), Volume XXV. 1974, XVIII + 144 pp. Also available as paperback.

76. John Emery Murdoch and Edith Dudley Sylla, *The Cultural Context of Medieval Learning. Proceedings of the First International Colloquium on Philosophy, Science, and Theology in the Middle Ages – September 1973*, Boston Studies in the Philosophy of Science (ed. by Robert S. Cohen and Marx W. Wartofsky), Volume XXVI. 1975, X + 566 pp. Also available as paperback.

77. Stefan Amsterdamski, *Between Experience and Metaphysics. Philosophical Problems of the Evolution of Science*, Boston Studies in the Philosophy of Science (ed. by Robert S. Cohen and Marx W. Wartofsky), Volume XXXV. 1975, XVIII + 193 pp. Also available as paperback.

78. Patrick Suppes (ed.), *Logic and Probability in Quantum Mechanics*. 1976, XV + 541 pp.

79. Hermann von Helmholtz: *Epistemological Writings. The Paul Hertz/Moritz Schlick Centenary Edition of 1921 with Notes and Commentary by the Editors.* (Newly translated by Malcolm F. Lowe. Edited with an Introduction and Bibliography, by Robert S. Cohen and Yehuda Elkana), Boston Studies in the Philosophy of Science (ed. by Robert S. Cohen and Marx W. Wartofsky), Volume XXXVII. 1977, XXXVIII+204 pp. Also available as paperback.

80. Joseph Agassi, *Science in Flux*, Boston Studies in the Philosophy of Science (ed. by Robert S. Cohen and Marx W. Wartofsky), Volume XXVIII. 1975, XXVI + 553 pp. Also available as paperback.

81. Sandra G. Harding (ed.), *Can Theories Be Refuted? Essays on the Duhem-Quine Thesis.* 1976, XXI + 318 pp. Also available as paperback.

82. Stefan Nowak, *Methodology of Sociological Research: General Problems.* 1977, XVIII + 504 pp.

83. Jean Piaget, Jean-Blaise Grize, Alina Szeminska, and Vinh Bang, *Epistemology and Psychology of Functions*, Studies in Genetic Epistemology, Volume XXIII. 1977, XIV+205 pp.

84. Marjorie Grene and Everett Mendelsohn (eds.), *Topics in the Philosophy of Biology*, Boston Studies in the Philosophy of Science (ed. by Robert S. Cohen and Marx W. Wartofsky), Volume XXVII. 1976, XIII + 454 pp. Also available as paperback.

85. E. Fischbein, *The Intuitive Sources of Probabilistic Thinking in Children.* 1975, XIII + 204 pp.

86. Ernest W. Adams, *The Logic of Conditionals. An Application of Probability to Deductive Logic.* 1975, XIII + 156 pp.

87. Marian Przełęcki and Ryszard Wójcicki (eds.), *Twenty-Five Years of Logical Methodology in Poland.* 1977, VIII + 803 pp.

88. J. Topolski, *The Methodology of History.* 1976, X + 673 pp.

89. A. Kasher (ed.), *Language in Focus: Foundations, Methods and Systems. Essays Dedicated to Yehoshua Bar-Hillel*, Boston Studies in the Philosophy of Science (ed. by Robert S. Cohen and Marx W. Wartofsky), Volume XLIII. 1976, XXVIII + 679 pp. Also available as paperback.

90. Jaakko Hintikka, *The Intentions of Intentionality and Other New Models for Modalities.* 1975, XVIII + 262 pp. Also available as paperback.

91. Wolfgang Stegmüller, *Collected Papers on Epistemology, Philosophy of Science and History of Philosophy*, 2 Volumes, 1977, XXVII + 525 pp.
92. Dov M. Gabbay, *Investigations in Modal and Tense Logics with Applications to Problems in Philosophy and Linguistics*. 1976, XI + 306 pp.
93. Radu J. Bogdan, *Local Induction*. 1976, XIV + 340 pp.
94. Stefan Nowak, *Understanding and Prediction: Essays in the Methodology of Social and Behavioral Theories*. 1976, XIX + 482 pp.
95. Peter Mittelstaedt, *Philosophical Problems of Modern Physics*, Boston Studies in the Philosophy of Science (ed. by Robert S. Cohen and Marx W. Wartofsky), Volume XVIII. 1976, X + 211 pp. Also available as paperback.
96. Gerald Holton and William Blanpied (eds.), *Science and Its Public: The Changing Relationship*, Boston Studies in the Philosophy of Science (ed. by Robert S. Cohen and Marx W. Wartofsky), Volume XXXIII. 1976, XXV + 289 pp. Also available as paperback.
97. Myles Brand and Douglas Walton (eds.), *Action Theory. Proceedings of the Winnipeg Conference on Human Action, Held at Winnipeg, Manitoba, Canada, 9-11 May 1975*. 1976, VI + 345 pp.
98. Risto Hilpinen, *Knowledge and Rational Belief*. 1979 (forthcoming).
99. R. S. Cohen, P. K. Feyerabend, and M. W. Wartofsky (eds.), *Essays in Memory of Imre Lakatos*, Boston Studies in the Philosophy of Science (ed. by Robert S. Cohen and Marx W. Wartofsky), Volume XXXIX. 1976, XI + 762 pp. Also available as paperback.
100. R. S. Cohen and J. J. Stachel (eds.), *Selected Papers of Léon Rosenfeld*, Boston Studies in the Philosophy of Science (ed. by Robert S. Cohen and Marx W. Wartofsky), Volume XXI. 1978, XXX + 927 pp.
101. R. S. Cohen, C. A. Hooker, A. C. Michalos, and J. W. van Evra (eds.), *PSA 1974: Proceedings of the 1974 Biennial Meeting of the Philosophy of Science Association*, Boston Studies in the Philosophy of Science (ed. by Robert S. Cohen and Marx W. Wartofsky), Volume XXXII. 1976, XIII + 734 pp. Also available as paperback.
102. Yehuda Fried and Joseph Agassi, *Paranoia: A Study in Diagnosis*, Boston Studies in the Philosophy of Science (ed. by Robert S. Cohen and Marx W. Wartofsky), Volume L. 1976, XV + 212 pp. Also available as paperback.
103. Marian Przełęcki, Klemens Szaniawski, and Ryszard Wójcicki (eds.), *Formal Methods in the Methodology of Empirical Sciences*. 1976, 455 pp.
104. John M. Vickers, *Belief and Probability*. 1976, VIII + 202 pp.
105. Kurt H. Wolff, *Surrender and Catch: Experience and Inquiry Today*, Boston Studies in the Philosophy of Science (ed. by Robert S. Cohen and Marx W. Wartofsky), Volume LI. 1976, XII + 410 pp. Also available as paperback.
106. Karel Kosík, *Dialectics of the Concrete*, Boston Studies in the Philosophy of Science (ed. by Robert S. Cohen and Marx W. Wartofsky), Volume LII. 1976, VIII + 158 pp. Also available as paperback.
107. Nelson Goodman, *The Structure of Appearance*, Boston Studies in the Philosophy of Science (ed. by Robert S. Cohen and Marx W. Wartofsky), Volume LIII. 1977, L + 285 pp.
108. Jerzy Giedymin (ed.), *Kazimierz Ajdukiewicz: The Scientific World-Perspective and Other Essays, 1931 - 1963*. 1978, LIII + 378 pp.

109. Robert L. Causey, *Unity of Science.* 1977, VIII+185 pp.
110. Richard E. Grandy, *Advanced Logic for Applications.* 1977, XIV + 168 pp.
111. Robert P. McArthur, *Tense Logic.* 1976, VII + 84 pp.
112. Lars Lindahl, *Position and Change: A Study in Law and Logic.* 1977, IX + 299 pp.
113. Raimo Tuomela, *Dispositions.* 1978, X + 450 pp.
114. Herbert A. Simon, *Models of Discovery and Other Topics in the Methods of Science,* Boston Studies in the Philosophy of Science (ed. by Robert S. Cohen and Marx W. Wartofsky), Volume LIV. 1977, XX + 456 pp. Also available as paperback.
115. Roger D. Rosenkrantz, *Inference, Method and Decision.* 1977, XVI + 262 pp. Also available as paperback.
116. Raimo Tuomela, *Human Action and Its Explanation. A Study on the Philosophical Foundations of Psychology.* 1977, XII + 426 pp.
117. Morris Lazerowitz, *The Language of Philosophy. Freud and Wittgenstein,* Boston Studies in the Philosophy of Science (ed. by Robert S. Cohen and Marx W. Wartofsky), Volume LV. 1977, XVI + 209 pp.
118. Tran Duc Thao, *Origins of Language and Consciousness,* Boston Studies in the Philosophy of Science (ed. by Robert S. Cohen and Marx. W. Wartofsky), Volume LVI. 1979 (forthcoming).
119. Jerzy Pelč, *Semiotics in Poland, 1894 - 1969.* 1977, XXVI + 504 pp.
120. Ingmar Pörn, *Action Theory and Social Science. Some Formal Models.* 1977, X + 129 pp.
121. Joseph Margolis, *Persons and Minds, The Prospects of Nonreductive Materialism,* Boston Studies in the Philosophy of Science (ed. by Robert S. Cohen and Marx W. Wartofsky), Volume LVII. 1977, XIV + 282 pp. Also available as paperback.
122. Jaakko Hintikka, Ilkka Niiniluoto, and Esa Saarinen (eds.), *Essays on Mathematical and Philosophical Logic. Proceedings of the Fourth Scandinavian Logic Symposium and of the First Soviet-Finnish Logic Conference, Jyväskylä, Finland, 1976.* 1978, VIII + 458 pp. + index.
123. Theo A. F. Kuipers, *Studies in Inductive Probability and Rational Expectation.* 1978, XII + 145 pp.
124. Esa Saarinen, Risto Hilpinen, Ilkka Niiniluoto, and Merrill Provence Hintikka (eds.), *Essays in Honour of Jaakko Hintikka on the Occasion of His Fiftieth Birthday.* 1978, IX + 378 pp. + index.
125. Gerard Radnitzky and Gunnar Andersson (eds.), *Progress and Rationality in Science,* Boston Studies in the Philosophy of Science (ed. by Robert S. Cohen and Marx W. Wartofsky), Volume LVIII. 1978, X + 400 pp. + index. Also available as paperback.
126. Peter Mittelstaedt, *Quantum Logic.* 1978, IX + 149 pp.
127. Kenneth A. Bowen, *Model Theory for Modal Logic. Kripke Models for Modal Predicate Calculi.* 1978, X + 128 pp.
128. Howard Alexander Bursen, *Dismantling the Memory Machine. A Philosophical Investigation of Machine Theories of Memory.* 1978. XIII + 157 pp.
129. Marx W. Wartofsky, *Models: Representation and Scientific Understanding,* Boston Studies in the Philosophy of Science (ed. by Robert S. Cohen and Marx W. Wartofsky), Volume XLVIII. 1979 (forthcoming). Also available as a paperback.
130. Don Ihde, *Technics and Praxis. A Philosophy of Technology,* Boston Studies in

the Philosophy of Science (ed. by Robert S. Cohen and Marx W. Wartofsky), Volume XXIV. 1979 (forthcoming). Also available as a paperback.

131. Jerzy J. Wiatr (ed.), *Polish Essays in the Methodology of the Social Sciences*, Boston Studies in the Philosophy of Science (ed. by Robert S. Cohen and Marx W. Wartofsky), Volume XXIX. 1979 (forthcoming). Also available as a paperback.

132. Wesley C. Salmon (ed.), *Hans Reichenbach: Logical Empiricist*. 1979 (forthcoming).

133. R.-P. Horstmann and L. Krüger (eds.), *Transcendental Arguments and Science*. 1979 (forthcoming). Also available as a paperback.

SYNTHESE HISTORICAL LIBRARY

Texts and Studies
in the History of Logic and Philosophy

Editors:

N. KRETZMANN (Cornell University)
G. NUCHELMANS (University of Leyden)
L. M. DE RIJK (University of Leyden)

17. Arpád Szabó, *The Beginnings of Greek Mathematics.* 1979 (forthcoming).

18. Rita Guerlac, *Juan Luis Vives Against the Pseudodialecticians. A Humanist Attack on Medieval Logic.* Texts, with translation, introduction and notes. 1978, xiv + 227 pp. + index.

SYNTHESE LANGUAGE LIBRARY

Texts and Studies
in Linguistics and Philosophy

Managing Editors:

JAAKKO HINTIKKA
Academy of Finland, Stanford University, and Florida State University (Tallahassee)

STANLEY PETERS
The University of Texas at Austin

Editors:

EMMON BACH (University of Massachusetts at Amherst)
JOAN BRESNAN (Massachusetts Institute of Technology)
JOHN LYONS (University of Sussex)
JULIUS M. E. MORAVCSIK (Stanford University)
PATRICK SUPPES (Stanford University)
DANA SCOTT (Oxford University)